ALLOCATING PUBLIC AND PRIVATE
RESOURCES ACROSS GENERATIONS

International Studies in Population

Volume 3

The International Union for the Scientific Study of Population (IUSSP)

The IUSSP is an international association of researchers, teachers, policy makers and others from diverse disciplines set up to advance the field of demography and promote scientific exchange. It also seeks to foster relations between persons engaged in the study of demography and population trends in all countries of the world and to disseminate knowledge about their determinants and consequences. The members of the IUSSP scientific group responsible for this book were chosen for their scientific expertise. This book was reviewed by a group other than the authors. While the IUSSP endeavours to assure the accuracy and objectivity of its scientific work, the conclusions and interpretations in IUSSP publications are those of the authors.

International Studies in Population (ISIP) is the outcome of an agreement concluded by IUSSP and Springer in 2004. The joint series covers the broad range of work carried out by IUSSP and includes material presented at seminars organized by the IUSSP. The scientific directions of the IUSSP are set by the IUSSP Council elected by the membership and composed of:

Allocating Public and Private Resources across Generations

Riding the Age Waves—Volume 2

Edited by

ANNE H. GAUTHIER
University of Calgary,
Alberta, Canada

C.Y. CYRUS CHU
Academia Sinica,
Taipei, Taiwan

and

SHRIPAD TULJAPURKAR
Stanford University,
California, U.S.A.

Springer

A C.I.P. Catalogue record for this book is available from the Library of Congress.

ISBN 978-1-4020-4790-9 (PB)
ISBN 978-1-4020-4480-9 (HB)
ISBN 978-1-4020-4481-6 (e-book)

Published by Springer,
P.O. Box 17, 3300 AA Dordrecht, The Netherlands.

www.springer.com

Printed on acid-free paper

CONTENTS

PREFACE

SHRIPAD TULJAPURKAR
C.Y. CYRUS CHU
ANNE H. GAUTHIER
NAOHIRO OGAWA
IAN POOL

Beginning in the last century and continuing into the 21st century, the populations of the world's nations have displayed large and long-lived changes in age structure. Many of these began with fertility change in the form of baby booms, busts, or declines, and are amplified by declining mortality and by migration within and between nations. These age-structural transitions have powerful effects on human affairs, from driving the demand for public and private goods and services for young and old, to determining the flow of resources across the different ages of the human life cycle. The consequences of age-structural transitions vary in emphasis and detail, but not in significance, across the spectrum of nations in different stages of economic development. Demography will matter in this century not by force of numbers, but by the pressures of waves of age-structural change.

In 1997 a committee of the International Union for the Scientific Study of Populations was charged with exploring age-structural transitions and their policy implications. The committee brought together distinguished scientists to examine the key demographic, social, economic, and policy aspects of age-structural change across a spectrum of nations at different stages of development. Readers will find a rich discussion of their work in *Riding the Age Waves*, a series of three books, "Population, Resources, and Development," "Allocating Public and Private Resources across Generations," and "Responses to Ageing in Advanced Industrial States." Volume 1 of the series contains discussions focused on developing countries and challenges from building human capabilities and creating jobs to creating industry, infrastructure, and institutions. Volume 3 of the series contains discussions focused on the highly industrialized countries and policy problems related to ageing and long run sub-replacement fertility.

The subject of this volume (volume 2) is intergenerational transfers, by which we mean transfers of all kinds (money, goods, services, time, culture, taxes and benefits, information, social capital) across generations. Some of the discussion also concerns intragenerational transfers that are often naturally intertwined with transfers across the age spectrum.

Why are intergenerational transfers important in an age of changing demographic structures? Intergenerational transfers are a key determinant of the well-being of individuals and families, and of aggregate human, social, and financial capital. Intergenerational transfers of time, money, goods, and attitudes from parents to young children contribute to children's educational achievement and later employment trajectories, and to intergenerational class mobility. Similarly, intergenerational transfers from grown-up children to elderly parents have been shown to contribute to parents' well-being. At a macro level, the direction and magnitude of intergenerational transfers has been proposed as a key factor determining fertility transitions and levels. In the reverse direction, patterns of intergenerational transfer are strongly affected by shifts in age structure, kinship structure, and social networks. The macroeconomics of populations undergoing fertility and mortality transitions can be usefully analysed in terms of intergenerational flows to illuminate questions ranging from savings rates to the need for public support to different age segments of the population.

In many industrialized countries, population age structures today reflect large fertility declines that occurred a generation or so ago, with resulting baby booms that have aged and are moving through the labour force. Fertility declines have had and continue to have significant social and economic effects, e.g., on the growth of individual savings and thus aggregate capital investment, on family sizes and the shift from traditional family structures toward nuclear families, and on the thinning of traditional kinship networks. Mortality declines have resulted in rapid growth of the elderly population and an increase in old-age dependency ratios in both population, and families. Thus, the joint effects of intergenerational transfers and age-structural transitions will have short- and long-run effects on policy with respect to education, social services, welfare, old-age support, and on macroeconomic and fiscal policy.

Several features distinguish these books from past writing on the subject. First, a joint examination of dimensions of age-structural change that have often been considered in isolation from each other (for example, in multidimensional measurement of transfers, and in cross-national comparisons); second, the papers here bring together the many policy implications of these dimensions; third, the use of case studies to examine policy consequences and options of particular dimensions of change; fourth, the development of qualitative and formal methods to analyse the long-term dynamic nature and consequences of age-structural change.

The committee (Shripad Tuljapurkar, C.Y. Cyrus Chu, Anne H. Gauthier, Naohiro Ogawa, Rafael Rofman, Ian Pool, Hassan Youssif) thanks the many people and agencies who made this work possible. In addition to the IUSSP, we thank the Asian Metacentre at the Singapore National University, the Academia Sinica in Taipei, and the Nihon University Population Research Institute in Tokyo. The editors of the individual volumes were C.Y. Cyrus Chu, Anne H. Gauthier, Naohiro Ogawa, Ian Pool, Vipan Prachuabmoh, and Shripad Tuljapurkar.

ACKNOWLEDGEMENTS

This book is the result of an IUSSP (International Union for the Scientific Study of Population) seminar on *Age Structure Transitions and Policy Dynamics: The Allocation of Public and Private Resources across Generations* held in Taipei, Taiwan December 6–8, 2001. The seminar was organized by the IUSSP Committee on Age Structure and Public Policy (Shripad Tuljapurkar, Chair, C.Y. Cyrus Chu, Anne H. Gauthier, Naohiro Ogawa, Ian Pool, Rafael Rofman, Hassan Yousif) and the Institute of Economics of Academia Sinica. An editorial committee reviewed a selected subset of the many excellent papers presented at the meeting. Following the usual practices in this and the previous series the revised set of chapters was subject to independent scrutiny through the IUSSP. The editors and external reviewers requested revisions and the final revised papers are included in this volume.

THE AUTHORS

Juha M. Alho is Professor of Statistics at the University of Joensuu and holds a Ph.D. from Northwestern University. His main research interests are statistical demography and the application of statistics to forestry and sociology. Alho is a Fellow of the American Statistical Association and member of the International Statistical Institute.

Sumon Kumar Bhaumik is a Lecturer of Economics at Brunel University in the U.K. He is also a Research Fellow of William Davidson Institute of Michigan, Ann Arbor and of IZA—Institute for the Study of Labour, Bonn. His research interests include, among other things, intergenerational relationships, impact of socioeconomic contexts on decisions of couples regarding childbirth, and decision-making in the context of labour-force participation.

C.Y. Cyrus Chu is Distinguished Research Fellow, Institute of Economics, Academia Sinica, Taiwan. His research interest covers family economics and economic analysis of law. Right now he is working on a new monograph entitled Understanding Chinese Families, to be finished by 2007.

Philip N. Cohen is an Associate Professor of Sociology at the University of California, Irvine. His research concerns gender, race/ethnic and class inequality, focusing on inequality within and between families across various social contexts, as well as micro–macro linkages in labour market inequality.

Eileen M. Crimmins is the Edna Jones Professor of Gerontology at the Andrus Gerontology Center and Director of the USC/UCLA Center on Biodemography and Population Health, University of Southern California. Her recent research has addressed issues surrounding historical trends in the health and functioning of the older population.

Marcus W. Feldman is Professor of Biological Sciences at Stanford University and Director of Stanford's Morrison Institute for Population and Resource Studies.

Anne H. Gauthier is the Canada Research Chair in Comparative Public Policy in the Department of Sociology at the University of Calgary, Canada. She obtained her doctorate degree from Oxford University. Her research interests include family policy, children's well-being, and the transition to adulthood. She is the author of *The State and the Family: A Comparative Analysis of Family Policies in Industrialized Countries* (Oxford 1996).

Noreen Goldman is Professor of Public and International Affairs at the Woodrow Wilson School and a Research Associate at the Office of Population Research at Princeton

University. Her primary research interests are the social determinants of health, biodemography, and survey design. She has been involved in major data collection efforts pertaining to the social environment and health in Guatemala and Taiwan.

Janet C. Gornick is Professor of Political Science and Sociology at the Graduate Center of the City University of New York, and Professor of Political Science at Baruch College. She is also Associate Director of the Luxembourg Income Study, an international social research institute. Her research primarily concerns cross-national variation in social welfare policy. Professor Gornick is co-author (with Marcia K. Meyers) of *Families That Work*: Policies for Reconciling Parenthood and Employment, published in 2003 by the Russell Sage Foundation. She is currently Guest Editor of two forthcoming issues of the Journal of Comparative Policy Analysis; both issues concern work-family reconciliation policies in cross-national perspective.

Mark D. Hayward is Professor of Sociology and Director of the Population Research Center at the University of Texas. His recent research has focused on health disparities of the older population stemming from race and life-cycle social disadvantage. Dr. Hayward is a long-time collaborator of Dr. Eileen Crimmins.

Xiaoyi Jin is a doctoral student at Population Research Institute, Xi'an Jiaotong University.

Stefan Hrafn Jonsson is a Ph.D. candidate for a dual degree in Sociology and Demography, Pennsylvania State University, USA.

Nan Li is a mathematical and statistical demographer. He has been a professor at Xi'an Jiaotong University in China before becoming a Senior Research Fellow at the Max Planck Institute for Demographic Research in Rostock, Germany. He is currently a Research Scientist in the Program on Population, Policy, and Ageing in the Terry Sanford Institute of Public Policy at Duke University.

Shuzhuo Li is Professor and Director at the Research Institute for Population and Economics at Xi'an Jiaotong University.

I-Fen Lin is Associate Professor of Sociology at Bowling Green State University. Her research investigates parent–child relations across the life span, with particular attention to child support, intergenerational exchanges, and the quality of survey data on separated families.

Yu-Hsuan Lin is Section Chief at Bureau of Health Promotion, Department of Health, Taiwan. Her current research is in the area of public health and epidemiology. She has taken part in several national surveys in Taiwan.

Diane Macunovich is currently Professor of Economics at the University of Redlands, California. She received her Ph.D. in Economics from the University of Southern California in 1989, after receiving her undergraduate degree from M.I.T. in 1966 and then working for 17 years as an economic and demographic consultant in the United States, United Kingdom, Iran, and Canada. She specializes in research on economic and demographic feedback effects: how population growth affects the economy, and how economic conditions affect population growth. Much of her work is presented in her recent book *Birth Quake: The Baby Boom and Its Aftershocks* (University of Chicago Press 2002).

Cem Mete is a senior economist at the World Bank. His current research focuses on the functioning of labour markets in transition economies and the evaluation of health sector reforms in Eastern Europe.

Robert Schoen is Hoffman Professor of Family Sociology and Demography at The Pennsylvania State University, University Park PA (USA). He holds a Ph.D. degree in Demography from the University of California at Berkeley and is an Associate of the Society of Actuaries. His current work explores issues in family and formal demography, including the analysis of population models with changing vital rates and the transition to first cohabitation, first marriage, and first birth.

T. Paul Schultz is Malcolm K. Brachman Professor of Economics and Demography, Yale University, USA.

Judith Treas is Professor of Sociology at the University California, Irvine, and Director of the Graduate Program in Demographic and Social Analysis. A specialist in family and ageing, she is currently engaged in cross-national research on the distribution of household labour and in a study of older people in America's immigrant families.

Shripad Tuljapurkar is Dean and Virginia Morrison Professor of Population Studies and Professor of Biological Sciences, Stanford University, USA.

Reijo Vanne is Chief Economist at the Central Pension Security Institute, Helsinki, Finland.

Alexander A. Weinreb is Lecturer in the Department of Sociology and Anthropology at Hebrew University. His research focuses on data collection methodologies in developing societies, and on the relationship between social networks (e.g., familial, political, social) and demographic behaviour. Most of his work, to date, has focused on fertility and HIV/AIDS in East Africa.

Maxine Weinstein is Distinguished Professor of Population and Health at Georgetown University where she has been a member of the faculty since 1987. She is a demographer whose primary interests lie in exploring the effects of the intersection of biological and behavioural factors on demographic phenomena.

Ruoh-Rong Yu is Associate Research Fellow, Center for Survey Research, Academia Sinica, Taiwan. Her main research interests include family economics, labor economics and survey design.

Zhenmei Zhang is assistant professor in Sociology at Bowling Green State University. Her research interests centre on health disparities arising from the family life course and gender.

INTRODUCTION

ANNE H. GAUTHIER
SHRIPAD TULJAPURKAR
C.Y. CYRUS CHU

Intergenerational transfers are a key determinant of individuals' and families' well-being, and of societies' aggregate level of human and financial capital. Intergenerational transfers of time, money, goods, and attitudes from parents to young children have been shown to contribute to children's educational achievement and employment trajectories in adulthood (Teachman 1987; Rindfuss 1999) and to intergenerational class mobility (Erikson and Goldthorpe 2002). Similarly, intergenerational transfers from grown-up children to elderly parents have been shown to contribute to parents' well-being (Chan 1996). At a macro level, the direction and magnitude of intergenerational transfers have been proposed as the key factor determining fertility transitions and levels (Caldwell 1976). The significance of intergenerational transfers is accentuated in populations undergoing age-structural transitions, which would include much of the world today. Age-structural transitions are driven by declines in fertility and mortality, and by migration, and result in shifts in age structure, kinship structure, and social networks, all of which can affect intergenerational transfers. The macroeconomics of populations undergoing fertility and mortality transitions can be usefully analysed in terms of intergenerational flows (Lee 2003) to illuminate questions ranging from the rapid increase in savings rates during a fertility transition to the sharing of risk and support in ageing societies. Rapid fertility declines have a large effect on dependency ratios both at the family level and at the national level, and can interact with cultural preferences to generate unusual shifts in the sex ratio of populations (see the chapter by Feldman et al., this volume). Rapid fertility declines also interact with traditional patterns of intergenerational transfers to pose challenges to rapidly industrializing countries in the provision of old-age support (Cameron and Cobb-Clark 2002; Zimmer and Kwong 2003).

In recent decades, there have been important theoretical and empirical studies on intergenerational transfers. Sociological and economic studies (e.g., Behrman, Pollack, and Taubman 1995; Harevan 1996; Mason and Tapinos 2000; Ythier, Kolm, and Gerard-Varet 2001) have contributed to our understanding of the complexities, multidimensionalities, and dynamics of intergenerational transfers. Yet, the nature of intergenerational transfers, their sources, direction, motives, mechanisms, and consequences are still not fully understood. As argued by Lee (2002), intergenerational studies are still under-researched.

Anne H. Gauthier et al. (eds.), Allocating Public and Private Resources across Generations, 1–11.

The papers included in this volume make a significant contribution to the understanding of the nature and measurement of transfers, their motives and mechanisms, their macro-level dimensions, and their policy consequences for populations undergoing demographic and economic transitions.

1. The Dimensions and Dynamics of Intergenerational Transfers

The complex nature of intergenerational transfers means that studies have tended to capture only a fraction of all such transfers. For example, studies have captured money transfers or time transfers but rarely both; they have captured transfers from grown-up children to elderly parents or from elderly parents to grown-up children but rarely both; they have captured parent–child transfers, but rarely transfers between siblings, members of the extended family, friends, and neighbours; and they have captured transfers during a very short period of time (usually 12 months) but not during a lifetime. These limitations most often reflect the limitations of the data itself. But they also suggest multidimensionalities that are often not discussed in studies.

In order to prepare the ground for the empirical chapters that follow, we summarize in Table 1 the main dimensions of intergenerational transfers in terms of their types, timings, periodicity, direction of net flow, and donors and recipients.

Table 1. Dimensions of intergenerational transfers.

Dimensions	Categories
Types	Time (including help, services, and visits) Money Goods (including shared space and goods through coresidence)
Timing	Inter-vivos Post-mortem
Periodicity	Regularly One-off Irregularly
Direction	Upward (e.g., from children to parents) Downward (e.g., from parents to children) Lateral (e.g., from siblings to siblings)
Donors and recipients	Children Parents Kins Friends Neighbours Community
Household units	Intrahouseholds Interhouseholds

Source: Adapted from National Research Council (2001), p. 173.

The papers in this volume include innovative approaches to the measurement, description, and dynamic modelling of transfers and their policy implications. Thus, Weinreb considers bidirectional transfers among all potential kin, not only parents and children. Treas and Cohen provide a cross-national comparison of transfers as measured by coresidence and visits. Feldman et al. examine vertical and oblique transfers of cultural norms and their interaction with resource transfers. Chu and Yu take a network approach to explaining transfer patterns in Taiwan. Explicit dynamic models are used by Hayward, Crimmins, and Zhang to study chronic health, and by Schoen and Jonsson to examine fiscal costs and benefits in populations with periodic changes in birth cohort size. Alho and Vanne provide one of the few stochastic projections of age-specific transfers and their attendant fiscal burdens.

An overview of the studies included in this volume in terms of country, dataset, intergenerational transfers, and focus is provided in Table 2 and attests to their international coverage and multidimensionality.

2. Motives, Mechanisms, and Micro-Level Patterns

Following the seminal work by Becker (1974), the motive of altruism has dominated the literature on intergenerational transfers. According to this theoretical perspective, individuals engage in intergenerational transfers because they care about the well-being of the other generations. In contrast, the exchange model suggested by Cox (1987) posits that generations exchange time and goods because both generations perceive benefits from such exchanges. Since these seminal studies, several other motives for intergenerational transfers have been suggested in the literature including the corporate model, insurance, access to credit, comparative advantage, jural obligations, structural model, and "warm glow."[1] It is beyond the scope of this introduction to review in detail these competing theoretical explanations and related studies. In general, the literature testing these different theoretical models has been inconclusive, showing support for the altruistic hypothesis in some studies, and for the exchange hypothesis in others (Cameron and Cobb-Clark 2002).

The paper by Lin et al. in this volume provides, in fact, support for both the altruistic and the reciprocal hypotheses—at least in the context of Taiwan. The paper also finds some evidence for a structural explanation by which parents who have more children are more likely to be recipients of intergenerational transfers. Their work adds to the growing body of evidence for multiple motivations underlying transfers in different countries (Schoeni 1997, Frankenberg, Lillard, and White 2002).

The other main issue that has been raised in the literature is the ways by which reciprocity is enforced in models that are based on the exchange model. For example, studies have raised the question of how can parents, who have invested in their children when they were young,

[1] The literature is inconsistent in the naming of these different motives, and in their number. For example, Altonji, Hayashi, and Kotlikoff (2000) identify six motives, while Lillard and Willis (1997) identify five (with only partial overlap between the two lists).

Table 2. Overview of studies included in this volume.

Panel A: Micro-level studies

Authors	Country of analysis	Data	Types of transfer	Direction of transfers	Reference period for transfers
Weinreb	Malawi	Malawi Family Transfers Project, 1999	Money and goods	Bi-directional transfers between kins	Last 9 months
Chu and Yu	Taiwan	Panel Study of Family Dynamics (PSDF), 1999–2000	Money and assets	Unidirectional transfers from children to parents	Monthly
Lin et al.	Taiwan	Survey of Health and Living Status of the Elderly 1989, 1999	Money	Unidirectional transfers from children to parents	Last year
Bhaumik	Germany	German Socio-Economic Panel (GSOEP) 1996, 1997	Money	Bidirectional transfers between kins (also unrelated persons)	Last year
Treas and Choen	Twenty industrialized countries	International Social Survey Program 1994	Coresidence and visits	Transfers between children and mothers	Last year
Feldman et al.	China	Survey of transmission of son preference and survey of marriage and old-age support	Money and goods	Unilateral transfers from children to parents	Last year

Panel B: Micro–macro and macro-level studies.

Authors	Country of analysis	Data	Focus
Mete and Schultz	Taiwan	Survey of Health and Living Status of the Middle Aged and Elderly, 1989, 1996	Impact of the introduction of national health insurance on labour-force participation
Gornick	Fourteen OECD countries	OECD Social Expenditures Database 1980–1996 and Luxembourg Income Study	Public transfers to elderly and children
Hayward et al.	USA	Health Retirement Survey 1992 and Assets and Health Dynamics Survey 1994	Impact of changes in aggregate educational attainment and mortality on active life expectancy and prevalence of functional problems
Macunovich	USA	State-level cross sections of personal consumption expenditures 1900–1982	Impact of changes in age structure on patterns of consumption and savings
Schoen and Jonsson	n/a	n/a	Impact of fertility changes and transfer systems on dependency ratios
Alho and Vanne	Finland	Official statistics on taxes, transfers, and expenditures	Impact of population ageing on intertemporal public liabilities

n/a: not applicable.

be assured (or can make sure) that their grown-up children will support them when they are old. Several mechanisms have been suggested as ways of enforcing such reciprocity including social norms, peer pressures, trust, and threat of disinheritance. For example, a study of intergenerational transfers by Lee (2000) in South Korea discusses the fact that in order to ensure that children comply with the expected reciprocity, "parents attempt to instill in their children such values as filial piety, respect for elders, and a felt obligation to support parents in old age" (p. 283). Very few papers, however, have empirically tested these mechanisms. In this regard, the paper by Chu and Yu is particularly interesting. The paper examines the fact that a large majority of parents in Taiwan divide their assets prior to their death. Moreover, and somewhat surprisingly, the child-to-parent transfer appears to be higher in families that have divided all their assets inter-vivos than among those who have not divided their assets or who have none. Chu and Yu explain this phenomenon through a theory of social network, which acts as a mechanism to enforce filial reciprocity.

Finally, another issue that has been raised in the literature is the intentional or unintentional nature of intergenerational transfers. Derived from a rational choice perspective, most of the literature has emphasized (or assumed) the intentional nature of transfers. The paper by Bhaumik challenges this assumption by examining the impact of predictable (e.g., marriage) and unpredictable (e.g., illness) events on intergenerational transfers. His analysis, based on the German Socio-Economic Panel Survey, suggests that events, such as marriage, divorce, and childbirth all increase the probability of private transfers and their magnitude.

These papers add useful empirical and theoretical dimensions to the current lively debate about the motives and structure of intergenerational giving. In particular they support the view that the standard economic arguments behind transfers are insufficient for complete understanding (Bawin-Legros and Stassen 2002; Cox and Soldo 2004).

Three other papers included in this volume do micro-level analyses that expose the multi-dimensional nature of transfers. The paper by Weinreb is particularly innovative in considering transfers among all potential kin—and not only between parents and children—and in considering both the giving and receipt of transfers. The paper also innovates in considering not only money transfers, but also goods. The paper shows that in Malawi a wide range of kin may be involved in inter- and intragenerational transfers, and that transfer behaviour is influenced by mortality within one's network. Recent work by Agree et al. (2001) takes a similarly inclusive look at transfers by considering social networks and transfers between multiple individuals or households. They too document complex patterns of transfers among older parents and their children, and suggest that network measures of resource flows would provide new understanding of the nature of social support and its effect on individual welfare.

The paper by Treas and Cohen provides an unusual cross-national dimension. The literature on intergenerational transfers has been overwhelmingly restricted to the analysis of single countries. Treas and Cohen instead examine intergenerational transfers in the form of coresidence and visits for 20 industrialized countries. Results reveal large variations across countries in the prevalence of intergenerational coresidence. Results also suggest that coresidence and visits are not substitutes but are instead complementary forms of intergenerational support.

The paper by Feldman et al. is original both in the scope and measurement of intergenerational transfers. They study cultural transmission between generations, of a preference for sons over daughters and of marriage patterns in rural China. They use a general formulation of transfers, allowing for complex (vertical, lateral, and oblique) cultural transmission; these are similar to the modes of transmission analysed for material resources by Weinreb in his chapter. Feldman et al. relate cultural transmission to the intergenerational transmission of resources, and combine the two types of transfers to examine the dynamics of sex ratio at birth.

3. Macro-Level Transfers and Dynamics

Intergenerational transfers have also important macro-level dimensions. Four such dimensions are addressed in the papers included in this volume: (1) the impact of public programs on micro-level transfers; (2) the nature of welfare state transfers in terms of their age bias or neutrality; (3) the impact of demographic changes on aggregate savings; and (4) the impact of intergenerational transfers on mortality. Each of these dimensions would deserve a long treatment. Here we provide only a brief summary and refer readers to specific chapters for a more detailed discussion.

3.1. THE IMPACT OF PUBLIC PROGRAMS ON MICRO-LEVEL TRANSFERS

One of the questions that has been asked in the literature is whether or not the introduction of public transfers (such as pensions) provides disincentive to savings or transfers, or in other words, whether or not public transfers crowd-out private transfers (Cox and Jimenez 1992). The literature provides mixed results, suggesting that the extent of displacement (or crowding-out) varies depending on the form and type of public and private transfers (Ezemenari 1997). The paper by Mete and Schultz in this volume addresses this issue by examining the impact of the introduction of a public health care insurance in Taiwan in 1995. Results suggest that the introduction of this insurance did not provide disincentives to save for one's future health care needs and to stay in the labour force at older ages. Their results are consistent with recent findings from other countries (Cameron and Cobb-Clark 2002).

3.2. THE NATURE OF WELFARE-STATE TRANSFERS IN TERMS OF THEIR AGE BIAS OR NEUTRALITY

In a widely cited paper, Preston (1984) pointed to the increasing gap in the financial situation of children and elderly in the United States. While poverty among elderly in the United States has declined since the 1960s, the opposite trend has been observed among children. Preston explained this increasing age gap as a result of the electoral power of the growing elderly population, and its ability to influence the adoption of policies that increased its well-being. This hypothesis has been tested cross-nationally in very few studies, and has resulted in mixed findings (see for example Esping-Andersen and Sarasa 2000; Lynch 2001). The paper by Gornick in this volume, which complements the micro-level cross-national comparison by Treas and Cohen in this volume, examines the trends in the allocation of public resources to children and elderly in 14 industrialized countries. Her results suggest that while public expenditures on elderly have risen since the 1980s, so did expenditures on children. The study, thus, raises some doubt as to the negative impact of population ageing on public investment in children. This is a potentially important finding given the concerns raised by Preston.

3.3. THE IMPACT OF DEMOGRAPHIC CHANGES ON AGGREGATE SAVINGS AND PUBLIC LIABILITIES

The phenomenon of age-structural transitions has been the subject of recent studies, including the volume by Tuljapurkar, Pool, and Prachuabmoh (2005, in this series). Demographic changes have in fact wide-ranging consequences, especially on the magnitude of the flow of private and public resources across generations. Lee (2003) shows how these consequences are likely to have shaped human affairs over long time periods, from the early evolution of human life histories to the modern creation of welfare states. Three papers in this volume examine this issue.

The paper by Schoen and Jonsson examines the impact of changing fertility on dependency ratios and on the costs and benefits of retirement systems. They construct a population renewal model that generates persistent cycles in cohort birth sizes, a stylized model for the kinds of birth cohort variation widely observed in the last century. They demonstrate that dependency burdens fluctuate cyclically by a cohort's year of birth, and they show how individuals' gains and losses fare under both defined contribution and defined benefit plans. Their paper is a useful contribution to the basic understanding of pension systems in unstable populations, extending the classic work of Keyfitz (1977).

The paper by Alho and Vanne is a sophisticated and major step forward in the analysis of the impact of population ageing on so-called intertemporal public liabilities, defined as the discounted future entitlements of all current and future generations (minus discounted taxes and public net wealth). Intergenerational transfers are the backbone of all public taxation and expenditure, and the proper accounting of demographic uncertainty is the key to making sound policy (Tuljapurkar 1992). Using the case of Finland, the paper constructs stochastic forecasts of population, taxes, and expenditures by age, and then evaluates the long-run balance of payments across generations. They demonstrate that although the current public wealth will be gradually consumed, "current and past generations are not leaving the future generations an inheritance of debt and misery."

Finally, the paper by Macunovich examines the impact of change in the age structure of the population on patterns of consumption and savings. Her paper is grounded in Easterlin's (2001) important work on the significance of cohort relative sizes and expectations. Using American historical data and a variety of models, her analysis supports the view that changes in the age structure of the population have had a major impact on the U.S. economy. The empirical analysis also suggests that children in high-income areas appear to motivate increased savings on the part of other members either through precautionary motives or to finance bequests.

3.4. THE IMPACT OF INTERGENERATIONAL TRANSFERS ON MORTALITY

As discussed at the beginning of this introduction, the downward intergenerational transfers of public or private resources (from the parent generation to the child generation) constitute a major form of investment in tomorrow's generations. This investment has consequences on the levels of human, financial, and social capital. The paper by Hayward, Crimmins, and Zhang takes an interesting approach in examining the impact of changes in the aggregate educational attainment on the health of generations. Based on a series of simulations involving a multistate life table, the paper reveals the positive impact of social investment programs on the health of the older population, including a decline in the prevalence rate of functional problems and an increase in the active life expectancy.

Overall, this collection of papers contributes to a better understanding of the micro-level dimensions of intergenerational transfers and to their micro–macro connections. By covering countries at different stages of their economic development, this collection of papers also contributes to the still under-analysed area of cross-national variations in intergenerational transfers.

4. The Structure of this Volume

The first part of the volume includes papers that examine intergenerational transfers at the micro level. Within these papers, we start with an in-depth analysis of Malawian transfers (the paper by Weinreb) and then move on to papers that vary in terms of their types of intergenerational transfers and country of analysis.

The second part of the volume starts with an analysis of the impact of public programs on micro-level behaviour (the paper by Mete and Schultz) and then moves to macro-level analyses. Based on various datasets and various analytical techniques, the papers unravel the complex links between age-structural transitions, transfers, and countries' well-being.

References

Agree, E.M., Biddlecom, A.E., Valente, T.W., Chang, Ming-cheng. (2001), "Social network measures of parent–child exchange: The applications in Taiwan and the Philippines." *Connections,* 24(2):59–75.

Bawin-Legros, B. and Stassen, J.F. (2002), "Intergenerational solidarity: Between the family and the state." *Current Sociology,* 50(2):243–262.

Becker, G.S. (1974), "A theory of social interactions." *Journal of Political Economy*, 82:1063–1093.

Behrman, J., Pollack, R. and Taubman, P. (1995), *From Parent to Child: Intrahousehold Allocations and Intergenerational Relations in The United States*. University of Chicago Press.

Cameron, L.A., Cobb-Clark, D. (2002), "Old-age support in developing countries: Labor supply, intergenerational transfers and living arrangements." Australia National University. On-line: http://econ.rsss.anu.edu/~dclarck/docs/papers/indo_draft19.pdf

Caldwell, J.C. (1976), "Toward a restatement of demographic transition theory." Population and development review, reprinted as Chapter 4 of John Caldwell (1982) *Theory of Fertility Decline*, Academic Press: New York. pp. 113–180.

Chan, A. (1996), "How do parents and children help one another?—Socioeconomic determinants of intergenerational transfers in peninsular Malaysia." *Journal of Population,* 2:1.

Cox, D. (1987), "Motives for private income transfers." *Journal of Political Economy*, 95:508–546.

Cox, D. and Jimenez, E. (1992), "Private transfers and public policy in developing countries: A case study for Peru." *World Bank Economic Review*, 6(1):155–169.

Cox, D. and Soldo, B.J. (2004), "Motivation for money and care that adult children provide for parents: Evidence from 'point-blank' survey questions." *Center for Retirement Research Working Papers* http://escholarship.bc.edu/retirement papers/11

Easterlin, R. (2001), "Income and happiness: towards a unified theory." *The Economic Journal*, 111:465–484.

Erikson, R., Goldthorpe, J.H. (2002), "Intergenerational inequality: A sociological perspective." *Journal of Economic Perspectives*, 16:31–44.

Esping-Andersen, G. and Sarasa, S. (2002), "The generational conflict reconsidered." *Journal of European Social Policy*, 12(1):5–21.

Ezemenari, K. (1997), "The link between public and private interhousehold transfers: Implications for the design of safety net programs in developing countries." *American Journal of Agricultural Economics*, 79(2): 666.

Frankenberg, E., Lillard, L. and Willis, R.J. (2002), "Patterns of intergenerational transfers in southeast Asia." *Journal of Marriage and Family*, 64:627.

Hareven, T.K. (ed.) (1996), *Aging and generational relations over the life course: A historical and cross-cultural perspective*. W. de Gruyter: Berlin; New York.

Lee, R. (2002), "Intergenerational research". Paper for the National Institute of Child Health and Human Development. On-line: http://www.nichd.nih.gov/cpr/dbs/sp.lee1.htm

Lee, R. (2003), "Demographic change, welfare, and intergenerational transfers: a global overview," *Genus*, LIX:(3–4) (July–December): 43–70.

Lee, Y.-J. (2000), "Support between rural parents and migrant children in a rapidly industrializing society: South Korea." In: *Sharing the Wealth: Demographic Change and Economic Transfers Between Generations,* E. Mason and G. Tapinos, eds. pp. 292–306. Oxford University Press.

Lynch, J. (2001), "The age-orientation of social policy regimes in OECD countries." *Journal of Social Policy*, 30(3):411–436.

Mason, E. and Tapinos, G. (eds.) (2000), *Sharing the Wealth: Demographic Change and Economic Transfers Between Generations*. Oxford University Press.

National Research Council (2001), *Preparing for an Aging World: The Case for Cross-National Research*. National Academy Press, Washington D.C.

Preston, S.H. (1984), "Children and the elderly: Divergent paths for America's dependents." *Demography*, 21:435–457.

Rindfuss, R. (1999), "Family inputs and the transition to adulthood." In: *Transitions to Adulthood in a Changing Economy: No Work, No Family, No Future*? A. Booth, A.C. Crouter, and M.J. Shanahan, eds. Praeger: Westport, CN.

Schoeni, R.F. (1997), "Private interhousehold transfers of money and time: New empirical evidence." *Review of Income and Wealth*, 43(4):423–448.

Teachman, D.D. (1987), "Family background, educational resources, and education attainment." *American Sociological Review,* 52:548–557.

Tuljapurkar, S. (1992), "Stochastic population forecasts and their uses." *International Journal of Forecasting*, 8:285–391.

Tuljapurkar, S., Pool, I. and Prachuabmob, V. (eds.). (2005). *Riding the Age Waves; Population Resources and Development*. Vol. 1. Springer: Dordrecht.

Ythier, J.M., Kolm, S.C. and Gerard-Varet, L.A. (eds.) (2001), *The Economics of Reciprocity, Giving and Altruism*. Palgrave Macmillan: UK.

Zimmer, Z. and Kwong, J. (2003), "Family size and support of older adults in rural China: Current effects and future implications." *Demography*, 40:23–44.

CHAPTER 1. SUBSTITUTION AND SUBSTITUTABILITY: THE EFFECTS OF KIN AVAILABILITY ON INTERGENERATIONAL TRANSFERS IN MALAWI

ALEXANDER A. WEINREB
Department of Sociology and Anthropology, Hebrew University

1. Introduction

Although increasing analytic attention has been focused on intergenerational support structures of late, little of that attention has, thus far, been focused on the lateral components of those structures, by which I refer to relations among uncles and nieces, aunts and nephews, and so on. This is problematic from a broad sociological perspective for two main reasons. The first is that a relatively narrow focus on vertical intergenerational links overlooks the extent to which these microstructures are embedded in multidirectional and multigenerational support networks involving the elderly's siblings, their nephews and nieces, other relatives, neighbours, friends, fictive kin, and so on. And the second is that, while the "structures of jural obligations" (Holy 1976:108)[1] which underlie intergenerational support structures may be associated with kin proximity measured in terms of closeness of blood ties—thus, for example, the common assumption that a given person has a greater obligation to provide assistance to his parents than to his uncles, aunts, or some unrelated individual—there is considerable ethnographic evidence that across many societies this is not the case; that individuals may have, and often do have, more intensive structured ties with second-order blood relations like uncles and aunts than with first-order blood relations like fathers; or indeed, that they may have more structured ties with non-kin (e.g., Whyte 1943; Wilmott and Young 1962; Stack 1974; Fox and Fox 1984; Lewis 1994; Griffiths 1997; Meriwether 1999; Feinberg and Ottenheimer 2002). In short, overlooking lateral intergenerational transfers relations implicitly privileges biology over culture and also ignores a key premise of multiple theoretical approaches to the analysis of exchange relations: the idea that these relations are both emergent product and producer of normative systems of social relations (e.g., Mauss 1925, 1990; Homans 1958; Blau 1964, 1986).

I have dealt with the first point—the effects of the relatively narrow focus on vertical intergenerational links—elsewhere, both by discussing the embeddedness of parent–child relations in broader networks, and by comparing the intensity of vertical and lateral

[1] Holy refers to a trend within structural–functional anthropology that would "associate corporate groups with systems of activities" (Holy 1976:108). What he refers to as "structures of jural obligations" is therefore a type of normative map for intrafamilial resource flows that shape actual transfer behaviours.

Anne H. Gauthier et al. (eds.), Allocating Public and Private Resources across Generations, 13–38.

intergenerational exchanges (Weinreb 2002). In this paper, I begin to address the second point. Specifically, I explore how individuals choose to shape their intergenerational support network out of a universe of people normatively defined as kin—a "potential kin network"—where prior and largely exogenous demographic events associated with both older generations' fertility and peers' migration and mortality have shaped the size and age structure of that network. I evaluate people's choices by looking for signs of apparent "substitution" of one type of kin for another in kin-specific transfer systems, as well as for signs of "substitutability" and "complementarity." Substitution refers to the extent to which individuals increasingly attempt to forge ties with existing kin in the absence of others. Substitutability refers to the extent to which these attempts are reciprocated. Complementarity is the inverse of substitution; it refers to the extent to which individuals reduce their transfers to others (I define these terms more systematically below).

These questions are of particular interest to social demographers since they deal first and foremost with the interaction between normative and demographic structures. Thus, at the microlevel they ask how individuals tend to structure their exchange networks in response to particular demographic constraints such as the lack of a father, maternal uncle, and so on. And at the macrolevel they ask how these aggregated individual actions signal emergent structures of jural obligations that are partly the product of these demographic constraints.

These are, of course, expansive questions, too expansive to be dealt with adequately in a single paper. But I pose them before fixing on more modest goals in order to both identify the ultimate empirical prize of this body of research—the identification of causal loops that bind demographic and cultural change—as well as, more immediately, to set a tone for this paper, which is, I think, largely empirical in content but speculative in overall design.

The source of this empirical *versus* speculative tension lies in both the research question, which is relatively new—to the best of my knowledge, no-one has yet explored these types of substitution effects in terms of a wide kin network—and in my general method. The latter is reflected in the two-section structure of the paper. In the first, I briefly and selectively review aspects of family theory that allow me to frame the notion of kin construction strategies in a way that I think is both credible and empirically useful. In the second section, I apply this framework to pilot data from Malawi in order to explore gross variation along multiple microlevel intergenerational dyads within a given kin network.

2. Theoretical Background

To facilitate the discussion, I introduce a visual schema in Figure 1. It portrays the key *potential* intergenerational relations associated with a given household, where the latter is, for conceptual ease, defined as a household constituted by a monogamous husband and wife.[2] Assuming that both the husband and wife have a full complement of parents

---

[2] I assume a simple ideal-type kin structure in which, for example, there are no step-parents or other fictive kin. It should be stressed, however, that the overall framework does not require such restrictions since, as described below, it is based on observed transfer patterns in a family network, where the latter can include both biological, affinal, and fictive kin.

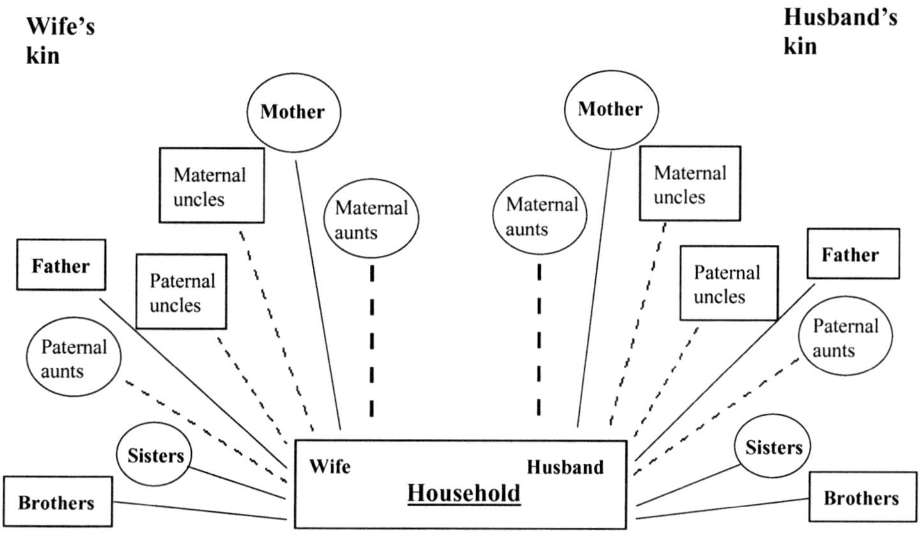

Figure 1. Potential intergenerational relational dyads in Malawi.

and uncles and aunts—a joint function of their grandparents' fertility and the cumulative mortality across their parents' generation's lifespan—decisions need to be taken about transfers with twelve types of relatives: four types of parents and eight types of uncles and aunts, divided equally among the husband's and wife's paternal and maternal uncles and aunts. In addition, decisions need to be taken about transfers with each set of brothers and sisters, each of whom is also embedded in a similar type of intergenerational network.

2.1. SPECIFYING THE PROBLEM

As implied above, the main problem with focusing solely on vertical intergenerational transfers is that, by doing so we ignore other potentially important intergenerational transfer routes, especially, as portrayed in Figure 1, transfers between uncles and aunts on the one hand, and nephews and nieces on the other. This in turn places us at odds with two observationally driven developments in family theory, especially as constructed by anthropologists.

First, following Schneider (1968), there has been an increasing acknowledgement of the flexibility of kinship terms, and an assertion of their importance as systems of symbols rather than as determinants of interactional patterns. Because of such assertions, researchers have (a) tended to maintain that in any given society there is only a minimal equation between structures of jural obligations and specific types of family relationships; and (b) they have also largely dissociated kinship terminology from deterministic patterns of social interaction, including transfers (see Peletz 1995 for a review). In the context of Figure 1, this means that an individual's transfer relationship with primary kin such as parents may, in any given society, be less important than those with secondary kin such as a type of uncle or aunt. Similarly, transfer relationships may be constructed with kin irrespective of the normative descent system (e.g., patri- or matrilineality) and post-marital residential rules (e.g., patri- or matrilocality) that predominate in a given context. A whole series of contingent factors (e.g., gender, residential proximity, wealth, likability, compatibility of forms of capital among the kin, rural or urban residence) may affect actual transfer relationships much more than such normative rules and preferences.

The second relevant development in family theory parallels the first. It is a product of qualitative and ethnographic studies from multiple contexts over several decades. These have shown that individuals forge and maintain intensive exchange relationships with both proximate and nonproximate kin in addition to, or as a substitute for, relationships with their parents. In both developed and developing world contexts, these lateral transfer patterns have been observed most frequently among the relatively impoverished. They are a staple in foundational urban ethnographies (e.g., Whyte 1943; Wilmott and Young 1962; Stack 1974). They are also implicit in every study of kinship, incorporating both rules and preferences related to marriage, remittances and other material resource flows, effects on employment opportunities, and so on (e.g., Grieco 1987; Massey et al. 1993).

In an earlier paper, I have attempted to evaluate some of the these ideas empirically by (a), as noted above, comparing the intensity of transfer relationships between prime-aged respondents and their older kin depicted in Figure 1; and (b) by exploring the effects of some of the contingent factors on estimated transfers (I discuss the data with which these estimations were conducted below). Results have, in general, supported these two developments (i.e., the increasing acknowledgement of the relative flexibility of kinship terms, and the observation that individuals forge and maintain intensive exchange relationships with a wide set of family members). Thus, lateral intergenerational transfer relationships with individual uncles and aunts can be as financially valuable as vertical relations with parents and, in the aggregate—that is, with grouped types of uncles and aunts—more valuable. Moreover, though this varies somewhat by ethnic group, factors unrelated to kin proximity such as the relative's residential, gender and marital status, tend to be more strongly associated with intergenerational transfer relationships than kin proximity itself, where the latter refers to the strength of the blood-tie (Weinreb 2002).

2.2. SYSTEMIC REGULARITIES AND PREFERRED TRANSFER ROUTES

Both these developments in family theory and prior work in Malawi serve as a springboard for a discussion of substitution and related effects for two reasons. The first is that they emphasize the constructed nature of kin networks, constructed, that is, within a universe of kin (wherein I assume that the costs and dangers of constructing such ties are much less than they would be among strangers).[3] And the second reason is that although they emphasize the importance of a social constructivist approach, they also imply that transfer systems have systemic regularities.

---

[3] In general I would assert that the costs are primarily moral—"why are you going outside the family?"—and the dangers financial—"what's to keep him from running off with our money?"—though obviously this reading could be complexified.

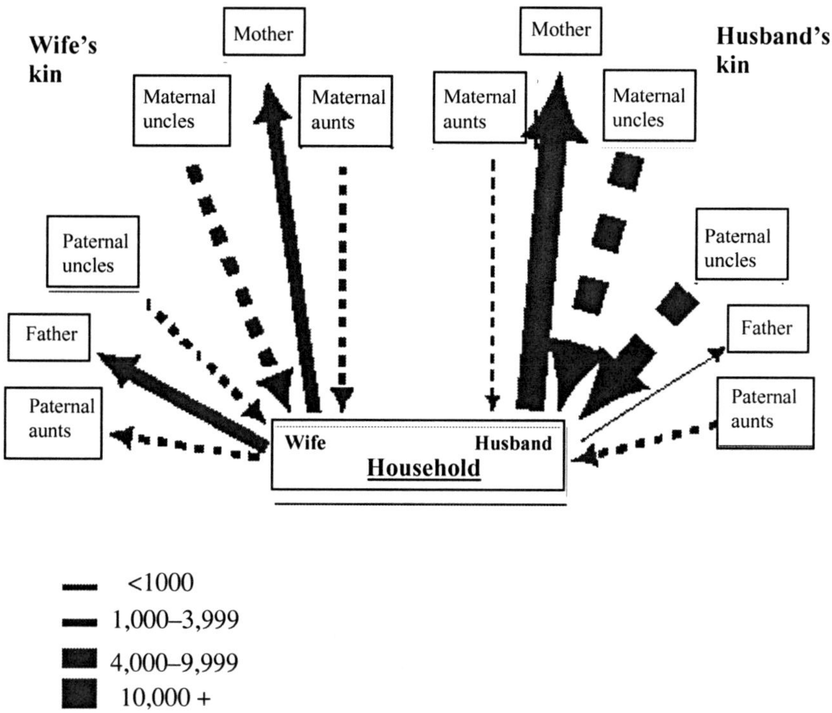

Figure 2. Sum of net transfers between respondent and kin, by type of kin, and direction and value of transfer.

In that prior research, for example, I distinguished between four directional types of transfer relationship that can exist between a respondent and kin: a zero transfer relationship, a unilateral transfer from a respondent to kin, a unilateral transfer from kin to the respondent, and an exchange of resources. Summing the value of transfers by the type of relative across these four types of possible transfers, then comparing the sum across all types of kin, provides an indicator of gross transfers by relational dyad. Figure 2 maps the product of this summation based on estimations (for a single period) reported in Weinreb (2002). It shows that, from a household's perspective there are net flows outwards to all four parents, and net flows inwards from all four uncles. Aunts are also primarily givers, though not on the same scale as the uncles.

Figure 2 serves as an empirical map of vertical and lateral intergenerational transfers since it implies that different types of older kin appear to play different roles in the overall transfer network (implicit in this is also an implication that shifts from being a net giver or receiver are specific to both the particular type of kin relation and to life-cycle). Parents, especially mothers, are the most frequent beneficiaries of prime-aged adult respondent's transfer decisions; and uncles, as a collective, are the most frequent donors to those same prime-aged adults. This does not mean, of course, that these types of kin cannot play other roles: that an elderly mother, for example, will not be a net donor to her prime-aged children rather than a net recipient of their transfers. But the systemic regularities identified by the distributions in Figure 2 signal preferred transfer routes which may, in turn, signal normative transfer rules (e.g., one such rule based on data presented in Figure 2 could assert that "prime-aged adults should give most to their mothers, but parents should give generously to nephews and nieces"). This increases the relevance of questions about substitution and related effects since it asks how people manipulate ties with members of their network when, for one reason or another, that network lacks the person to whom they ideally ought to transfer resources, and/or the person from whom they can expect resources.

2.3. TYPES OF SUBSTITUTION, SUBSTITUTABILITY, AND COMPLEMENTARITY, AND THEIR ANALYTIC INDICATORS

The core notion underlying kin substitution, substitutability, and complementarity—that individuals increase or reduce their relations with existing kin in order to minimize the deleterious effects of a dearth of other types of kin—can be disaggregated along two main dimensions.

The first disaggregation refers to the fact that data on transfers between a respondent (R) and his kin (K) reflect the decisions of *two sets of actors*, each of who's perception of relations between the other and a third party of unavailable kin (Q) may be different. From R's perspective, for example, K, in this case a paternal uncle, may be a useful substitute to Q, a recently deceased maternal uncle. But from K's perspective, Q may have been an affinal relation with whom he had limited contact (assuming that K and Q are not related in some other way). In short, from R's perspective K and Q may be substitutes, indicated by R's attempt to increase transfer relations with K in the absence of Q. But from K's perspective, R and Q are, if anything, complements, since the absence of Q in R's immediate network reduces the attractiveness of K investing in a transfer relationship with R.

Table 1 attempts to identify the six possible changes in transfer relationship between R and K in the absence of Q in terms that are consistent with the dual perspectives of each of the actors in the transfer dyad. It also defines the analytic indicators that will be used in the following analysis, and which are derived from this disaggregation. It should be recognized that the six possible changes also follow from the distinction between unilateral and bilateral transfers made above. Thus, for each of the three possible transfer directions—unilateral transfers from R to K, from K to R, and bilateral—there may have been either an increase or decrease in transfers in response to the absence of Q.

The key premise underlying the identification of analytic indicators in Table 1— marked in the final two columns from R's and K's perspective, respectively—is that individuals' investments in transfers and exchanges, measured jointly by both the likelihood of making an investment and by its value, index the value they assign to a given relationship (I describe the relative benefits of looking at likelihood and value below). This is an extremely simple behavioural premise. It is also one that, in one form or another, is deeply rooted in social science, underlying both the Maussian and social exchange theories mentioned above, neo-Durkheimian approaches which focus on material transfers as a ceremonial focus of

Table 1. Six possible changes in transfer relationship between respondent R and kin K in the absence of deceased kin Q, and their respective analytic labels.

No.	Change in transfer relationships	Description	Relationship between K and Q from R's perspective	Relationship between R and Q from K's perspective
1.	Increase in R→K	R is *more* interested in forging ties; no change in K's behaviour	Substitutes	No relationship between R and Q
2.	Decrease R→K	R is *less* interested in forging ties; no change in K's behaviour	Complements	No relationship between R and Q
3.	Increase R↔K	Both R and K are *more* interested in forging ties	Substitutable	Substitutable
4.	Decrease R↔K	Both R and K are *less* interested in forging ties	Reduced complementarity	Reduced complementarity
5.	Increase in K→R	K is *more* interested in forging ties; no change in R's behaviour	No relationship between K and Q	Substitutes
6.	Decrease K→R	K is *less* interested in forging ties; no change in R's behaviour	No relationship between K and Q	Complements

emotional exchanges (e.g., Collins 2004), quasi-ethical microeconomic approaches, and so on. I do not intend to discuss linkages to these approaches any further here. But it is important to at least briefly note their relationship to the current framework.

Table 1 suggests the following with respect to the identification of substitution and other strategies in response to kin availability: where R has heightened transfers to K in the absence of Q, we can assume a greater motivation to build a transfer relationship with K (case #1 in Table 1). *Ceteris paribus*, the desire to substitute one kin for another is a likely origin of this greater motivation. Diminished transfers in this situation, in contrast, reflect R's view that K and Q are complements, as per the situation described above (case #2 in Table 1). Similarly, where both R and K have heightened transfers with each other in the absence of Q, this indicates a reciprocally augmented motivation to use the other as a substitute for Q. This is what I refer to as "substitutability" (case #3 in Table 1). Diminished transfers in this situation imply an equally reciprocated perspective in which both R and K think of the other as a complement to Q (case #4 in Table 1). Finally, where R *receives* either heightened or diminished transfers from K in the absence of Q, this tells us nothing whatsoever about R's motivation, for here R is simply the subject of K's action (cases #5 and 6 in Table 1).

The second disaggregation refers to group-level patterns that can be derived from Figures 1 and 2 and then applied to any of the six possible changes described in Table 1. Specifically, it refers to the extent to which, within a given person's natal kin network, these changes in transfer behaviour occur within and across lineages (e.g., paternal *versus* maternal), within and across generations (e.g., from siblings to parents), and within and across gender (e.g., from paternal uncles to paternal aunts). Similarly, to these subcategories of lineage, generation and gender, we can also add substitution of affinal for natal kin. For example, to what extent do women who lack certain kin in their natal network attempt to fill those relational gaps by forging ties with their husband's kin?

It is also useful to identify three conditioning effects on these types of substitution. The first deals with the source of lack of kin. Simply, in any given case, a person may not have a given type of kin D for one of two main reasons. Either Q never existed in that kin network—e.g., a maternal grandmother never gave birth to sons so R has never had any maternal uncles—or else Q was lost to mortality. (One can expand these categories to take into account symbolic or "social" deaths such as occur after a serious family feud. In the first case, an individual would never have been made aware of the kin by his elders, and in the second he himself would have voluntarily ceased to map them as part of the family.) It is reasonable to suggest that these sources of nonavailability of kin should have distinct effects on substitution strategies since, in the former case, the person's kin transfer strategies have never had to take account of the presence of that type of kin, whereas in the latter, clearly, this is not the case.

The second of the conditioning effects refers to the type of transfer. It may be the case, for example, that R may consider K a substitute for Q with respect to, say, monetary transfers, but not with respect to the proffering of nonmonetary assistance. I would expect this conditioning to be especially relevant to transfer substitutions where R and K are of different genders, or vary in wealth, since such kin may vary in their access to, and control over, monetary resources.

The third conditioning effect refers to the relations between R and K where, for example, multiple types of kin are not available in the network (e.g., Q_1, Q_2, and Q_3). This clearly reduces the number of choices available to R with respect to ways in which to structure the transfer network, at least to the extent that R wants to stay within a kin network.

3. Data and Methods

The analysis proceeds in two main sections. The first is a necessary detour: I describe the research setting, the data, the estimation procedures and some data-related issues specific to this type of analysis. I then proceed to a description and discussion of two sets of results. The first explores the effects of differential kin availability on the relative likelihood of being in one of the six types of transfer relationships delineated above. And the second explores the effects of differential kin availability on the total net value of transfers exchanged along given relational dyads.

3.1. RESEARCH SETTING

As mentioned above, I use data from Malawi, a relatively small sub-Saharan African country with a population of almost 11 million, about 86 percent of whom reside in rural areas (World Bank 2001; United Nations 2001). Malawi's economic, demographic, health and political profile make it well suited for a study of kin transfer systems. First, it is a poor country, even by African standards: its GNI per capita is $190 USD, in comparison to a sub-Saharan mean of $480 (World Bank 2001). Second, there are very limited market- and policy-based alternatives to kin support networks. While there has been no prior large-scale systematic study of these networks, missionaries, ethnographers and more contemporary qualitative researchers have confirmed their importance in day-to-day family interactions across all the major ethnic groups (e.g., Frazer 1914; Johnson 1922; Coudenhove 1926; Young 1932, 1970; Ntara 1949, 1973; Mitchell 1956, 1962; Hirschmann 1990; Mtika 2001). Third, a series of long-term changes in Malawi have been threatening to compromise the capacities of those kin networks by reducing the proximity of family members or else undermining individuals' motivations to invest in and maintain them (e.g., Malawi 1998; Mtika 2001). These include high labour migration rates, both rural-to-rural, rural-to-urban, and from Malawi to South Africa (Segal 1985; Hirschmann 1990; Kalipeni 1996), and increasing levels of education (adult literacy rates, 54 percent for women and 75 percent for men, are slightly highly than the average for sub-Saharan Africa (UNESCO 2005)—education, by allowing for greater social mobility, is said to reduce the motivations to remain entrenched in transfer networks (Caldwell 1982), although recent school enrollment has increased faster than the sub-Saharan African mean [World Bank 2001]); they also include, most recently and destructively, AIDS. HIV/AIDS prevalence among adults is of the order of 14 percent (Republic of Malawi 2005).

More generally, Malawi has a relatively high dependency ratio, consistent with both high, albeit falling, fertility of 6.7 children per woman, and with the impact of AIDS on the middle generation of prime-aged adults. Although there has been a threefold increase in adult mortality at ages 20–59 since the 1980s (Doctor and Weinreb 2005), adult mortality at ages above 60 has continued to fall (Doctor 2001). Thus, roughly 5 percent of the total population is over 60, and that proportion is increasing rapidly (World Bank 2001; United Nations 2001).

3.2. DATA

The specific data which I use come from the Malawi Family Transfers Project (FTP), a pilot project fielded in three rural areas between June and August, 1999. The FTP drew on a cluster sample developed the previous year for the first wave of the Malawi Diffusion and Ideational Change Project (MDIC), an ongoing social networks study focused on AIDS and Family Planning related behaviour. It targeted about 50 percent of those listed in the original MDIC sample list, yielding interviews with 723 ever-married women aged less than 50 and 532 of their husbands, an 84.5 and 77.9 percent response rate, respectively: the lower response rate for men reflects both higher male labour migration and separation/divorce—in which case women, but not their ex-spouses, were interviewed; the lower number of men also reflects polygyny.

The sample is drawn more or less equally from three areas—Balaka, Mchinji, and Rumphi districts—which are, respectively, in Southern, Central and Northern Region (the ethnic composition of the dataset is therefore more or less equally divided between Yao, Tumbuka, and Chewa). In socioeconomic terms, the three areas are similar and broadly representative of rural areas in Malawi in general (for example, on these dimensions they are equivalent to areas described in Hirschmann 1990 and Mtika 2000). In particular, formal sector employment opportunities are equally limited, and almost all the families in the areas therefore make their living through petty trade, usually subsidized by the production of staples and cash crops on small patches of land, by remittances from family members, and especially at harvest times, by selling labour to local farmers who can afford to employ others, a system known as "*ganyu* labour." In addition, village-level data collected in combination with the main survey data indicate that there is also little variation between the areas in terms of access to commercial activities and institutions (daily markets, supermarkets, and banks), to health providers (hospitals, health centers, maternal, and child health clinics), and to other state-related offices (post offices and police stations). Similarly, almost every village in each of the sampled areas is within a few kilometres of a road on which public transport—publicly owned buses as well as privately owned minivans and pick-up trucks—is available to take residents to the nearest large town. In fact, the most notable difference between the areas is in the ethnicity of the dominant group. (I have addressed ethnic differences in the structure of transfer networks elsewhere [Weinreb 2002].)

Respondents were asked three types of questions in the survey: questions about themselves, questions about kin survivorship with respect to their father, mother, paternal uncles, paternal aunts, maternal uncles, maternal aunts, and brothers and sisters, then questions specific to each of the surviving relatives.[4] These latter included questions about these relatives' basic sociodemographic characteristics and about two possible types of transfers made between the respondent and that relative. The first referred to material goods and money that (a) could be assigned a monetary value, and (b) had occurred "since the beginning of the last growing season," roughly 9 months before data collection, and the second referred to in-kind assistance over the last month, which we defined as the gift (or receipt) of time. Since the latter was asked differently with respect to parents and uncles and aunts, this paper explores the first of these only.[5] Also note that our specification of different reference periods reflect the fact that while many types of monetary transfers are infrequent, informal assistance is often provided daily, or at least a few times per week. Lengthening the reference period for the latter would have diluted the intensity of the informal assistance.[6]

---

[4] Questions were also asked about the respondents' siblings. These data are discussed in Weinreb, Behrman, and Mtika (2001).

[5] Note that most of the material transfers were not monetary. The most common examples were assorted quantities of agricultural products like maize (shelled and ground), groundnuts (shelled and unshelled), rice, and cassava, and other goods, especially clothes, shoes, and soap. Where the respondent did not report the cost of these gifts, a monetary value was assigned based on the reported quantity of goods and local market prices. Where the respondent did not specify the amount of a monetary gift (in less than 10 percent of cases), those gifts were assigned the value of 50 Malawi Kwacha, equivalent to the value of monetary gifts whose value had been specified by the respondent. It should also be noted that "since the beginning of the last growing season" was specified because prior research by one of the investigators had indicated that this reference period would be more easily understood than the more open-ended "since this time last year."

[6] I have described the distribution of both the likelihood and value of material transfers and the likelihood and time-use values of informal assistance at some length in Weinreb (2002), including a description of important ethnic variation in these patterns. As mentioned earlier, the sum of net transfers depicted in Figure 2 derives from that description. For the purposes of this paper, all that needs to be derived from that discussion is that: (1) uncles and aunts appear to play an important role in the intergenerational transfer system; (2) there are some systemic regularities in these arrangements; and (3) among these regularities is the fact that men play larger donor roles—according to both men's and women's reports—in the transfer of monetary goods.

Table 2. Number of surviving kin (and transfer dyads) associated with the 494 households (HH's) in which both the husband and wife were interviewed, and percentage of HH's without any kin, by type of kin.

Type of kin	Husband interviewed	Percent of HH's with zero K
Wife's kin		
Father	277	34.3
Mother	353	20.4
Paternal uncles	540	33.6
Paternal aunts	556	29.0
Maternal uncles	616	22.9
Maternal aunts	665	22.5
Brothers	*n/a*	12.7
Sisters	*n/a*	14.3
Husband's kin		
Father	223	38.7
Mother	281	27.4
Paternal uncles	538	31.4
Paternal aunts	489	31.2
Maternal uncles	623	23.5
Maternal aunts	552	30.2
Brothers	*n/a*	14.0
Sisters	*n/a*	14.1
N = Number of surviving kin in older generation	5,713	

Prior to analysis I restructured the data in two ways. First, I merged the women's and men's data in order to create a "household" data file in which, from the woman's perspective (at least where the husband was successfully interviewed so we were able to collect information on his kin), there was information on transfers to both natal and affinal kin. Information was collected from both the wife and her husband in 494 out of the 723 cases. Second, I then reshaped this household dataset to nest data on each relative/transfer dyad within the data on that particular relative. As presented in Table 2, this procedure yielded data on 5,713 elderly natal and affinal relatives of the interviewed wives and their husbands—the same set of kin depicted in Figure 1—or an average of 11.6 intergenerational dyads per household.[7]

[7] A further 203 women reported themselves as currently married: but we could not interview their husbands (in most cases because they were labour migrants); and 26 were currently unmarried. Although previous research in both Malawi and elsewhere in sub-Saharan Africa has shown that women-headed households have different income bases and subsistence practices (e.g., Berheid and Segal 1994; Brouwer, Hoorweg and van Liere 1997), which has implications for intergenerational transfer relationships, this paper leaves comparisons between these male- and female-headed households to a later paper.

The data quality aspects of these reported survival data are discussed in Weinreb (2002). In brief, reported parents' survival of both fathers and mothers by both men and women is consistent with the UN model life table (West), with life expectancy set to 50. Reported survival of uncles and aunts is slightly less robust. In particular, as implied in the second column of Table 2, male respondents appear to have underreported maternal aunts; and, to a lesser extent, women respondents may also have slightly underreported paternal aunts' survival. These differences in reported survivorship are related to ethnicity—specifically, there are relatively few maternal aunts among the patrilineal Tumbuka and relatively few paternal uncles among the matrilineal Yao (Weinreb 2002, pp. 111–113). Insofar as the forthcoming analysis does not explore ethnic-specific substitution patterns, this apparent reporting bias is of minimal concern.

3.3. ESTIMATIONS

Table 1 described the six analytic indicators used in this analysis. I now discuss the procedures used to estimate those relations.

Since, as noted above, it is important to explore these effects in relation to both (a) the relative likelihood of having one of three types of nonzero transfer relationships with a given type of kin (i.e., unilateral transfers from R to K, unilateral from K to R, and bilateral transfers), and (b) the net value of all transfers with that type of kin, I estimated two series of models, each series involving 16 separate estimations (i.e., one for each type of kin). The first series specified multinomial logit models that identified covariates of the directions of transfer relationships seen in Table 3 relative to "no reported transfers" (i.e., unilateral transfers from the respondent to kin, unilateral from kin to respondent, and bilateral transfer). The second series of models specified censored regressions of the net value of a transfer between a respondent and his/her parent, uncle, or aunt. Censoring was defined as a zero-transfer relationship. More formally, a linear approximation of the estimated models is

$$T_{\mathrm{rk}} = aR_{\mathrm{r}} + bP_{\mathrm{rk}} + cK_{\mathrm{rk}} + dK_{\mathrm{rk}}Q_{\mathrm{rk}} + u_{\mathrm{rk}},$$

where T_{rk} is the vector of transfers between the respondent and each member of the kin network—taking the form of a 4-category variable in the logits and a continuous variable in the regressions—R_{r} refers to a matrix of observed respondent characteristics; P_{rk} refers to a kin member's residential proximity to the respondent's household; K_{rk} to a series of dummy identifiers for the 12 different types of older kin; Q_{rk} refers to zero surviving kin of a given type k; u_{rk} is a vector of stochastic shocks; and four coefficient matrices—a, b, c, and, d—are estimated in the model.

Within each of these two series of models, discrete effects of the presence or absence of one of the 16 types of kin Q in a given household's kin network on a husband's and wife's relations with other kin K (i.e., where $K \neq Q$) was estimated. The specific estimate appeared in an interaction term measured in the coefficient matrix d, because, where a respondent reported that they had no surviving relatives in those categories, Q_{rk} was coded as "1." (Column 2 in Table 2 shows the proportion of respondents without kin in these

Table 3. Cell-sizes for multinomial logit models used to estimate substitution and related effects, as presented in Appendix Tables 1a and b.

	No transfers reported: Reference category	Transfers reported and *has* surviving kin Q; therefore used to estimate cK_{rk} main effect in model			Transfers reported but *does not have* surviving kin Q; therefore used to estimate $dK_{rk}Q_{rk}$ interaction effect in model			Total
		R→K	K→R	R↔K	R→K	K→R	R↔K	
	(1)	(2)	(3)	(4)	(5)	(6)	(7)	(8)
Wife's kin								
Father	85	35	41	50	18	22	26	277
Mother	71	70	44	111	18	11	28	353
Paternal uncles	405	38	30	21	19	15	11	540
Paternal aunts	389	54	26	39	22	10	16	556
Maternal uncles	407	48	73	39	14	22	12	616
Maternal aunts	438	63	63	49	18	18	14	665
Husband's kin								
Father	51	45	9	51	29	6	32	223
Mother	52	70	9	87	27	3	33	281
Paternal uncles	345	59	41	33	27	19	15	538
Paternal aunts	294	67	28	39	30	13	18	489
Maternal uncles	383	65	63	56	20	19	17	623
Maternal aunts	336	64	40	47	28	17	20	552

16 groups). Owing to the large number of models, I restrict the tabular presentation of results—in Tables A.1, A.2, and A.3—to coefficient matrix d. Table A.1 explores the effects of "no wife's kin" of type Q on the relative *likelihood* of wife's and husband's transfer relations with other kin; Table A.2 explores equivalent effects in relation to "no husband's kin." In both of these, results are presented as relative risk ratios, where relative is defined in relation to the likelihood of an equivalent transfer to kin K where other kin Q are not absent from the network (i.e., $Q = 0$). Substantive results from these 16 models are summarized in Table 4.

Similarly, Table A.3 presents results in relation to the effects of both "no wife's" and "no husband's" kin on the *value* of such transfers. Here, the predicted value of a transfer associated with a given type of relative who has surviving kin of type Q is presented in the top left corner of each cell; and the estimated difference for those who do not have such kin—with an associated indicator for a test for significance of that difference—is presented in the lower right corner of each cell. Substantive results with respect to changes in the value of these flows are summarized in Table 5.

Table 4. Summary of substitution, complement, and substitutability effects in response to the lack of kin of type Q. All reported effects significant at least at the 5 percent level (bold font) and 10 percent level *(italicized non-bold font)*.

Missing kin of type X	*Type and direction of transfer*						Total num. of effects of lack of kin Q by type of kin Q
	Substitute R→K increase	Complement R→K decrease	Substitute R↔K increase	Reduced complementarity R↔K decrease	Substitute (K's perspective) K↔R increase	Complement (K's perspective) K↔R decrease	
Wife's kin							
Father	HPU^{kx}	—	*WMU; HMA*	—	—	*WM; HPA*	1 *(4)*
Mother	WMA	—	**WMA**	**WF**	**WPU**^{gx,lx}	—	4 *(0)*
Paternal uncles	**WF; HPA**^{gx,kx}	—	*HF; HPU*	—	—	—	2 *(2)*
Paternal aunts	—	—	*HMA*	—	—	**HMU**^{gx,lx}	1 *(1)*
Maternal uncles	—	*HPU*	—	—	—	*HPU*	0 *(2)*
Maternal aunts	*WM*	—	—	—	—	—	0 *(1)*
Brothers	—	**HF**^{lx}	—	**WF; HF**^{kx}	*HPA*	—	3 *(1)*
Sisters	—	**WF**^{gx}	—	**WF**^{gx}**; WM**	—	**WF**^{gx}; *WPU*	4 *(1)*
Husband's kin							
Father	**HMU**^{lx}**; HPU**	—	—	*WPA*^{gx,kx}	—	**WM**^{gx,kx}	3 *(1)*
Mother	—	—	**WPU**^{gx,kx}**; HMA**	—	**WMU**^{gx,kx}	—	3 *(0)*
Paternal uncles	*HPA*	—	**WMU**^{kx}**; HPA**^{gx}	—	—	—	2 *(1)*
Paternal aunts	*HPU*	—	—	—	—	*HMU; HMA*	0 *(3)*
Maternal uncles	**HPA**^{gx,lx}	—	—	—	—	—	1 *(0)*
Maternal aunts	—	—	—	**HPA**^{lx}	—	*HPU*	1 *(1)*
Brothers	**WF**^{kx}**; HPA**^{gx}**; HMA**^{gx}	—	**WF**^{kx}	HM^{gx}	**WF**^{kx}; *HMU*	*WPU*	6 *(2)*
Sisters	**HPU**^{gx}; *HM*	**WF**^{gx,kx}	**WMA**^{kx}**; HM**	**WF**^{gx,kx}**; WM**^{kx}	—	*HPU*	6 *(2)*
Total number of effects by type of transfer	10 *(5)*	3 *(1)*	8 *(5)*	10 *(0)*	3 *(2)*	3 *(9)*	37 *(22)*

Notes. Estimated coefficients from which this summary table is derived are presented in Appendix Tables 1a and b. Abbreviations denote the following: W = Wife; H = Husband; F = Father; M = Mother; P = Paternal; M = Maternal; U = Uncle; A = Aunt. Transfer response to lack of kin is: ^{gx} = cross-gender; ^{lx} = cross-lineage; ^{kx}= cross-kin (i.e., from natal to affinal kin).

Table 5. Summary of changes in value of transfers associated with the lack of kin of type Q. All reported effects significant at least at the 5 percent level (bold font) and 10 percent level *(italicized non-bold font).*

Missing kin of type X	**Increase in net value of transfers received**	**Decrease in net value of transfers received**	**Total**
Wife's kin:			
Father	—	—	—
Mother	*HPU*	*WPU*	*(2)*
Paternal uncles	—	—	—
Paternal aunts	**HMU**^{gx,kx}	**HF**^{gx,kx}	2
Maternal uncles	*HPU*	—	*(1)*
Maternal aunts	—	*HMA*	*(1)*
Brothers	—	**HPU**^{kx}**;**	2
Sisters	HPU^{gx,kx}	**HPA**^{gx,kx}	1
Husband's kin:			
Father	—		1
Mother	—	*HMA*	—
Paternal uncles	**WF**^{gx,kx}**; HF**	—	1*(1)*
Paternal aunts	**HMU**^{gx,kx}**;**	—	1(1)
Maternal uncles	HPU	—	—
Maternal aunts	—	—	—
Brothers	—	—	2
Sisters	—	**WF**^{kx}**; HPA**^{gx}	2
	WPU^{gx,kx}**; HF**	—	
Total number of effects by type of transfer	6 *(4)*	5 *(3)*	11 *(7)*

Notes: Estimated coefficients from which this summary table is derived are presented in Appendix table 1c. Abbreviations denote the following: W = Wife; H = Husband; F = Father; M = Mother; P = Paternal; M = Maternal; U = Uncle; A = Aunt. Transfer response to lack of kin is: ^{gx} = cross-gender; ^{lx} = cross-lineage; ^{kx} = cross-kin (i.e., from natal to affinal kin).

Consistent with the model described above, these results are net of the effects of controls for women's age and schooling (other women's characteristics such as parity, household characteristics including wealth, and equivalent husband's characteristics had no independent effects and no effects on the variables of interest), the residential proximity of relatives, and dummy identifiers for type of kin. These variables are described in Table A.4. In addition, the estimated logits have modified standard errors to take account of clustering on the respondent (subscript r).

Finally, although I mentioned "conditioning effects" on substitution and related strategies in the prior discussion, it should be noted that I do not incorporate any of these effects—that is, the source of kin loss, the type of transfer, and the possible loss of multiple types of kin—in these estimations. The reasons, respectively, are that we have no data on the first of these; both the cross-sectional nature of the data and their size (discussed below) inhibit our dealing with the second; and straightforward sample size problems impede the third.

3.4. CELL SIZES

The estimation procedure described above is sensitive to sample size since it demands the specification of both dummy identifiers for each type of kin (cK_{rk}), and interaction effects that explore the difference in transfers between R and K in the absence of a given type of kin (coefficient matrix d). Moreover, within a multinomial logit model—which is the best theoretical match, since it simultaneously estimates the likelihood of different types of transfer relationships in comparison to zero-transfer relationships—sample-size becomes an even more significant issue insofar as this previous product of kin-specific samples and survival must be disaggregated into the relevant transfer categories.

In order to highlight some of the structural weaknesses of this approach with respect to data demands in general, and these pilot data in particular, Table 3 presents the cell sizes used in the multinomial logit estimations described above. As implied above, these are the product of (a) kin-specific sample sizes (column 1 in Table 2), (b) variation in the proportion of that type of kin who have survived (column 2 in Table 2), and (c) kin-specific probabilities of making a transfer of a given type—unilateral to kin; unilateral from kin; bilateral—(not shown in this paper). Varying (b) enables us to estimate kin- and transfer-type-specific cell sizes. Thus, in columns (2)–(4) of Table 3, the product of (a) and (c) is multiplied by (b) in the form 1–(% zero kin/100) in order to estimate the number of K in cells used to produce the main effect cK_{rk}; and in columns (5)–(7) the product of (a) and (c) is multiplied by (c) in the form (% zero kin/100) in order to estimate the number of K in cells used to produce the interaction effect $dK_{rk}Q_{rk}$. Results highlight both the relative inequality of cell sizes across different types of transfer behaviour, which has some consequences for the validity of across-kin comparisons, and the weakness of particular columns.

In general, the three columns in which the interaction effect of interest $dK_{rk}Q_{rk}$ is estimated—columns (5)–(7)—are the weakest: 22 of the 36 cells in these columns contain less than 20 kin, and 5 of these cells contain less than 12 kin. The weakest column within this area is column (4), which refers to unilateral transfers from K to R. As noted in Table 1, these estimates are used as indicators of K's motivation to establish and/or maintain ties with R. Four of the five cells with less than 12 kin are in column (4). Two of these, representing cells for husband's parents, jointly represent only 9 kin (this is too little to produce meaningful estimates so I discount these cells in my subsequent discussion). In short, to statistically differentiate themselves from the main effects dK_{rk}, relative risk ratios in columns (5)–(7) must, on average, be either less than 0.4 or greater than 2.1.

This high bar needs to be borne in mind as we review the results. But it also needs to be borne in mind in relation to potential future studies, since it implies that we need to significantly increase sample sizes over those used in this study before we can reliably estimate a broad array of substitution and related effects, including disaggregating them by, say, ethnic subpopulations.

4. Results

4.1. EFFECTS OF DIFFERENTIAL KIN AVAILABILITY ON RELATIVE LIKELIHOOD OF BEING IN A TRANSFER RELATIONSHIP

Three general results are worth noting. First, lack of kin in each of the 16 categories is associated with a significant change in the relative likelihood of having given, received, or exchanged resources with at least one other member of one's intergenerational kin network. In fact, notwithstanding cell-size effects, there are 37 such changes significant at the 5 percent level, and a further 22 at the less reliable 10 percent level out of a possible total of 180. (In the following comments I only discuss those significant at the 5 percent level, but given the cell-size constraints, it is worth at least presenting the latter.) Second, there is a considerable difference among kin in their impact on these intergenerational transfers when they are missing from the kin network. The most widespread changes in these patterns were observed in relation to the lack of brothers and sisters. Together, these 4 models produced 19 of the 37 changes in reported transfers. By contrast, the 4 parent models produced a more modest 11 changes, and the 8 uncle and aunt models produced the remaining 7 changes. Third, in terms of the six-part disaggregation of the types of substitution and related effects identified in Table 1, the most frequent changes were in apparent substitution and reduced complementarity—accounting for 10 each of the 37 changes—and substitutability effects—accounting for a further 8 changes. In contrast, the lower likelihood of transfers from R to K (the indicator of K and Q complementarity) and of K to R accounted for a mere 3 each of the 37 changes.

Other patterns are also worth noting. First, there are no cases in either the substitution or substitutability categories where one type of parent appears to functionally replace another. Rather, all heightened transfer relationships in the absence of a parent are with an uncle or aunt, and in only one case, a wife's paternal uncle's (WPU) substitutability for the husband's mother (HM), is that relationship with an uncle or aunt of different gender to the deceased parent. In other analytic categories, on the other hand, the absence of one parent is associated with different transfer relationships with one's other parent or parents-in-law. Thus, a deceased wife's mother is associated with both reduced bilateral exchanges with the wife's father and reduced unilateral transfers from the husband's father.

Second, the absence of siblings primarily affects transfer relations with parents rather than with uncles and aunts. It accounts for all 7 of the changes associated with the lack of a wife's siblings, and 8 of the 13 changes associated with the lack of a husband's siblings. A specific example: the absence of the wife's brothers is associated with reduced transfers to, and bilateral exchanges with, the husband's father, as well as reduced bilateral exchanges with the wife's father. Similarly, the absence of any wife's sisters is associated with reduced transfers to, bilateral exchanges with, and receipts of transfers from, the wife's father. It is also associated with reduced bilateral exchanges with the wife's mother.

Third, responses to the lack of kin can be observed within and across gender, within and across lineages (i.e., within one's own natal kin), and within and across ones affinal and natal kin. Cases that cross these boundaries are indicated in Table 4 by the superscripts ^{gx}, ^{lx} and, ^{kx}, respectively. These indicators fall in all sections of the table, showing that all three types of kin strategies, those that involve substitution, substitutability, and complementarity, occur across the gamut of kin, irrespective of social group identities. Indeed, this raises a more general issue with respect to the results in Table 4—the fact that there are few systematic patterns that exist across all relatives. I return to this after discussing apparent substitution and related responses with respect to the value of transfers.

4.2. EFFECTS OF DIFFERENTIAL KIN AVAILABILITY ON THE TOTAL NET VALUE OF TRANSFERS

Differences in the value of transfers between those with surviving kin Q and those without such kin are much less in evidence than are differences in the likelihood of having made or received a transfer, as just described. Only 11 differences significant at the 5 percent level are identified in Table 5, out of a total possible 180 intergenerational dyads (the total number of cells in Table A.3). Nor can this relative lack of significant results be blamed on small cell size since in relation to the value of transfers, the smallest cell size includes 45 kin. Before attempting to explain these impoverished associations as part of a larger pattern, it is useful to review some of them in greater detail.

I begin with the statistically significant results. Consistent with Table 4, the bulk of the significant effects—7 of the 11—stem from the absence of siblings. Thus, a sisterless wife tends to receive more from her husband's paternal uncle than a wife who has sisters. Similarly, a sisterless husband receives more from his wife's paternal uncles. In contrast, both wives and husbands who lack brothers tend to receive less from a number of older relatives—husband's paternal uncles and aunts in the case of the wife lacking brothers—and the wife's father and husband's paternal aunt in the case of the husband lacking brothers. These differences fit with the complementarity story described above insofar as the lack of brothers—who are mostly adults and, therefore, among the most active labour force participants—makes the overall network much less attractive to potential kin investors.

A second set of significant results reflects an apparent gender difference in the effects. Of the 11 significant changes in the value of an exchange with kin, 9 involve exchange with a male relative, and 8 involve exchange with husband's rather than wife's kin (the latter includes all 5 of the changes associated with an absence of given types of wife's kin). Both of these are consistent with the greater male role in monetary transfers, a role that may extend to his own kin's responses to his wife's familial constraints.

It is also worth noting dyads where there was no association. In particular, the absence of parents has no effect on the net value of transfers exchanged with other kin. Thus, the remaining four effects on the value of transfers are related to the absence of uncles and aunts. Two of these effects arise out of the absence of wife's kin (her paternal aunt) and two out of the absence of husband's kin (his paternal uncle and paternal aunt). This difference in the relative effects of absent parents *versus* uncles and aunts on the value of transfers contrasts with the equivalent associations depicted in Table 4 in relation to their effects on the likelihood of transfers.

5. Conclusion

The results imply that individuals' transfer relationships are at least in part contingent on the overall structure of the kin network, including the availability of certain types of kin. In response to the absence of selected kin, transfer relationships with some kin are heightened, and with others they are diminished. In some of these cases, only one of the two parties involved in the relational dyad appears to have changed transfer behaviour. In others, both parties have modified their behaviour. In short, though weakened by some data problems, the analysis has demonstrated the array of transfer-related responses to various kin-specific absences. Some of these imply active substitution of surviving kin for deceased. Others, in which transfers are diminished in the absence of a given kin member, imply that the two types of surviving kin are complementary. These effects can be observed both within and across boundaries demarcating gender, lineage, generation, and natal kin. Overall, they are consistent with the post-Schneider constructivist approaches to the study of kinship in which people's actual kin networks are limited much less by normative structures that affect lineage and descent, than by contingent patterns of social relations.

That said, the very breadth of these observed effects, in combination with the data limitations, mean that it is difficult to isolate systematic pathways for either of the three main transfer strategies. For example, those which figure most prominently in Table 4 tend to be the ones which have the largest cell sizes in Table 3. This makes it difficult to follow up on the intriguing map of vertical and lateral intergenerational transfers presented in Figure 2—in which, for example, paternal uncles appear to play such an important role—because analytic leverage is correlated with the key variables of interest.

Similarly, kin availability appears to affect the likelihood of transfers much more than it does their value. The data do not allow us to isolate the reason for this difference. It could stem from differential quality in measurement to the extent that people may be less likely—intentionally or not—to give accurate responses to questions about specific transfers even if they remember that a transfer occurred. Otherwise, measurement problems could arise because relational dyads in which there was one transfer in the reference period—say, from a migrant kin—were indistinguishable from relational dyads where there were multiple transfers over an extended period. Alternatively, the difference between changes in the likelihood and value of transfers may reflect an important theoretical distinction between these two transfer currencies. In Maussian terms, for example, it is less the value of a given transfer that is important than the fact that it was made in the first place. Transfers bind individuals together at a symbolic level.

As implied by these remarks, this analysis could be modified and improved in several ways. Most of the weaknesses, I think, stem from the nature of these pilot data. I have already noted some of them—which I referred to as "conditioning effects"—in relation to the estimation procedures. It is worth noting at least two others, however, since this is a relatively open and new area of empirical research.

First, in order to portray an ideal-type household with kin networks involving both the husband's and wife's kin, I have restricted the analysis to currently married couples where both the husband and wife were interviewed. This implicitly does two things. First, it skirts over issues of husband–wife competition and negotiations about transfers to their respective natal kin. And second, it ignores some of the more impoverished households in sub-Saharan Africa, those headed by women, which are increasing in number in the wake of AIDS and broader cultural changes. Specifically, to the extent that the married women in our analysis "lineage hop" within their own natal kin, as well as into spouse's kin networks in transfer relationships, and to the extent that such tie-building strategies successfully forge reciprocal exchanges; the substitution, substitutability, and complementarity effects identified in this paper beg questions about equivalent effects and strategies among single women (among men too, but as women are more usually burdened with children and restricted to domestic labour spheres, theirs is generally considered a more problematic status). Indeed, in the most general terms, the analysis points to one of the potential reasons that unwed women and men are at some disadvantage in general, and in societies with limited alternatives to kin networks in particular. Simply, larger networks allow for a larger number of potential transfer relationships. Smaller networks, in contrast, are more impoverished.

More generally, and on a more qualitative level, we also know little about the extent to which these individuals are aware of both the larger systemic regularities and preferred transfer routes implied in Figure 2 and, therefore, little about these individuals' motivations with respect to given transfers. I have therefore simply assumed a general equivalence between changes in a transfer relationship and changes in motivation with respect to that transfer relationship.

For all of these limitations, however, I think that addressing substitution and related effects is a promising and useful enterprise. This analysis has confirmed the overall embeddedness of vertical and lateral intergenerational transfer relations in broader kin networks (hence the effects of the lack of respondent's siblings on intergenerational relations), which supports general constructivist approaches to intergenerational and kin relations (i.e., modified transfers over an array of kin). In turn, this provides at least one framework for exploring the effects of ongoing demographic and ideational change in much of the developing, and recently developed, world. As mentioned at the start, this is particularly useful from a social demographic perspective since the grand theoretical objective of this paper is to contribute to the exploration of links between normative and demographic structures. To the extent that there are such linkages, the consequences of demographic transitions with respect to kin networks will play out normatively over the next few generations. Identifying strategies that individual's use in response to such changes—including transfer strategies—may, in this context, be a useful way to chart those developments.

Acknowledgement

Alexander A. Weinreb acknowledges comments on a previous version of this paper from Jere Behrman and Anne Gauthier.

Appendix

Table A.1. Estimated effects of having no surviving wife's kin of type K_X on the likelihood of having a transfer relationship with other wife's and husband's kin (i.e., kin $K_{\sim X}$), by direction of reported transfer. The effects were estimated in a series of multinomial logits in which the omitted category was "no reported transfer." The effects are net of controls, as described in the text, and are reported as risk ratios relative to the likelihood of an equivalent transfer to kin $K_{\sim X}$ where respondent *has* surviving kin of type K_X. A summary of these effects is presented in Table 4.

	Wife has no surviving:																							
	father			mother			paternal uncles			paternal aunts			maternal uncles			maternal aunts			brothers			sisters		
	R→K	*K→R*	*2way*	*R→K*	*K→R*	*2way*	*R→K*	*K→R*	*2way*	*R→K*	*K→R*	*2way*	*R→K*	*K→R*	*2way*	*R→K*	*K→R*	*2way*	*R→K*	*K→R*	*2way*	*R→K*	*K→R*	*2way*
Wife's kin																								
Father				0.62	0.51	0.44**	2.5**	1.5	1.0	0.84	0.54	0.69	0.78	0.69	0.60	0.61	0.72	0.76	0.54	0.57	0.32**	0.27**	1.1	0.36**
Mother	1.5	0.48*	1.5				1.4	0.97	0.86	1.6	1.4	1.2	0.90	0.50	0.86	2.0*	1.1	0.92	1.1	0.53	0.57	0.56	0.45	0.26***
Paternal uncle	1.1	1.6	1.5	0.83	2.5**	1.1				0.66	0.83	1.8	0.43*	0.97	1.0	0.96	1.0	1.4	0.56	0.88	0.33	0.51	0.88	1.1
Paternal aunt	1.5	0.60	1.0	1.3	0.61	0.81	0.96	1.1	1.8				1.3	1.4	0.58	1.5	0.60	1.2	0.93	0.51	0.47	0.86	0.67	1.5
Maternal uncle	1.5	0.97	2.0*	1.3	0.76	1.5	1.4	0.99	1.5	0.69	1.0	1.0				1.3	0.67	1.4	0.50	1.6	0.66	1.2	1.1	0.84
Maternal aunt	1.5	0.83	1.6	2.2**	1.1	2.3**	1.0	0.98	1.6	0.71	0.94	1.1	1.2	1.5	1.2				0.45	0.72	0.62	0.76	0.90	0.65
Husband's kin																								
Father	1.0	<0.1***	1.3	0.53	<.1***	0.62	1.8	1.4	2.0*	0.59	0.32	0.65	1.8	0.61	1.4	0.67	0.16*	0.84	0.34**	0.23	0.40**	0.96	<.1***	0.67
Mother	1.7	0.78	1.7	0.68	0.76	0.66	1.4	0.63	1.4	0.76	0.99	0.82	0.72	0.63	0.93	0.69	1.7	1.3	0.52	0.40	0.59	1.5	1.3	1.4
Paternal uncle	1.8**	1.0	1.5	0.95	0.87	0.72	1.1	0.90	2.0*	1.6	1.5	0.99	1.1	0.50*	1.8	1.4	0.94	1.2	0.71	1.3	1.2	1.4	0.39*	0.36
Paternal aunt	1.2	0.48*	0.91	0.90	1.5	0.72	1.8**	0.86	1.6	1.5	1.2	0.78	0.81	1.2	0.73	1.1	1.0	1.1	0.87	2.7*	1.2	1.3	0.74	0.57
Maternal uncle	1.2	0.84	1.4	1.0	0.69	0.93	0.96	1.1	1.3	0.72	0.44**	0.62	1.3	0.65	1.0	0.73	0.61	0.82	0.86	0.63	0.59	1.3	1.3	0.70
Maternal aunt	1.5	1.1	1.9*	1.1	0.73	1.1	1.2	0.79	1.2	1.1	1.2	1.8*	0.68	1.3	0.72	0.94	0.99	1.8	1.6	0.98	1.2	0.77	1.6	0.83
$n = 5,713 - N_{kX}$		5,436			5,360			5,173			5,157			5,097			5,048			5,713			5,713	
Pseudo R^2		0.100			0.087			0.099			0.102			0.103			0.105			0.100			0.100	

*Significant at the 0.10 level.
**Significant at the 0.05 level.
***Significant at the 0.01 level.

Table A.2. Estimated effects of having no surviving husband's kin of type K_X on the likelihood of having a transfer relationship with other wife's and husband's kin (i.e., kin $K_{\sim X}$), by direction of reported transfer. The effects were estimated in a series of multinomial logits in which the omitted category was "no reported transfer". The effects are net of controls, as described in the text, and are reported as risk ratios relative to the likelihood of an equivalent transfer to kin $K_{\sim X}$ where respondent *has* surviving kin of type K_X. A summary of these effects is presented in Table 4.

	Husband has no surviving:																							
	father			mother			paternal uncle			paternal aunt			maternal uncle			maternal aunt			brother			sister		
	R→K	*K→R*	*2way*	*R→K*	*K→R*	*2way*	*R→K*	*K→R*	*2way*	*R→K*	*K→R*	*2way*	*R→K*	*K→R*	*2way*	*R→K*	*K→R*	*2way*	*R→K*	*K→R*	*2way*	*R→K*	*K→R*	*2way*
Wife's kin																								
Father	1.5	1.2	0.88	1.3	1.5	1.3	1.4	0.66	1.0	0.57	0.71	0.67	0.64	1.0	1.0	0.89	1.1	1.1	5.7**	5.4**	9.0***	0.27**	1.1	0.36**
Mother	0.95	0.44**	0.73	0.86	0.78	0.61	0.93	0.66	0.76	0.66	0.69	0.78	1.6	0.73	1.1	0.98	0.64	0.91	0.87	1.4	1.3	0.56	0.46	0.26***
Paternal uncle	0.85	0.60	1.3	0.58	1.1	2.7**	0.90	1.2	1.8	1.0	0.98	1.4	0.55	0.97	0.81	0.85	0.49*	0.92	1.7	1.0	0.18*	0.57	0.14*	0.64
Paternal aunt	0.80	0.56	0.46**	0.92	1.4	0.62	0.88	1.8	1.2	0.65	0.52	1.2	0.91	1.0	0.72	0.80	0.89	0.35**	1.5	1.3	0.49	0.66	0.87	0.51
Maternal uncle	1.6	0.83	0.95	0.96	1.9**	1.1	1.6	1.2	2.6***	0.95	0.70	0.94	1.3	0.81	1.5	1.2	1.2	0.79	1.8	0.87	0.87	0.94	0.68	0.67
Maternal aunt	0.91	0.89	0.79	0.74	0.96	0.90	0.65	1.1	1.6	1.4	1.3	1.7	1.1	0.90	1.2	1.1	1.1	1.5	1.5	0.51	1.8	0.75	0.87	2.4**
Husband's kin																								
Father				0.73	2.3	0.75	0.64	0.86	0.65	1.4	0.75	1.4	0.50	0.20	0.82	1.3	1.1	0.84	0.93	2.3	0.61	1.1	0.10***	0.52
Mother	1.5	0.79	1.8				0.99	0.60	0.97	1.0	0.60	0.58	0.77	0.37	0.48	1.2	0.92	0.90	0.93	0.10***	0.81	4.4*	2.3	4.7**
Paternal uncle	1.7*	0.94	1.5	1.0	1.3	0.86				1.8*	0.52	0.60	0.87	1.2	0.60	1.1	1.5	1.1	1.5	1.1	0.61	2.3**	1.4	1.5
Paternal aunt	1.2	1.1	0.75	0.91	1.3	1.0	1.7*	0.79	2.6***				1.9**	1.1	0.79	1.6	1.2	0.61	3.2***	1.8	1.4	1.4	1.5	1.6
Maternal uncle	2.5***	1.1	1.5	1.1	0.93	1.5	1.5	0.41	1.6	1.0	0.50*	0.70				1.4	1.1	1.1	1.4	0.69	2.1*	1.1	1.6	0.92
Maternal aunt	1.2	0.93	1.2	1.1	0.97	2.5**	1.6	0.86	1.7	0.97	0.52*	0.77	1.4	0.96	0.74				2.2**	2.0	1.3	1.8	1.1	0.50
$n = 5,713 - N_{kX}$		5,490			5,432			5,175			5,224			5,090			5,161			5,713			5,713	
Pseudo R^2		0.093			0.085			0.106			0.106			0.106			0.106			0.102			0.102	

*Significant at the 0.10 level.
**Significant at the 0.05 level.
***Significant at the 0.01 level.

Table A.3. Estimated effects of having no surviving wife's or husband's kin of type K_X on the value of transfers with other wife's and husband's kin (i.e., kin $K_{\sim X}$). Effects were estimated in 16 censored regressions (with censoring defined as no transfer having been reported in a given dyad). The effects are net of controls, as described in the text. Two estimates are reported in this table. In the top left of each cell is the estimated value of transfers exchanged with a given relative where R has kin of type X ($\alpha + \beta_1$); in the bottom right of each cell is the estimated value of transfers exchanged with a given relative where R has no kin of type X (β_2). Significance tests examine the difference between these estimated coefficients (i.e., $\beta_1 = \beta_2$). Where the $p(\beta_1 = \beta_2) < .1$, β_2 is displayed in bold type. A summary of these effects is presented in Table 5.

	Wife has no surviving								Husband has no surviving							
	father	mother	pat uncle	pat aunt	mat uncle	mat aunt	brother	sister	father	mother	pat uncle	pat aunt	mat uncle	mat aunt	brother	sister
	(1)	(2)	(3)	(4)	(5)	(6)	(7)	(8)	(9)	(10)	(11)	(12)	(13)	(14)	(15)	(16)
Wife's kin																
Father		130	148	107	100	125	115	124	105	149	78	130	137	133	149	125
		50	−63	91	10	−31	109	11	55	−70	**132****	−26	−18	39	**−158***	10
Mother	70		66	79	39	70	81	70	66	85	87	94	85	−45	84	71
	1		29	5	53	−13	−14	61	33	−8	−18	-61	-6	31	−59	60
Pat uncle	439	462		417	387	422	404	401	381	402	372	404	393	279	406	386
	−90	**−139***		29	27	−42	50	52	28	−18	−32	3	75	68	−5	**242****
Pat aunt	387	402	414		369	388	379	383	338	376	359	367	386	251	380	373
	−12	29	−69		−4	−6	62	5	83	0	−50	41	21	83	7	92
Mat uncle	316	307	312	309		294	298	300	300	309	280	273	300	194	285	282
	−70	22	−47	−17		−11	−58	−64	−41	−73	−60	39	−11	−25	24	67
Mat aunt	332	351	336	323	314		306	314	297	300	280	334	329	229	315	322
	−50	−69	−39	18	−64		86	9	21	28	0	−57	−23	−54	−15	−42
Husband's kin																
Father	86	93	60	162	26	27	31	47		45	39	46	48	−24	58	42
	−68	−60	22	**−245*****	75	130	**319*****	164		136	**123***	56	123	−70	53	**241****
Mother	226	268	237	225	234	212	221	220	226		238	218	227	125	235	243
	6	−94	−6	38	−74	44	137	89	36		7	30	55	−3	−19	54
Pat uncle	200	216	194	231	174	200	249	196	238	233		190	244	137	209	229
	57	**125***	81	11	**121***	88	**−165****	**183****	−67	−66		**127***	−69	−66	87	35
Pat aunt	233	281	287	276	237	257	286	255	249	266	244		280	167	281	268
	65	−5	−60	−10	22	10	**−163****	39	12	−32	−57		−43	−48	**−163****	43
Mat uncle	243	237	233	204	206	204	220	229	241	227	185	186		123	237	234
	−41	44	1	**116****	12	96	63	0	−53	−8	47	**124****		−8	−74	−32
Mat aunt	299	295	291	332	257	312	284	282	306	284	267	254	289		284	287
	−50	13	−40	−81	13	**−107***	−29	−20	**−88***	−38	−44	60	−8		−52	−53

*Significant at the 0.10 level. **Significant at the 0.05 level. ***Significant at the 0.01 level.

Table A.4. Descriptive statistics for control variables in models reported in Tables 4 and 5.

Variable	Mean	Effect on:[2] i. likelihood of transfers	Effect on:[2] ii. value of transfers
Women's age (measured in years)	29.53 (8.5)	+	+
Women's schooling (measured in years)	3.75 (3.3)	+	−
Kin residential proximity to respondent's HH (%)			
- same village	18.9	ref	ref
- same Traditional Authority (TA)[1]	29.9	ref	ref
- other rural area	27.4	ref	ref
- city outside TA	17.5	−	+
- abroad	6.3	−	+

[1] Traditional authority is roughly equivalent to a U.S. county.
[2] These effects are uniform across the models and the vast majority of them (i.e., across each set of 16 models) are statistically significant at the 5 percent level.

References

Berheide, C.W. and Marcia, T. Segal. (1994), "Controlling less land, producing less food: The fate of female-headed households in Malawi." In: *Women, the Family, and Policy: A Global Perspective*, Esther Ngan-ling Chow and Catherine White Berheide, eds. pp. 145–62. State U of New York Press: Albany.

Blau, P.M. (1964/1986), *Exchange and Power in Social Life*. Wiley and Sons: New York.

Brouwer, I. D., Hoorweg, J. C., and Marti J. van Liere. (1997), "When households run out of fuel: Responses of rural households to decreasing fuelwood availability, Ntcheu District, Malawi." *World Development*, 25(2):255–266.

Caldwell, J.C. (1982), *A Theory of Fertility Decline*. New York: Academic Press.

Collins, R. (2004), *Interaction Ritual Chains*. Princeton: Princeton University Press.

Coudenhove, Hans. (1926), *My African Neighbors: Man, Bird, and Beast in Nyasaland*. Little, Brown, and Company: Boston.

Doctor, Henry V. (2001), "*Changes in Life Expectancy in Malawi, 1977–1998*." Population Studies Center, University of Pennsylvania: Mimeo.

Doctor, H.V. and Weinreb, A. A. (2005), "Mortality among married men in rural Kenya and Malawi." *Journal of African Population Studies*. 20(2):165–177.

Feinberg, R. and Ottenheimer, M. (ed.). (2002), *The Cultural Analysis of Kinship: The Legacy of David M. Schneider*. University of Illinois Press: Chicago.

Fox, R. and Fox, R. (1984), *Kinship and Marriage*. Cambridge University Press: Cambridge.

Frazer, Donald. (1914), *Winning a Primitive People: Sixteen Years' Work Among the Warlike Tribe of the Ngoni and the Senga and Tumbuka Peoples of Central Africa*. E.P. Dutton and Company: New York.

Grieco, M. (1987), *Keeping It in the Family: Social Networks and Employment Chance*. Tavistock: London.

Griffiths, Anne M.O. (1997), *In the Shadow of Marriage: Gender and Justice in an African Community*. University of Chicago Press: Chicago.

Hirschmann, D. (1990), "Malawi's 'captured' peasantry: An empirical analysis." *Journal of Developing Areas*. 24(4):467–488.

Holy, Ladislav. (1976), "Kin groups: Structural analysis and the study of behavior." *Annual Review of Anthropology*, 5:107–131.

Homans, George C. (1958), "Social behavior as exchange." *American Journal of Sociology*, 63:597–606.

Johnson, W.P. (1922), *Nyasa, the Great Water: Being a Description of the Lake and the Life of the People*. Oxford University Press: London.

Kalipeni, E. (1996), "Demographic response to environmental pressure in Malawi." *Population and Environment*, 17(4):285–308.

Kaspin, D. (1995), "The politics of ethnicity in Malawi's democratic transition." *Journal of Modern African Studies*, 33(4):595–620.

Lewis, Ioan M. (1994), *Blood and Bone: The Call of Kinship in Somali Society*. Red Sea Press.

Malawi, Government of, and World Bank. (1998), *Malawi AIDS Assessment Study*. Blantyre and Washington: Government of Malawi and World Bank.

Massey, Douglas S., Joaquin Arango, Graeme Hugo, Ali Kouaouci, Adela Pellegrino, J. Edward Taylor. (1993), "Theories of international migration: A review and appraisal." *Population and Development Review*, 19(3):431–466.

Mauss, M. (1925/1990), *The Gift: The Form and Reason for Exchange in Archaic Societies*. W.W.Norton: New York.

Meriwether, M.L. (1999), *The Kin Who Count: Family and Society in Ottoman Aleppo, 1770–1840*, University of Texas Press: Austin.

Mitchell, J.C. (1956), *The Yao Village: A Study in the Social Structure of a Nyasaland Tribe*. University of Manchester Press: Manchester.

Mitchell, J.C. (1962), "Marriage, matriliny, and social structure among the Yao of southern Nyasaland." *International Journal of Comparative Sociology*, 3(1):29–42.

Mtika, Mike M. (2001), "The AIDS epidemic in Malawi and its threat to household food security." *Human Organization*, 60(2):178–188.

Ntara, S.J. (1949/1973), *The History of the Chewa*. Franz Steiner Verlag: Weisbaden.

Peletz, M.G. (1995), "Kinship studies in late twentieth-century scholarship." *Annual Review of Anthropology*, 24:343–372.

Republic of Malawi (2005), *Malawi HIV and AIDS Monitoring and Evaluation Report 2005*. Lilongwe: Office of the President and the Cabinet. http://data.unaids.org/pub/Report/2006/2006_country_progress_report_malawi_en.pdf

Schneider, D.M. (1968), *American Kinship: A Cultural Account*. Prentice Hall: Englewood Cliffs, NJ.

Segal, E.S. (1985), "Projections of internal migration in Malawi: Implications for development." *Journal of Modern African Studies*, 23(2):315–329.

Stack, C. (1974), *All Our Kin: Strategies for Survival in a Black Community*. Harper and Row: New York.

United Nations. (2001) *Women's Indicators and Statistics Database, Version 4 (CD-Rom)*. United Nations: New York.

UNESCO (2005). Malawi Country Page, www.uisunesco.org/profiles/EN/EDU/CountryProfile_en.aspx?code=4540

Weinreb, A.A. (2002), "Lateral and vertical intergenerational exchange in rural Malawi." *Journal of Cross-Cultural Gerontology*, 17(2):101–138.

Whyte, W.F. (1943), *Street Corner Society: The Social Structure of an Italian Slum*. University of Chicago Press: Chicago.

Wilmott, P. and Young, M. (1962), *Family and Kinship in East London*. Harmondsworth: London.

World Bank. (2001), *African Development Indicators*. World Bank: Washington.

Young, T.C. (1932/1970), *Notes on the History of the Tumbuka-Kamanga Peoples in the Northern Province of Nyasaland*. Frank Cass and Company: London.

CHAPTER 2. KINSHIP NETWORKS AND INTERGENERATIONAL TRANSFERS

C. Y. CYRUS CHU
Institute of Economics, Academia Sinica

RUOH-RONG YU
Survey Research Centre, Academia Sinica

1. Introduction

Most existing literature on intergenerational transfers in the West focuses on downward transfers from parents to children. Two strands of studies have been developed in the past decade. One strand emphasizes *inter vivos* transfers while the other strand looks at *post-mortem* bequests. For a related discussion, references can be found in Rosenzweig and Stark (1997).

As described in various sources, in oriental societies child-to-parent transfers may be more essential than that from parent to child. The Taiwan Elderly Survey conducted in 2000 shows that more than 47 percent of the elderly rely on financial support from their children to sustain their living.[1] Another island-wide survey entitled Panel Study of Family Dynamics (PSFD) shows a similar pattern.[2] Among interviewees (aged 36–65) surveyed in 1999–2000 by PSFD, 53 percent reported that they *provided* financial transfers to their parents regularly, while only less than 5 percent *received* financial transfers from their parents.[3]

As to the timing concerning when parents transfer their assets, some phenomena found in Taiwan also deserve special attention. According to the PSFD survey, among respondents with at least one living parent 23 percent of them answered that their parents had already transferred *all* their assets. As we can see from Table 1, although the ratio of families, which had transferred all their assets is a bit higher for samples with only one parent alive, all these ratios are significantly larger than zero.[4]

---

[1] The Elderly Survey is conducted by the Ministry of Interior Affairs of Taiwan.

[2] Detailed information of the PSFD survey is provided later in Section 3.

[3] These figures are calculated based on respondents with at least one living parent.

[4] Since there are no mandatory social security pensions in Taiwan, the proportion of bequeathable assets relative to total assets should be higher in Taiwan than in the US.

Anne H. Gauthier et al. (eds.), Allocating Public and Private Resources across Generations, 39–61.

Table 1. Asset transfer status in different types of families (percent).

Samples	All transferred	Partly transferred	Not yet transferred	Nothing to be transferred
All	22.96	6.30	36.76	33.98
With father alive only	25.79	5.26	32.63	36.32
With mother alive only	36.40	4.50	19.02	40.08
With both parents alive	12.08	7.95	51.22	28.75

The relationship between asset transfer and child-to-parent remittance constitutes the main theme of this chapter. A preliminary investigation on the PSFD data set shows that with respect to those parents who had already divided and transferred their assets completely, the average income remittance provided by their children is higher than is the case for families with assets undistributed or with no assets. For family heads who had distributed all their assets, the average amount of child-to-parent transfers received per month is about 9,374 NT dollars, the highest among the four asset categories. As we shall see later, after controlling for other explanatory variables (including the children's and parents' socioeconomic status variables), the above relationship between asset status and child-to-parent remittance still holds.[5]

Two main competing motives are proposed concerning intergenerational transfers in the past literature: altruism and exchange. In the altruistic framework advanced by Becker (1974), transfers are made since a benevolent individual cares about the well-being of other individuals. As to the exchange model (Bernheim et al. 1985; and Cox 1987), transfers are made by one party in exchange for certain responses from another party. These two motives have attracted much attention in the past, and various empirical studies had been exercised to disentangle these two motives.[6] Though these two hypotheses are quite successful in explaining transfer behaviour in the West, none of them provide satisfactory explanations with regard to the phenomenon observed in Taiwan.

According to the exchange-motive hypothesis, adult children give transfers to their parents in exchange for something (services or future bequests) in return. However, as explained above, both financial transfers and visits are largely unilateral (from child to parent) in Taiwan. Influenced by the custom of filial piety, adult children (especially males) often provide monetary transfers regularly to their parents, and take care of their parents' old age. In any event, for families in which parents have already distributed *all* their assets, there is no financial return for the children to expect from their parents. As such, the exchange motive seems to fail in explaining the child-to-parent transfers in this direction.

Altruistic children of course may choose to make more transfers to their parents when the latter have distributed all the assets and hold nothing in their hands. However, the

---

[5] The results are presented in Tables 3a and 3b, which will be explained later in the context.

[6] See Altonji et al. (1997); Bernheim et al. (1985); Cox (1987); Perozek (1998); and Wilhelm (1996) for further discussion.

altruism hypothesis cannot explain why children with different backgrounds show different feedback behaviours. Specifically, a preliminary investigation on the PSFD data set shows that children with tighter kinship networks transfer more to their parents than children with looser kinship networks.

Since neither the altruistic nor exchange motives can fully explain the phenomenon observed in Taiwan, we are forced to find some other possible explanations. The purpose of this paper is to propose a theory of *social network* to explain the feedback of children to their parents, and then study empirically whether the data of Taiwan can support such a theory. We argue that within a kinship network, the major driving force behind the children's behaviour may be the *norm*, instead of the lure of assets held by the parents, as proposed by Bernheim et al. (1985).

In Section 2 of this paper we shall propose a theoretical argument explaining why a complete transfer of assets may increase the children's feedback to their parents. Roughly speaking, within a tightly knit kinship network, if parents were to complain that their children's feedback toward them was not sufficient enough after they distributed all their assets, then the condemnation received by the children (from relatives within the network) for their lack of filial devotion would be strong. By contrast, children surrounded by a loose kinship network would not face such pressure even when their parents' assets are all transferred. Thus, the transfer of assets can be seen as triggering a test of the children's filial behaviour. Since such a trigger effect is more significant within tighter kinship networks, the econometric setup should take such concerns into account.

The remainder of this paper is arranged as follows. Following the theoretical discussion in Section 2, the empirical model together with the data set are explained in Section 3. Estimation results are presented in Section 4, and the final section provides the conclusion.

2. Theories of Child-To-Parent Transfers

In this section we propose a model of intergenerational transfers by taking the role of kinship networks into account. For simplicity, we consider a family structure consisting of only one parent (p) and one child (k), and we assume that each individual lives for three periods. The first period of a person's life is childhood, the second is adulthood, and the third is old age. The parent and the child overlap for two periods, with the parent's old-age corresponding to the child's adulthood, and the overlapping periods are named as "period 1" and "period 2," respectively, in the following analysis. In period 1, the parent decides whether to distribute his assets. In period 2, the child decides the amount of transfers to his parent.

As described by Sung (1974) and Chan (1990), the practice and timing of asset transfer in Taiwan is largely dominated by custom, with the transfer more likely to take place when the family head (usually father) dies, or when children grow up to establish a new family or develop their own career. For the time being we treat the asset-transfer status in period 1 as exogenous and focus on the intergenerational interaction in period 2.

In the case where the parent distributes all his assets in period 1, the parent may have to rely on financial support from the child or from the government in period 2. Since mandatory social security pensions are not enforced in Taiwan yet, as suggested by Kotlikoff and Spivak (1981) and Laitner (1997, pp. 231–233) families may function as substitutes for market annuities. If the parent distributes all assets to his child in period 1, the parent and the child may agree, perhaps implicitly, that the child will support the parent in period 2.

How are such agreements enforced? Laitner (1997, pp. 232) offers two possible answers. One possibility arises from the altruism of the child toward his parent. Another possibility is that the two generations value each other's affections and they fear that any default on a family-line agreement would automatically jeopardize such ties. As denoted by Laitner (1997, pp. 232), this fear could ensure that the child obeys such an agreement, facilitating implicit contracts which would otherwise be impractical. In the following analysis, we will explore these two possibilities, respectively.

2.1. PURE ALTRUISM HYPOTHESIS

Suppose that the child cares about the well-being of the parent. In period 2, the child maximizes his own utility with respect to the amount of transfers made to his parent (T):

$$U_\mathrm{k} = u_\mathrm{k}(c_\mathrm{k}) + \delta u_\mathrm{p}(c_\mathrm{p}),$$

where u_k and u_p denote the well-being of the child and the parent respectively; c_k and c_p are the child's and parental consumption, respectively; and δ is the degree of the child's altruism toward the parent. In the above expression, c_k equals $y_\mathrm{k} + (1+r)D - T$, where y_k is the child's earnings in period 2; r is the interest rate; and D is the transferred asset from the parent in period 1. In the expression of u_p, c_p is the parent's consumption, which equals $(1+r)(A_\mathrm{p} - D) + T$, where A_p is the total assets owned by the parent in period 1. The problem faced by the child can thus be rewritten as follows:

$$\max_T \; U_\mathrm{k} = u_\mathrm{k}(y_\mathrm{k} + (1+r)D - T) + \delta u_\mathrm{p}((1+r)(A_\mathrm{k} - D) + T).$$

Maximizing the above object function with respect to T yields the following child-to-parent transfers function:

$$T = T(D, A_\mathrm{p}, y_\mathrm{k}, r).$$

If both the utility functions of the child and the parent are well behaved, then the following property of the transfers function is expected:

$$\frac{\partial T}{\partial D} > 0,$$

which means that an altruistic child will transfer more to his parent if more assets are distributed in period 1.

A disadvantage of the altruism hypothesis is that it cannot explain the possible connection between the size of child-to-parent transfers and the structure of the kinship network of the child. This connection, however, does exist in Taiwan according to our empirical evidence.

2.2. KINSHIP PRESSURE HYPOTHESIS

As mentioned earlier in this section, the fear of defaulting on a family-line agreement could ensure that the child feeds back to his parent even if the parent holds nothing in period 2. According to Wellman (1990) and Bott (1968, p. 122), the degree of fear or the pressure felt by the child may be related to the tightness of kinship networks faced by the child, which will be explained later in the context.

Following the setup of Kandel and Lazear (1992), Lindbeck (1997), and Lindbeck et al. (1999), we incorporate the cost of kinship pressure into the utility function of the child. Assume that the child's utility function can be written as

$$\tilde{U}_{\mathrm{k}} = u_{\mathrm{k}}(c_{\mathrm{k}}) - v(T, \Omega), \tag{1}$$

where v is the cost of peer pressure when T is chosen; and Ω is the parameter vector characterizing the environment which affects the network-pressure effect. It is usually assumed that $\partial v/\partial T < 0$, meaning that more child-to-parent transfers correspond to less norm pressure from the kinship network.

Now we try to explore the factors behind Ω. After the parent divides all his assets, the child is expected to receive "norm discipline" of which the strength hinges upon the closeness of the network members (Wellman 1990) as well as the economic ties among kin (Bott 1968, p. 122). Bott (pp. 202, 206) also pointed out that most spontaneous expressions of norms take the form of gossip from friends, neighbours, and relatives, whereby "gossip is one of the chief means by which norms are stated and reaffirmed." Thus, the transfer of assets provides a reference point in time for the kinship network to observe a child's subsequent feedback behaviour. If the child does not feedback properly after the asset transfer, gossip may arise and thereby force the child to behave differently.

The norm discipline would in fact be stronger if the kinship network were stronger, for two reasons. First, kin closely related to the child in question are more likely to comment on the child's alienating behaviour; and second, parents within a closer kinship network are more likely to complain to relatives and friends about an estranged parent–child relationship. As such, the feedback of children toward their parents differs according to the tightness of the kinship.

Let N be an index characterizing the tightness of the respondent's kinship network. According to our previous discussion, we assume that N and D are the two major variables in Ω. Suppose the pressure of social norm is strong enough, such that for families with tighter kinship networks, the following result can be expected:

$$\frac{\partial T}{\partial D} > 0 \quad \text{if } N \text{ is large,}$$

while for families with looser kinship networks, we expect the following result:

$$\frac{\partial T}{\partial D} = 0 \quad \text{if } N \text{ is small.}$$

To sum up, for families with tighter kinship networks, the children are likely to transfer more to their parents if the latter have distributed all their assets. As to families with a looser kinship network, the above relationship between asset disposition and child-to-parent transfers may not hold. The above argument forms the major hypotheses to be tested in the following section.[7]

3. Empirical Analysis

3.1. ECONOMETRIC FRAMEWORK

Let N be a dummy variable characterizing the tightness of the kinship network. According to the value of N, two kinds of families can be observed: families with a tighter kinship network ($N = 1$) and families with a looser kinship network ($N = 0$). Thus, we expect to observe the following relationship:

$$\frac{\partial T}{\partial D} > 0, \quad \text{if } N = 1,$$
$$\frac{\partial T}{\partial D} = 0, \quad \text{if } N = 0.$$

The dependent variable, child-to-parent transfers (T), is measured by the amount of transfers (per month) provided by the children. Since coresidence of parents with their adult children is quite common in Taiwan,[8] and since the distance of a child's residence from his parents (denoted by L) may be correlated with T,[9] we construct the following two equations to allow for the simultaneity of child-to-parent transfers and coresidence decisions:

$$T_i^* = X_i\alpha_1 + D_i\beta_1 + \varepsilon_{1i}, \tag{2}$$
$$L_i^* = X_i\alpha_2 + D_i\beta_2 + \varepsilon_{2i}, \tag{3}$$

where T_i^* and L_i^* denote the latent variables characterizing the desired transfers and residence distance, respectively, of which the relationship with the *observed* T_i and L_i will be discussed later. Term D_i contains a vector of dummies used to characterize the asset-division status of the family. The vector X_i denotes other demographic and economic variables for family i. The error terms ε_{1i} and ε_{2i} are allowed to be correlated, revealing the decision simultaneity between T^* and L^*.

[7] The above hypothesis only asserts that the marginal impact of changing the asset disposition status on child transfers should be different within different kinship networks; it does not imply that the kinship network effect per se has no direct effect on child transfers.

[8] In the PSFD data set, about 25 percent of the interviewees' parents coreside with their children.

[9] For more details see Weinstein et al. (1994); Yeh (1994); and Freeman et al. (1994).

Since the observed values of T_i^* are nonnegative, we treat T_i as a censored variable. The relationship between the observed transfers (T_i) and T_i^* is as follows:

$$T_i^* = \begin{cases} T_i^*, & \text{if } T_i^* \geq 0, \\ 0, & \text{otherwise.} \end{cases} \tag{4}$$

A further problem arises from the residence distance variable (L_i). In the PSFD questionnaire, the answers to the question "How far away is your parents' residence from yours?" are categorical, counted from 1 to 7 as the distance (measured by the interviewees' most frequent transportation time) increases. Given that the observed L_i ranges from 1 to 7, its relationship with L_i^* is assumed to be as follows:

$$L_i = \begin{cases} 1, & \text{if } -\infty < L_i^* \leq c_1, \\ m, & \text{if } c_{m-1} < L_i^* \leq c_m, m = 2, \ldots, 6 \\ 7, & \text{if } c_6 < L_i^* \leq \infty. \end{cases} \tag{5}$$

The error terms in the above two equations, ε_{1i} and ε_{2i} are assumed to be bivariate normally distributed with mean zero and a correlation coefficient ρ_{12}. To help identify the system, we normalize the variance of ε_{2i} to be 1.

Our econometric model is composed of equations (2) to (5). Given the assumption of a bivariate normal distribution for (ε_{1i},ε_{2i}), we adopt the maximum likelihood estimation method to estimate the model. The derivation of the likelihood function is specified in Appendix A.

According to our theoretical framework, we divide the whole sample into two subgroups: samples with tighter kinship networks ($N_i = 1$); and samples with looser kinship networks ($N_i = 0$). We estimate equations (2)–(5) for these two subsamples separately, and then examine the null hypotheses established in the beginning of this section.

3.2. DATA

The data is taken from the PSFD survey conducted in Taiwan.[10] The PSFD survey is an island-wide interview exercised yearly since 1999. The main respondents are randomly sampled adults born between 1934 and 1963. Since one of the major purposes of the PSFD survey is to investigate the familial intergenerational interactions in Taiwan society, detailed demographic and economic characteristics of the respondents and their parents are collected in this survey. Furthermore, various measures of intergenerational transfers are

[10] This data set was collected with the support of the CCK foundation and the National Science Council of Taiwan, under the guidance of Gary Becker, Angus Deaton, Robert Hauser, James Heckman, Cheng Hsiao, Ronald Lee, William Parish, George Tiao, Jim Vaupel, Arthur Wolf, Cyrus Chu, and other local collaborators. The website http://psfd.sinica.edu.tw contains detailed information (including the contents of questionnaires) on this survey project.

also available in this data set. The above information helps us construct relevant variables in the empirical framework.

The data used in this paper are taken from the first-year survey collected during 1999 to 2000. The number of respondents in the original data set was 2,959. Since *inter vivos* child-to-parent intergenerational transfers are the major concern of this paper, we confine the sample to those respondents with at least one living parent. We then delete those samples with missing values, with the final sample size turning out to be 1,333.[11]

The definitions and basic statistics (including means and standard errors) of the variables are summarized in Table 2. The third column contains the basic statistics for all samples, while columns 4–7 contain the basic statistics corresponding to samples with a specific asset status. In the questionnaire, the respondents were asked whether their parents had distributed their assets to their children. Possible answers to this question include: 1. All the assets have been distributed; 2. Only part of the assets have been distributed; 3. The assets have not been distributed yet; 4. Parents do not have any assets. Corresponding to the four categories, four dummies (D_1, D_2, D_3, and D_4) are constructed, whereby D_k(k $= 1, \ldots, 4$) equals one if the respondent chooses category k, and zero otherwise.

The child-to-parent transfers (T_k) are defined as the average monthly remittance flow provided by the child. The measurement of the residence distance variable refers to the following question in the questionnaire: "How far is your parents' residence from yours?" Possible answers to this question include: 1. coresiding; 2. neighbouring; 3. (interviewees' most frequent) traveling time is within 10 minutes; 4. traveling time is between 10–30 minutes; 5. traveling time is between 30–60 minutes; 6. traveling time is between 60–120 minutes, and 7. traveling time is more than 2 hours. This classification is so designed in that Taiwan is a small island and most populated areas can be reached within 120 minutes; and we leave the rest in category 7. This "distance of residence" variable is therefore coded as an ordinal index with discrete values ranging from 1 to 7. Detailed definitions of other variables can be found in the PSFD website given in footnote 10.

The tightness of the kinship network (N) is measured by the frequency of ancestral worship and attending the Chinese New Year family gathering. Those who answered "frequently" are coded as 1, and 0 otherwise.[12] If the respondent has only one living parent, then the average education and average age refer to this parent's years of education and age. If there are two living parents, then the usual meaning of average applies. All other variables in Table 2 are self-explanatory and we shall not repeat the explanations here. To facilitate a comparison, most of the regressors used are similar to those adopted by Cox (1987), Cox et al. (1998), and Perozek (1998). One exception is the mainlander ethnicity dummy, which is used to capture the possible behavioural difference of recent (those coming to Taiwan from Mainland China during and after World War II) versus earlier immigrants to Taiwan.

---

[11] Among the 2,595 observations, 1,199 of them do not have living parents, 250 do not remember their father's exact age, 76 of them do not know the education of their parents, and 46 of them are reluctant to provide the data of T, the size of transfers.

[12] It is a Chinese custom to have a kinship union at the Chinese New Year eve dinner; most members in the network will attend the gathering unless they are occupied by other important commitments, or feel alienated.

Table 2. Means (standard deviations) of variables.

Variable	Definition	Means (std. dev.) All	Means (std. dev.) $D_1 = 1$	Means (std. dev.) $D_2 = 1$	Means (std. dev.) $D_3 = 1$	Means (std. dev.) $D_4 = 1$
T_k	Remittance per month (in NT thousand dollars)	5.725	9.374	4.587	5.951	3.228
		(28.574)	(49.760)	(7.468)	(23.902)	(9.561)
L	Distance of residence	4.038	3.569	4.214	4.114	4.238
		(2.249)	(2.131)	(2.157)	(2.291)	(2.261)
N	1 if child attended tomb visiting, ancestral worshipping, and Chinese New Year gathering Frequently (0 otherwise)	0.680	0.680	0.738	0.690	0.660
		(0.466)	(0.467)	(0.442)	(0.463)	(0.474)
D_1	1 if parents had transferred all of their assets (0 otherwise)	0.229	—	—	—	—
		(0.421)	—	—	—	—
D_2	1 if parents had transferred part of their assets (0 otherwise)	0.063	—	—	—	—
		(0.243)	—	—	—	—
D_3	1 if parents had not transferred their assets yet (0 otherwise)	0.367	—	—	—	—
		(0.482)	—	—	—	—
D_4	1 if parents do not have any assets (0 otherwise)	0.340	—	—	—	—
		(0.474)	—	—	—	—
Child level information						
AGEC	Age (in years)	47.364	50.810	44.631	44.655	48.472
		(7.040)	(7.403)	(5.155)	(5.910)	(6.912)
MALEC	1 if male	0.491	0.516	0.548	0.056	0.448
		(0.500)	(0.501)	(0.501)	(0.500)	(0.498)
EDUC	Education (in years)	10.005	8.814	12.119	11.069	9.267
		(4.201)	(4.437)	(3.639)	(3.768)	(4.169)

(continued)

Table 2. (*continued*)

Variable	Definition	Means (std. dev.) All	Means (std. dev.) $D_1 = 1$	Means (std. dev.) $D_2 = 1$	Means (std. dev.) $D_3 = 1$	Means (std. dev.) $D_4 = 1$
INCOMEC	Income per month (in NT thousand dollars)	65.433	61.195	94.857	76.345	51.035
		(244.283)	(290.108)	(150.663)	(320.958)	(54.202)
MARRIEDC	1 if married	0.862	0.869	0.845	0.847	0.876
		(0.345)	(0.338)	(0.364)	(0.360)	(0.330)
CHILDC	1 if has children	0.931	0.967	0.893	0.894	0.954
		(0.201)	(0.178)	(0.311)	(0.308)	(0.210)
SIBC	Number of siblings	4.474	4.797	4.226	4.128	4.673
		(2.002)	(2.039)	(2.147)	(1.809)	(2.090)
RURALC	1 if resident in rural area	0.767	0.804	0.702	0.778	0.742
		(0.500)	(0.400)	(0.460)	(0.416)	(0.438)
Parent level information						
ALIVEP	1 if both parents alive	0.491	0.258	0.619	0.684	0.415
		(0.500)	(0.438)	(0.489)	(0.466)	(0.493)
AGEP	Average age of alive parents (in years)	75.020	78.536	73.310	72.166	76.235
		(8.016)	(8.041)	(7.179)	(7.217)	(7.738)
MAINP	1 if father mainlander	0.105	0.052	0.048	0.110	0.146
		(0.307)	(0.223)	(0.214)	(0.313)	(0.353)
EDUP	Average education of alive parents (in years)	3.907	2.708	5.345	4.702	3.591
		(3.993)	(3.786)	(3.936)	(3.931)	(3.950)
PHEALTHP	1 if one or both of the alive parents had very poor health	0.032	0.042	0.012	0.018	0.042
		(0.175)	(0.202)	(0.109)	(0.134)	(0.201)
Number of observations		1,333	306	84	490	453

4. Results

4.1. PRELIMINARY FINDINGS

A key premise of this paper is that child-to-parent transfers are sensitive to the tightness of kinship networks. Before pursuing this issue, we first use the whole sample to estimate equations (2)–(5) without taking kinship networks into account. The results are listed in Tables 3a and b, where Table 3a includes only one asset-transfer dummy variable (D_1) while Table 3b includes D_1–D_3.[13] Table 3a shows that for families with assets completely transferred, the child-to-parent remittance is significantly higher than families with another asset disposition status. Similar results can be found in Table 3b, which shows that the dummy D_1 has a distinct effect on children's transfers, while the effects of other asset-transfer dummies are statistically indistinguishable. These results are consistent with the statistics listed in the previous section and the expression made in the beginning of this paper.

Another interesting feature in Tables 3a and b is that the asset-transfer dummies show opposite signs in the equation of residence distance as compared with the signs shown in the transfers equation. This means that children will enact a feedback either by providing monetary transfers or by residing closer to their parents. The ρ_{12} parameters in Tables 3a and b are not significantly different from zero, implying that we could have estimated the two equations separately.

With regard to other exogenous variables, Tables 3a and b demonstrate that most of the significant variables have signs as predicted by existing literature. With regard to the variables on the children's side, we found that younger children tend to provide more transfers to their parents. This is probably due to younger children having less family commitments of their own and hence are capable of transferring more income to their parents. Comparing with females, male respondents provide a similar amount of transfers, yet they live closer to their parents. These results are largely consistent with that in Weinstein et al. (1990), which indicates the role expectation of a male child's responsibility to look after his parents.

Children with a higher education tend to transfer more money to their parents, but live farther away. More educated children have generally received more educational support from parents in the past and have larger wealth now, and therefore tend to feed back more to their parents. The fact that more educated children tend to live farther away is probably because their jobs are more professional and harder to find in towns or cities near their parents.

Married children and children having their own kids also live farther away from their parents. This is the case because married couples are much more constrained by the joint

[13] To simplify the presentation, the estimates for parameters c_1–c_6 are omitted in these tables. The purpose of including D_1 is to identify the difference between the complete-distribution status and other scenarios; the purpose of including D_1–D_3 is to show that the complete-distribution status is distinct even under a more refined classification of asset disposition.

Table 3a. MLE estimates: D_1 included.

	Remittance per month Eq. (2)	Distance of residence Eq. (3)
Constant	−16.368	0.361
	(0.707)	(0.965)
Child level information		
AGEC	−0.865**	0.007
	(2.202)	(1.083)
MALEC	3.284	−0.720***
	(1.036)	(10.552)
EDUC	2.054***	0.018*
	(3.724)	(1.863)
INCOMEC	0.003	0.000
	(0.136)	(1.408)
MARRIEDC	4.022	0.339***
	(0.605)	(3.401)
CHILDC	1.814	0.530***
	(0.215)	(4.052)
SIBC	2.173**	0.044***
	(2.302)	(2.657)
RURALC	−1.289	−0.088
	(0.343)	(1.158)
Parent level information		
ALIVEP	6.056*	0.031
	(1.838)	(0.468)
AGEP	0.066	−0.009
	(0.204)	(1.619)
MAINP	−4.346	−0.084
	(0.597)	(0.771)
EDUP	−0.172	0.001
	(0.331)	(0.086)
PHEALTHP	−0.512	−0.334*
	(0.044)	(1.809)
D_1	10.443***	−0.274***
	(2.742)	(3.465)
ρ_{12}		0.025
		(0.559)
Number of observations	1,333	
Log likelihood	−6,252.58	

Note: Absolute *t* values are in parentheses.
*Significant at the 10 percent level.
**Significant at the 5 percent level.
***Significant at the 1 percent level.

Table 3b. MLE estimates: D_1, D_2, D_3 included.

	Remittance per month Eq. (2)	Distance of residence Eq. (3)
Constant	−21.507	0.410
	(0.917)	(1.072)
Child level information		
AGEC	−0.823**	0.007
	(2.094)	(1.024)
MALEC	3.238	−0.720***
	(1.017)	(10.525)
EDUC	1.988***	0.018*
	(3.547)	(1.917)
INCOMEC	0.002	0.001
	(0.128)	(1.441)
MARRIEDC	4.024	0.339***
	(0.606)	(3.400)
CHILDC	1.272	0.524***
	(0.279)	(3.977)
SIBC	2.273**	0.044***
	(2.354)	(2.623)
RURALC	−1.585	−0.085
	(0.417)	(1.108)
Parent level information		
ALIVEP	5.299	0.039
	(1.601)	(0.583)
AGEP	0.076	−0.009
	(0.236)	(1.633)
MAINP	−3.578	−0.089
	(0.486)	(0.811)
EDUP	0.187	0.001
	(0.354)	(0.091)
PHEALTHP	0.108	−0.342*
	(0.009)	(1.847)
D_1	13.013***	−0.298***
	(2.771)	(3.458)
D_2	2.951	−0.008
	(0.347)	(0.060)
D_3	5.617	−0.061
	(1.249)	(0.802)
ρ_{12}		0.359
		(0.599)
Number of observations	1,333	
Log likelihood	−6,252.58	

Note: Absolute *t* values are in parentheses.
*Significant at the 10 percent level.
**Significant at the 5 percent level.
***Significant at the 1 percent level.

location choice of the families or the location of their children's schools, and therefore are unable to live closer to their parents even if they wanted to. Children with more siblings provide more transfers and live farther away. The effect found in the residence distance equation can be explained by the free-rider effect, whereas the effect found in the transfers equation seems to reveal a demonstration effect.[14] Since monetary transfer is a more convenient (less time-consuming) way of feeding back than living closely by, siblings may choose the more convenient way and avoid the costly visits.

Concerning variables on the parent's side, as conjectured, children tend to transfer more to their parents when both parents are alive. Moreover, for parents who have very poor health conditions, their children tend to live closer. Notice that the asset-transfer variable D_l is very significant in both Tables 3a and b.

4.2. THE KINSHIP NETWORK EFFECT

In this subsection we consider the effect of kinship networks and estimate the econometric model constructed in Section 3. We first separate the sample into two groups according to the tightness of kinship networks. We then estimate the econometric model for these two groups separately and present the results in Tables 4a and b. Tables 4a and b are only different in the choice of asset-transfer dummy variables.

Both Tables 4a and b show that for children who have tighter kinship networks ($N = 1$), their money transfers are significantly higher when their parents have transferred all the assets. However, such an effect does not exist in the sample with looser kinship networks ($N = 0$). The above results are consistent with our previous conjecture that asset transfer may trigger gossip among relatives if the child's behaviour does not fit the norm.

In the residence-distance equations, similar results are found. Children who have tighter kinship networks ($N = 1$) will live closer to their parents' residence when the latter's assets are exhausted. As to the sample with looser kinship networks ($N = 0$), only a minor effect exists for the D_l variable in Table 4b.[15] The ρ_{12} parameters in Tables 4a and b are insignificant, again implying that the residence-distance decision and child-to-parent transfers decision are uncorrelated.

Some exogenous variables behave differently with respect to the two sample groups (those with $N = 1$ and those with $N = 0$). As we can see from Tables 4a and b, kinship networks impose more social pressure cost on children with certain characteristics (such as younger children, richer children, or children with more siblings) and induce these children to provide more transfers, whereas such pressures are not significant for samples with $N = 0$.

---

[14] Similar results are found in Perozek (1998). The free-rider effect reflects a person's moral hazard of "letting other siblings do the visits," whereas the demonstration effect means that other siblings' transfer sizes set up the standard for a particular child's transfer amount.

[15] In Table 4b, the D_l variable is significant at the 10 percent level.

Table 4a. MLE estimates: D_1 included.

	$N = 1$		$N = 0$	
	Remittance per month	**Distance of residence**	**Remittance per month**	**Distance of residence**
Constant	−4.134	0.513	−47.758	−0.192
	(0.139)	(1.122)	(1.144)	(0.274)
Child level information				
AGEC	−1.063**	0.005	−0.473	0.013
	(2.094)	(0.643)	(0.681)	(1.067)
MALEC	5.651	−0.788***	−2.371	−0.510***
	(1.223)	(9.389)	(0.362)	(4.043)
EDUC	1.879***	0.018	2.064**	0.010
	(2.607)	(1.504)	(2.091)	(0.579)
INCOMEC	0.026	0.002**	0.003	0.000
	(0.707)	(2.302)	(0.109)	(0.986)
MARRIEDC	5.495	0.303**	0.181	0.330**
	(0.611)	(2.221)	(0.018)	(2.085)
CHILDC	1.770	0.613***	0.853	0.545***
	(0.156)	(3.425)	(0.070)	(2.595)
SIBC	3.494***	0.061***	−0.627	0.011
	(2.746)	(3.013)	(0.360)	(0.340)
RURALC	−1.113	−0.047	−0.321	−0.096
	(0.219)	(0.487)	(0.044)	(0.703)
Parent level information				
ALIVEP	3.907	−0.001	12.758*	0.042
	(0.829)	(0.013)	(1.841)	(0.345)
AGEP	−0.084	−0.013*	0.465	−0.001
	(0.204)	(1.914)	(0.781)	(0.116)
MAINP	−2.510	−0.186	−4.668	−0.005
	(0.264)	(1.402)	(0.391)	(0.024)
EDUP	0.036	0.004	−0.741	−0.002
	(0.054)	(0.389)	(0.731)	(0.087)
PHEALTHP	−3.298	−0.381*	5.642	−0.189
	(0.214)	(1.663)	(0.293)	(0.592)
D_1	12.290**	−0.331***	7.910	−0.188
	(2.379)	(3.245)	(1.002)	(1.333)
ρ_{12}		0.019		0.030
		(0.329)		(0.309)
Number of observations	907		426	
Log likelihood	−4,278.28		−1,939.09	

Note: Absolute t values are in parentheses.
*Significant at the 10 percent level.
**Significant at the 5 percent level.
***Significant at the 1 percent level.

Table 4b. MLE estimates: D_1, D_2, D_3 included.

	$N = 1$		$N = 0$	
	Remittance per month	**Distance of residence**	**Remittance per month**	**Distance of residence**
Constant	−7.951	0.471	−53.382	−0.075
	(1.012)	(0.264)	(1.268)	(0.103)
Child level information				
AGEC	−1.048**	0.005	−0.387	0.011
	(2.060)	(0.681)	(0.543)	(0.889)
MALEC	5.658	−0.789***	−2.355	−0.514***
	(1.212)	(9.353)	(0.355)	(4.070)
EDUC	1.831**	0.018	1.906*	0.016
	(2.477)	(1.438)	(1.877)	(0.854)
INCOMEC	0.028	0.001**	0.003	0.000
	(0.730)	(2.228)	(0.099)	(1.037)
MARRIEDC	5.594	0.307**	−0.363	0.352**
	(0.622)	(2.235)	(0.035)	(2.214)
CHILDC	1.851	0.615***	2.607	0.487**
	(0.162)	(3.409)	(0.202)	(2.294)
SIBC	3.526***	0.062***	−0.544	0.009
	(2.768)	(3.022)	(0.301)	(0.272)
RURALC	−1.226	−0.049	−0.494	−0.086
	(0.235)	(0.505)	(0.067)	(0.619)
Parent level information				
ALIVEP	3.108	−0.005	12.039*	0.065
	(0.657)	(0.057)	(1.703)	(0.532)
AGEP	−0.064	−0.013	0.437	−0.000
	(0.154)	(1.887)	(0.716)	(0.025)
MAINP	−2.214	−0.173	−2.770	−0.064
	(0.232)	(1.292)	(0.220)	(0.303)
EDUP	0.050	0.004	−0.794	−0.000
	(0.074)	(0.350)	(0.772)	(0.004)
PHEALTHP	−2.996	−0.370	6.205	−0.209
	(0.193)	(1.608)	(0.313)	(0.659)
D_1	14.088**	−0.310***	11.513	−0.286*
	(2.297)	(2.801)	(1.362)	(1.847)
D_2	−2.344	0.137	11.982	−0.385
	(0.219)	(0.832)	(0.824)	(1.483)
D_3	4.687	0.027	7.286	−0.212
	(0.789)	(0.290)	(0.928)	(1.518)
ρ_{12}		0.020		0.033
		(0.341)		(0.336)
Number of observations	907		426	
Log likelihood	−4,276.82		−1,935.65	

Note: Absolute t values are in parentheses.
*Significant at the 10 percent level.
**Significant at the 5 percent level.
***Significant at the 1 percent level.

In order to test whether there is a significant network effect, we combine the two groups with $N = 0$ and $N = 1$, and run a regression including both N and $N \times D_1$ as explanatory variables. As one can see from Table 5, the kinship network does play a significant role, either independently or through the interaction with D_1, in children's decisions on their living locations as well as the size of their transfers. Children who have a tighter kinship network tend to live closer to their parents, and for those whose parents have divided and transferred all their assets, the kinship network has a statistically significant magnifying effect in child-to-parents filial piety.

4.3. FURTHER DISCUSSION

A further complication to the empirical setting is to consider the possible endogeneity of asset-transfer variables (D). Suppose the parent knows the interaction effect of the kinship network on the asset transfer status. Rational parents may then choose a certain asset status in order to take advantage of the network pressure on their children, thereby increasing the latter's feedback. Alternatively, if there are some variables which may help determine the child's remittance as well as the parent's asset transfer, but which are unknown to the econometrician, then simultaneity may also arise. In either case, we should take into account this endogeneity.

We perform endogeneity tests based on the procedure proposed by Rivers and Vuong (1988) and Vella (1993). The testing procedure and some of the testing results are listed in Appendix B. The test result shows that the hypothesis of endogeneity is rejected, which suggests that the asset-transfer variables can be treated as exogenous in our empirical framework.[16]

Another query, which might be raised is that our sample includes respondents of both sexes instead of just males. As noted by Chan (1990) and other scholars, Taiwan is a society dominated by male lineage. By custom the adult sons are responsible for their parents' old-age support. According to the above argument, confining the sample to male respondents seems to be more reasonable. We reestimate our empirical model by restricting the sample to males only. The main results found in the previous section still hold true after this manipulation.[17] These efforts indicate that our results are robust with respect to different econometric settings and different sample selection rules.

5. Conclusion

Previous literature on parent-to-child transfers focuses on two kinds of behaviour: *inter vivos* and *post-mortem*. Some recent surveys conducted in Taiwan show phenomena different from what have been established in the literature. First, concerning the interaction between adult children and their parents, child-to-parent transfers are more common than

[16] Consistent with our findings, Sung (1974) noted that the practice and timing of asset distribution in Taiwanese families is dominated by custom. In addition, assets are often shared evenly by the male descendants in most families.

[17] The relevant tables are not listed in this paper, but are available from the authors.

Table 5. MLE estimates: N, $D_1 \times N$ included.

	Remittance per month Eq. (2)	Distance of residence Eq. (3)
Constant	−16.621	0.446
	(0.710)	(1.178)
Child level information		
AGEC	−0.867**	0.008
	(2.220)	(1.270)
MALEC	3.105	−0.701***
	(0.855)	(10.248)
EDUC	2.025***	0.020**
	(3.712)	(2.065)
INCOMEC	0.003	0.001*
	(0.190)	(1.865)
MARRIEDC	3.564	0.355***
	(0.533)	(3.512)
CHILDC	1.947	0.564***
	(0.228)	(4.275)
SIBC	2.150**	0.045***
	(2.231)	(2.700)
RURALC	−1.419	−0.073
	(0.369)	(0.955)
Parent level information		
ALIVEP	5.863*	0.023
	(1.708)	(0.349)
AGEP	0.085	−0.010*
	(0.259)	(1.808)
MAINP	−4.035	−0.104
	(0.552)	(0.954)
EDUP	−0.238	0.003
	(0.462)	(0.323)
PHEALTHP	−0.264	−0.328*
	(0.023)	(1.787)
N	0.907	−0.218***
	(0.231)	(3.242)
$D_1 \times N$	11.421**	−0.347***
	(2.560)	(3.454)
ρ_{12}		0.026
		(0.581)
Number of observations	1,333	
Log likelihood	−6,241.22	

Note: Absolute t values are in parentheses.
*Significant at the 10 percent level.
**Significant at the 5 percent level.
***Significant at the 1 percent level.

parent-to-child transfers. Second, the ratio of parents who have distributed all their assets is quite high in Taiwan, contrary to the practice in the West. A preliminary investigation on the relationship between the asset-transfer status and child-to-parent transfers show that, compared with those families whose assets remained undistributed or who have no assets, child-to-parent transfers are relatively higher for families whose assets have been distributed completely. The existing literature does not explain this unique practice in Taiwan, and therefore we propose a theory of social networks to explain the phenomena, and examine this theory empirically using the Taiwan PSFD data set.

The empirical results show that for children who belong to tighter kinship networks, their transfers (the residence distance) will be higher (closer) in the case when their parents' assets are already completely distributed. However, the relationship between children's feedback and parents' asset-transfer status no longer holds for children with looser kinship networks. These results are consistent with our theoretical argument that asset-transfer may trigger gossip among relatives if the child's feedback behaviour does not fit the norm.

Considering the possibility that the asset-transfer decision may be manipulated by the parents, we perform an endogeneity test. Our results show that the hypothesis of endogenous asset-transfer is rejected. This finding is consistent with Sung's (1974) argument that the practice and timing of asset transfer in Taiwan is largely dominated by custom.

Appendix A: Derivation of the Likelihood Function

We first rewrite equations (2) and (3) as follows:

$$\begin{aligned} T_{ki}^{*} &= X_i\alpha_1 + D_i\beta_1 + \varepsilon_{1i} \\ &\equiv Z_i\gamma_1 + \varepsilon_{1i}, \\ L_{ki}^{*} &= X_i\alpha_2 + D_i\beta_2 + \varepsilon_{2i} \\ &\equiv Z_i\gamma_2 + \varepsilon_{2i}, \end{aligned}$$

where

$$\begin{bmatrix} \varepsilon_{1i} \\ \varepsilon_{2i} \end{bmatrix} \sim N\left(\begin{bmatrix} 0 \\ 0 \end{bmatrix}, \begin{bmatrix} \sigma_{11} & \sigma_{12} \\ \sigma_{12} & 1 \end{bmatrix} \right).$$

Defining $c_0 = -\infty$ and $c_7 = \infty$, the likelihood function can be written as follows:

$$\begin{aligned} L = &\prod_{j=1}^{7} \prod_{\substack{T_{ki} > 0 \\ L_i = j}} \int_{c_{j-1}-Z_i\gamma_2}^{c_j - Z_i\gamma_2} f(\varepsilon_{1i}, \varepsilon_{2i})\, d\varepsilon_{2i} \\ &\times \prod_{j=1}^{7} \prod_{\substack{T_{ki} = 0 \\ L_i = j}} \int_{-\infty}^{-Z_i\gamma_1} \int_{c_{j-1}-Z_i\gamma_2}^{c_j - Z_i\gamma_2} f(\varepsilon_{1i}, \varepsilon_{2i})\, d\varepsilon_{2i}\, d\varepsilon_{1i}. \end{aligned}$$

Appendix B: Endogeneity Test of Asset-Division Status

For ease of demonstration, consider the case that D_i contains D_{1i} only. To allow for the endogeneity of the asset-division status variable (D_1), we add the following equation in addition to equations (2) and (3):

$$D^*_{1i} = W_i\delta + \varepsilon_{3i}, \tag{A1}$$

where D^*_{1i} is a latent variable and only the sign of D^*_{1i} is observed:

$$D_{1i} = \begin{cases} 1, & \text{if } D^*_{1i} > 0, \\ 0, & \text{otherwise.} \end{cases}$$

As suggested by Rivers and Vuong (1988) and Vella (1993), equations (2) and (3) can be rewritten as follows

$$T^*_{ki} = X_i\alpha_1 + \beta_1 D_{1i} + \lambda_1\varepsilon_{3i} + \varepsilon'_{1i} \tag{A2}$$

$$L^*_{ki} = X_i\alpha_2 + \beta_2 D_{1i} + \lambda_2\varepsilon_{3i} + \varepsilon'_{2i}. \tag{A3}$$

The null hypothesis for the endogeneity test can be specified as

$$H_0 : \lambda_1 = \lambda_2 = 0.$$

To perform the endogeneity test, we first estimate equation (A1) by the probit method and then obtain the generalized residuals

$$\hat{\varepsilon}_{3i} = (D_{1i} - \hat{\Phi}_i)\,\hat{\phi}_i(1 - \hat{\Phi}_i)^{-1}\,\hat{\Phi}_i^{-1},$$

where $\hat{\Phi}_i$ and $\hat{\phi}_i$ are the CDF and PDF of the standard normal distribution evaluated at the probit estimates of equation (A1).

The regressors ε_{3i} in equations (A2) and (A3) are then replaced by the residuals $\hat{\varepsilon}_{3i}$ and these new equations are called (A2′) and (A3′). The maximum likelihood estimation of (A2′)–(A3′) produces consistent estimates of α_1, α_2, β_1, β_2, λ_1, and λ_2. The Wald test is performed to examine the endogeneity of the asset-division status.

For brevity, we only report the results corresponding to the setup of Table 3a here. The results for equations (A2′)–(A3′) and equation (A1) are presented in Tables B1 and B2, respectively. The Wald statistics listed in Table B1 show that the endogeneity of D_1 is rejected.

Table B1. MLE estimates: endogeneity test of D_1.

	Remittance per month Eq. (2)	Distance of residence Eq. (3)
Constant	−15.810	0.361
	(0.608)	(0.964)
Child level information		
AGEC	−0.860**	0.007
	(2.182)	(1.082)
MALEC	3.251	−0.720***
	(1.025)	(10.533)
EDUC	2.057***	0.018*
	(3.724)	(1.864)
INCOMEC	0.003	0.001
	(0.154)	(1.515)
MARRIEDC	3.979	0.339***
	(0.597)	(3.395)
CHILDC	1.624	0.530***
	(0.193)	(4.049)
SIBC	2.136**	0.044***
	(2.240)	(2.657)
RURALC	−1.211	−0.088
	(0.320)	(1.158)
Parent level information		
ALIVEP	6.862**	0.031
	(2.083)	(0.442)
AGEP	0.038	−0.009
	(0.117)	(1.599)
MAINP	−3.853	−0.084
	(0.533)	(0.776)
EDUP	−0.155	0.001
	(0.298)	(0.085)
PHEALTHP	−0.734	−0.334*
	(0.062)	(1.809)
D_1	15.418***	−0.275**
	(2.264)	(2.074)
$\hat{\varepsilon}_3$	−33.310	0.006
	(1.014)	(0.009)
ρ_{12}		0.025
		(0.562)
Number of observations	1,333	
Log likelihood	−6,251.70	
Wald statistic for $\lambda_1 = \lambda_2 = 0$	1.031	

Note: Absolute t values are in parentheses.
*Significant at the 10 percent level.
**Significant at the 5 percent level.
***Significant at the 1 percent level.

Table B2. Probit estimates for asset division status (D_1).

	Coefficient Estimates	Absolute *t* values
Constant	−0.627***	3.067
Child level information		
Living in rural area before 16 (1 = yes)	0.120	0.895
Parent level information		
Both parents alive (1 = yes)	−0.384***	3.146
Father alive (1 = yes)	−0.263**	2.219
Father mainlander (1 = yes)	−0.557***	3.339
Average education (in years)	−0.021*	1.802
Father born before 1921 (1 = yes)	0.536***	4.186
Father born in 192130 (1 = yes)	0.161	1.334
Number of sons	−0.028	0.966
Number of daughters	0.001	0.051
Number of observations	1,333	
Log likelihood	−644.964	

Note: Absolute *t* values are in parentheses.
*Significant at the 10 percent level.
**Significant at the 5 percent level.
***Significant at the 1 percent level.

Acknowledgement

An earlier version of this paper was presented at the 2001 IUSSP International Meeting on Age-Structure Transitions and Policy Dynamics. We have benefited from comments and suggestions from Anne Gauthier, Robert Schoen, Shripad Tuljapurkar, and Oded Stark.

References

Altonji, J.G., Hayashi, F. and Kotlikoff, L.J. (1997), "Parental altruism and *inter vivos* transfers: Theory and evidence." *Journal of Political Economy,* 105:1121–1166.

Becker, G.S. (1974), "A Theory of social interactions." *Journal of Political Economy,* 82:1063–1093.

Bernheim, B.D., Shleifer, A. and Summers, L.H. (1985), "The strategic bequest motive." *Journal of Political Economy,* 93:1045–1076.

Bott, E. (1968), *Family and Social Network: Roles, Norms and External Relationships in Ordinary Families.* Tavistock Publications: London.

Chan, C.-N. (1990), *Families and Society.* Linking Publications (in Chinese): Taipei.

Cox, D. (1987), "Motives for private income transfers." *Journal of Political Economy,* 95:508–546.

Cox, D., Eser, Z. and Jimenez, E. (1998), "Motives for private transfers over the life cycle: An analytical framework and evidence for Peru." *Journal of Development Economics,* 55:57–80.

Freeman, R., Thornton, A. and Yang, L.-S. (1994), "Determinants of coresidence in extended households." In: *Social Changes and the Family in Taiwan*, A. Thornton and H.-S. Lin, eds. University of Chicago Press: Chicago.

Kandel, E. and Lazear, E.P. (1992), "Peer pressure and partnership." *Journal of Political Economy,* 100:801–817.

Kotlikof, L.J. and Spivak, A. (1981), "The family as an incomplete annuities market." *Journal of Political Economy,* 89:372–391.

Laitner, J. (1997), "Intergenerational and intrahousehold economic links." In: *Handbook of Population and Family Economics,* Vol. 1, A. M. R. Rosenzweig and O. Stark, eds. North-Holland: Amsterdam.

Lindbeck, A. (1997) "Incentives and social norms in household behavior." *American Economic Review,* 87:370–377.

Lindbeck, A., Nyberg, S. and Weibull, J.W. (1999), "Social norms and economic incentives in the welfare state." *Quarterly Journal of Economics,* 114:1–35.

Perozek, M.G. (1998), "A reexamination of the strategic bequest motive." *Journal of Political Economy,* 106:423–445.

Rivers, D. and Vuong, Q.H. (1988), "Limited information estimators and exogeneity tests for simultaneous probit models." *Journal of Econometrics,* 39:347–366.

Sung, L.-S. (1974), *Inheritance and Kinship in North Taiwan*. Ph.D. dissertation, Department of Anthropology, Stanford University.

Vella, F. (1993), "A simple estimator for simultaneous models with censored endogenous regressors." *International Economic Review,* 34:441–457.

Weinstein, M., Sun, T.-H., Chang, M.-C. and Freedman, R. (1990), "Household composition, extended kinship, and reproduction in Taiwan: 1965–1985." *Population Studies,* 44:217–239.

Weinstein, M., Sun, T.-H., Chang, M.-C. and Freedman, R. (1994), "Co-residence and other ties linking couples and their parents." In: *Social Changes and the Family in Taiwan*, A. Thornton and H.-S. Lin, eds. University of Chicago Press: Chicago.

Wilhelm, M.O. (1996), "Bequest behavior and the effect of heirs' earnings: Testing the altruistic model of bequests." *American Economic Review,* 86:874–892.

Wellman, B. (1990), "The place of kinfolk in personal community networks." *Marriage and Family Review,* 15:195–228.

Yeh, K.-H. (1994), "Living arrangement of elderly parents in Taiwan: A psychological perspective." *Bulletin of the Institute of Ethnology, Academia Sinica,* 83:121–168 (in Chinese).

CHAPTER 3. STABILITY AND CHANGE IN PATTERNS OF INTERGENERATIONAL TRANSFERS IN TAIWAN

I-FEN LIN
Department of Sociology, Bowling Green State University

NOREEN GOLDMAN
Office of Population Research, Princeton University

MAXINE WEINSTEIN
Centre for Population and Health, Georgetown University

YU-HSUAN LIN
Department of Health, Bureau of Health Promotion, Survey Research Centre for Population and Health

1. Introduction

Many researchers have documented patterns of intergenerational exchanges (e.g., Cox and Rank 1992; Eggebeen 1992; Hermalin, Ofstedal, and Chi 1992; Lee, Parish, and Willis 1994; Soldo and Hill 1995; Logan and Spitz 1996; Lillard and Willis 1997; Agree et al. 2005), but, by and large our understanding of the helping behaviours between generations reflects a static view. This limitation results from the reliance of most of these previous studies on a snapshot approach to understanding monetary or time transfers. Far fewer studies provide a dynamic view of what happens in the *same* family over a longer period of time (but for exceptions, see Whitbeck, Simons, and Conger 1991; Whitbeck, Hoyt, and Huck 1994; Silverstein 1995; Henretta et al. 1997; Chang 1999; Silverstein et al. 2002). In this study, we use a unique data set that follows a cohort of more than two thousand Taiwanese older adults over a 10-year period to assess stability in intergenerational transfers. Specifically, we raise two questions in this paper: first, do patterns of transfers in the same family change over time? Second, how are ageing parents' demographic, social, and economic characteristics associated with changes in the patterns of transfers?

Historically, Taiwan has been a patriarchal society that places a great emphasis on obedience to, and respect for, older family members. Traditionally, adult sons have been expected to bear the major responsibility for regular care and support of their parents. Ageing parents' coresidence with a married son has been a dominant and preferred living arrangement.

Anne H. Gauthier et al. (eds.), Allocating Public and Private Resources across Generations, 63–87.

During the past few decades, however, Taiwan has been undergoing rapid demographic and social transitions. In particular, increases in life expectancy, declines in fertility to below replacement level, increased out-migration among younger generations, and weakening norms regarding filial support have affected economic and social aspects of the lives of older adults. For instance, although coresidence with a married son is still a common practice, a growing number of older persons today are living alone or with their spouse only (Weinstein et al. 1990; Sun and Liu 1994; Hermalin, Ofstedal, and Chang 1996; Chang 1999). At the same time, independent living has gradually become a preferred living arrangement (Sun and Liu 1994; Lee, Lin, and Chang 1995; Chattopadhyay and Marsh 1999; Hsu, Lew-Ting, and Wu 2001).

In the face of these rapid demographic and social changes, we know little about whether patterns of intergenerational transfers in Taiwan are stable or changing over time. On the one hand, it is plausible that as filial norms weaken, the prevalence of provision of support from adult children to their parents may also decrease. On the other hand, adult children may provide financial support to compensate for not living with their parents, resulting in an increase in the prevalence of financial transfers.

To date, we know of only three published studies in Taiwan that have looked at changes in financial support over time, each of which is a repeated cross-sectional study based on reports of children of ageing parents. Chattopadhyay and Marsh (1999) found that the percentage of respondents who think sons *must* provide financial support to their older parents declined from 34.8 in 1963 to 8.6 in 1991, but the percentage of respondents who actually provided financial support to their older parents increased from 69.6 to 79.6 during the same period. The increase in financial support, the authors explained, is likely to be attributable to the decline in coresidence. Consistent with these results, Weinstein et al. (1990) and Sun and Liu (1994) also found that while coresidence declined between 1965 and 1986, the percentage of respondents who reported giving money to their husbands' parents was higher in the 1980s than in the 1970s.

Although these findings are informative, generalization of these studies is limited in three respects. First, studies based on repeated cross-sectional surveys cannot tell us much about the familial processes that are associated with changes in intergenerational transfers. Second, all of these studies rely on adult children's reports, but previous research has shown that children acknowledge more frequent financial transfers to their parents than their parents do (Roan, Hermalin, and Ofstedal 1996). This tendency to "over-report" may be exacerbated by children's guilt associated with permitting parents to live alone. Finally, these studies are based mainly on data collected in the 1960s, 70s, and 80s. New public programs designed to enhance the economic well-being of older adults, such as the national health insurance program initiated in 1995, may have affected private transfers. Thus, it is important to use recent data to examine patterns of intergenerational support.

In this study, we address these limitations by examining financial transfers provided by adult children to their parents in the *same* family between 1989 and 1999. The study is based on older parents' reports as opposed to children's. Our main objective is to explore the association between patterns of transfers and change in the parents' resources and needs over time. Because we are unable to identify the relative timing of various changes and we cannot rule out that these changes are caused by some common, unmeasured family characteristics, we are cautious about making causal inferences between major life transitions and changes in financial support.

In the following section of the paper, we review previous research and lay out three mechanisms that explain why adult children provide support to their parents. Next, we describe the data, measures, and the analytic strategy used in the study. We then present the results of the statistical analysis. Finally, we discuss the findings and their implications.

2. Theoretical Background

Researchers have proposed at least three primary mechanisms explaining why adult children provide support to their parents: the altruistic explanation, the reciprocal explanation, and the structural explanation. In the following section, we introduce each explanation and summarize previous findings concerning monetary support based on each mechanism.[1] Since financial transfers from children to their ageing parents are less common in Western than in Asian societies[2], most of the empirical findings discussed below are based on studies in Asia.

2.1. ALTRUISTIC EXPLANATION—AGEING PARENTS' NEEDS

The altruistic explanation argues that children's provision of support is contingent upon the needs of their parents (Becker 1991). Intergenerational transfers are given because of altruistic feelings by one family member for another. Riley (1983) also suggests that families are embedded in a "latent matrix" of social support and that this latent matrix can be achieved in terms of needs. In other words, adult children are likely to help their parents even if they had minimum contact with their parents or their relationships were antagonistic.

Consistent with this explanation, previous research in Asian societies has shown that parents who are not working are more apt than those who are working to receive financial support from their children (Taiwan: Hermalin, Ofstedal, and Lee 1992; China: Lee and Xiao 1998). Parents with lower incomes and with less education are also more likely than their counterparts to receive monetary assistance from their children (Taiwan: Hermalin, Ofstedal, and Lee 1992; Lee, Parish, and Willis 1994; China: Lee and Xiao 1998). Taiwanese parents are

---

1 Here we limit the literature review to studies examining financial support because the underlying dynamic that affects the propensity to receive financial support differs from that associated with other types of support, such as help with activities of daily living (Lin et al. 2003).

2 For instance, the estimated flow of financial support from adult children to their parents is quite low in the United States. The 1992 Health and Retirement Study shows that 9.2 percent of the respondents gave $500 or more to their nonresident parents in the past 12 months; the 1988 Panel Study of Income Dynamics reports 5.2 percent of children providing monetary support to their nonresident parents in the last year (McGarry and Schoeni 1995); and the 1987–1988 National Survey of Families and Households estimates that 4 percent of children gave a gift or loan worth more than $200 to their nonresident parents during the past 5 years (Eggebeen and Hogan 1990).

likely to receive support when they have already divided all property or when they have no property to divide (Lee, Parish, and Willis 1994). In China, parents' pensions reduce the likelihood and the amount of money that children provide to their parents (Yang 1996; Lee and Xiao 1998). Parents living in rural areas are more likely to receive monetary assistance than parents living in urban areas (Taiwan: Hermalin, Ofstedal, and Lee 1992; China: Lee and Xiao 1998).

However, not all research results are consistent with the altruistic explanation. For instance, although parents without a spouse may need more support from their children than parents with a spouse, Lee and her colleagues (Lee, Parish, and Willis 1994) find that Taiwanese sons are *not* more likely to provide financial support to widowed parents than to married parents. This may occur, they explain, because the total amount of support required for a widowed parent may be less than that for both parents together. Empirical results concerning the relationship between health and provision of financial support are mixed. Parents in poor health may receive more support than parents in better health. Some researchers find that parents with difficulty in activities of daily living are more likely than parents without difficulty to receive financial support (China: Lee and Xiao 1998); while others find that parents' health is unrelated to the likelihood and the amount of monetary transfers that children give to their parents (Taiwan: Hermalin, Ofstedal, and Lee 1992; China: Yang 1996).

2.2. RECIPROCAL EXPLANATION—PARENTS' INVESTMENT IN OR HELP WITH CHILDREN

In contrast to the altruistic explanation is the idea that intergenerational support is contingent upon the particular history of the parent–child relationship. The reciprocal explanation (Thibaut and Kelley 1959; Gouldner 1960; Greenberg 1980) suggests that giving support to their children legitimates parents' expectations of receiving support from their children and that the extent to which adult children help their parents is based on the parents' past contribution to them. Using data from Malaysia, Lillard and Willis (1997) find that adult children are more likely to pay monetary transfers and pay a larger amount when their parents provide childcare than when their parents do not. In Japan, ageing parents "earn" the support of their children through past sacrifices and by making contributions to their children's households (Hashimoto 1996).

In the face of increases in marital disruption over recent decades, American scholars have paid considerable attention to the effects of marital disruption on the bonds between parents and their children. New research explores the long-term consequences of marital disruption on intergenerational transfers during divorced parents' old age. This line of research has shown that children of divorced parents are less likely to provide financial support to their parents than children growing up in intact families and that the negative effect of divorce seems to be stronger for ageing fathers than for ageing mothers (Furstenberg, Hoffman, and Shrestha 1995; Pezzin and Schone 1999). Although the rate of marital disruption is still quite low in Taiwan, we examine whether parents' divorce or separation is associated with intergenerational transfers in this analysis because, as far as we know, no study has yet examined this issue.

2.3. STRUCTURAL EXPLANATION—AVAILABILITY OF POTENTIAL HELPERS

Older parents' needs cannot be completely fulfilled if they do not have family support networks, particularly when they have no children or can rely on only one child to provide support. Researchers have proposed two hypotheses concerning the relationship between number of children and the provision of support from children to their ageing parents. The additive model (Logan and Spitze 1996) hypothesizes that various potential providers in the family are functionally equivalent. Thus, parents are likely to receive support and will receive more support if they have a larger number of children. By contrast, the substitution model (Litwak 1985) suggests that there is a hierarchy of potential helpers based on the recipient's preferences and on the capacities of helpers. Only when the preferred helper is not available, do other children step in. Therefore, the number of children is unrelated to the provision of support. In both Taiwan and China, researchers have shown that, consistent with the additive model, the number of living children has a positive association with the propensity to receive financial support (Taiwan: Hermalin, Ofstedal, and Lee 1992; China: Lee and Xiao 1998).

3. Data and Methods

Data for this analysis come from the 1989 and 1999 waves of the Survey of Health and Living Status of the Elderly in Taiwan (for a detailed description of the survey design, see TPIFP et al. 1989). Baseline interviews for this longitudinal study were conducted in 1989 with a national probability sample of 4,049 persons aged 60 and over. Only one older family member from each household was interviewed. The response rate for the baseline survey was 92 percent. Respondents interviewed at baseline were followed up in 1993, 1996, and 1999. Our analysis is restricted to ageing parents who participated in the baseline study and the 1999 follow-up, a period that is long enough to examine changes in the provision of support.

Of the 4,049 persons interviewed at baseline, 2,310 (57 percent) were re-interviewed in 1999. Among nonrespondents, 85.5 percent ($N = 1{,}486$) died during the 10-year period, 13 percent ($N = 226$) did not complete the 1999 interview, and 1.6 percent ($N = 27$) were lost to follow-up. We further restrict the analysis to 2,187 parents who had at least one living child in both years. This number differs slightly from the total number of parents who were re-interviewed in 1999 ($N = 2{,}310$) because 85 parents did not have any living children in both years, 20 parents had a living child in 1999 only, and 18 parents had a living child in 1989 only.[3]

[3] Although we could include these 18 parents in the analysis and examine the extent to which death of children is associated with loss of financial support, we are unable to obtain complete information about the death of children over the 10-year period. In the 1993 and 1999 follow-up surveys, parents were asked whether or not any of their children passed away since the last interview (5.6 percent of the 3,155 respondents in 1993 and 4 percent of the 2,310 respondents in 1999 reported losing a child since the last interview). However, in the 1996 survey, parents were asked whether or not they had *ever* lost a child. Because of the lack of time frame in the 1996 survey question, as many as 27.8 percent of the 2,669 respondents reported that at least one of their children had passed away.

The data are richer than most other data on older adults in three respects. First and foremost, the surveys are the only panel study that provides a national sample of older adults in Taiwan.[4] Second, the surveys include detailed information on all household members and ask an extensive set of questions on instrumental and financial exchanges in which the ageing parent is the provider or the recipient. Finally, the surveys collect a wide range of detailed information about marital, residential, and occupational histories, household composition, social and economic exchanges, emotional and instrumental supports and demands, and physical and mental health.

Table 1 shows the distribution and mean of characteristics of parents at baseline and of parents who were interviewed in both years (i.e., follow-up sample). Since the attrition of this sample results mostly from mortality, it is not surprising to find that the follow-up sample is more likely than the baseline sample to be female, younger (between ages 60 and 69), and in excellent or very good health. Marriage, education, and socioeconomic status have been shown to be associated with better health outcomes (e.g., Beckett et al. 2002). This is reflected in the follow-up sample, which is more likely than the baseline sample to be educated, married, employed, and to own property and assets. As to financial transfers, parents who had died prior to follow-up were more likely to have received financial support from their children at baseline than parents who were interviewed at follow-up (76 versus 69 percent, data not shown). Parents who survived but did not complete the follow-up interview and parents who were lost to follow-up were less likely to receive financial support from their children at baseline than parents who were re-interviewed (48 and 56 percent versus 69 percent, data not shown).

3.1. DEPENDENT VARIABLE

Information was obtained for four types of intergenerational transfers—help with activities of daily living, help with household chores, material support, and financial support. In this paper, we focus only on financial support because it is one of the most frequent transfers from children to their parents in Taiwan (Lin et al. 2003).[5] In the 1989 survey, the participants were asked, "Is there anyone who gives money to help you now?" If the participants said yes, they were asked, "Who provides this support to you? Anyone else?" Interviewers then marked a check next to each provider recorded on the coresident and nonresident household rosters. In the 1999 survey, the participants were asked, "In the past year, has your child or relative given you or your spouse money for your daily living expenses?" If the participants said yes, they were asked, "Who gave you money for

[4] Although a new panel study, the Panel Study of Family Dynamics, started its first wave of data collection in 1999 and expects to follow the families over time, the study is intended to represent a national sample of a younger population (i.e., aged 35–64 in 1999) than the one analysed here. For details about the PSFD study, see http://psfd.sinica.edu.tw.

[5] Using the 1989 survey data, we found that 2.7 percent of older adults received help with activities of daily living (ADL), 13.7 percent received help with household chores, 35.9 percent received material support, and 71.8 percent received financial support from their children (Lin et al. 2003). Our preliminary analysis of the 1999 survey data also suggests a low prevalence of receiving help with ADL activities, help with household chores, and material support from children, although question wordings differ between the two surveys.

Table 1. Parents' characteristics.

Characteristics	Percentage*/Mean	
	Baseline sample	Follow-up sample
Receipt of financial support from children, 1989–1999		
Receives support in 1989 only		22.0
Receives support in 1999 only		13.5
Receives support in both years		47.6
Does not receive support in either year		16.8
Sex		
Male	57.1	53.4
Female	42.9	46.6
Age in 1989		
60–69 years old	65.1	76.8
70–79 years old	28.7	20.9
≥ 80 years old	6.2	2.2
Ethnicity		
Fukienese	60.5	62.2
Mainlander	22.2	20.5
Hakka	14.9	15.5
Other ethnicity (Aboriginal, Foreign)	1.8	1.5
Missing	0.6	0.4
Education completed by 1989		
Received no formal education (illiterate or can read)	50.2	46.5
Primary school	30.6	32.7
Junior high school	8.1	9.2
Senior high school	5.6	5.6
College or higher	5.1	5.5
Missing	0.4	0.5
Marital status in 1989		
Married	64.2	73.9
Cohabiting	0.4	0.2
Separated or divorced	3.0	1.8
Widowed	28.5	23.8
Never married	3.8	0.2
Missing	0.1	0.0
Self-reported health status in 1989		
Excellent or very good health	37.7	43.5
Good health	36.9	38.4
Fair or poor health	21.8	17.2
Missing	3.5	0.9
Number of living sons in 1989	2.3	2.5
Number of living daughters in 1989	2.3	2.4

(continued)

Table 1. (*Continued*)

Characteristics	Percentage*/Mean	
	Baseline sample	Follow-up sample
Coresiding with a child in 1989		
Yes	71.0	76.7
No	29.0	23.3
Area of residence in 1989		
Rural township	34.8	35.8
Urban township	36.4	35.8
City	28.8	28.4
Work status in 1989		
Working, full time	20.4	24.4
Working, part time	7.4	8.8
Not working	72.0	66.6
Missing	0.2	0.2
Property division status in 1989		
Has divided all or part of property	24.2	24.5
Has not divided property	42.3	49.1
Has no property	31.3	26.3
Missing	2.2	0.2
Assets status in 1989		
Has assets	63.7	69.0
Does not have assets	36.2	30.9
Missing	0.1	0.1
Pension status in 1989		
Has a pension	19.9	17.7
Does not have a pension	79.9	82.2
Missing	0.3	0.1
Death of a spouse, 1989–1999		
Death of a spouse		16.2
No death of a spouse		83.8
Change in health status, 1989–1999		
Same health status		30.2
Health declines		51.4
Health improves		17.5
Missing		0.9
Change in coresidence status, 1989–1999		
Lives with a child in 1989 only		18.1
Lives with a child in 1999 only		7.1
Lives with a child in both years		58.6
Does not live with a child in either year		16.2
Change in work status, 1989–1999		
Works in 1989 only		26.8
Works in 1999 only		1.7

Table 1. (*Continued*)

Characteristics	Percentage*/Mean	
	Baseline sample	Follow-up sample
Works in both years		6.2
Does not work in either year		64.8
Missing		0.6
Property divided between 1989 and 1999		
Yes		10.0
No		90.0
Change in assets status, 1989–1999		
Has assets in 1989 only		25.9
Has assets in 1999 only		9.6
Has assets in both years		43.1
Does not have assets in either year		21.3
Missing		0.1
Change in pension status, 1989–1999		
Has a pension in 1989 only		7.5
Has a pension in 1999 only		10.3
Has a pension in both years		10.2
Does not have pensions in either year		71.8
Missing		0.2
Help children with household chores and childcare, 1989–1999		
Provides help in 1989 only		9.6
Provides help in 1999 only		10.2
Provides help in both years		1.1
Does not provide help in either year		79.1
Number of cases (percentage)	4049 (100 percent)	2187 (100 percent)

*Column totals in each panel may deviate slightly from 100 percent due to rounding errors.

your living expenses?" Interviewers then wrote down each provider's relationship to the respondent, including coresident and nonresident providers.[6]

The question wording is not identical for the two surveys. Two differences in wording, which are likely to have counterbalancing effects, may affect our measure of the stability of transfers. First, the surveys refer to different time frames. The 1989 survey refers to current transfers, whereas the 1999 survey explicitly indicates transfers made in the past year. Because the time frame is longer in the more recent survey, we would expect the frequency

[6] In the 1999 survey, parents were further asked about the total amount of financial transfers provided by their children and relatives. Since the same question was not asked in the 1989 survey, we are not able to compare the size of transfers over the decade.

of transfers to be higher in 1999 than in 1989. Second, the definition of financial transfers is more restrictive in the 1999 survey (i.e., for "living expenses") than in the 1989 survey. Thus, the frequency of reports of financial support may be lower in 1999 than in 1989.

Table 1 shows the percentage of parents who received financial support from their children in 1989 and 1999. Of the 2,187 parents interviewed in both years, 47.6 percent received support in both years; 16.8 percent received no support in either year; 22.0 percent received support at baseline but not at follow-up; and 13.5 percent received no support at baseline but had support at follow-up. In sum, the data indicate that about two-thirds of the older adults have a stable pattern of financial transfers, in contrast to one-third of the older adults who gain or lose support between 1989 and 1999. Because of the differences in question wording noted above, the estimate of two-thirds with a stable pattern of financial transfers is likely to be conservative. Unlike previous findings based on repeated cross-sectional studies, our results suggest a higher percentage of ageing parents losing than gaining support over time.

3.2. EXPLANATORY VARIABLES

Our explanatory variables include the parents' characteristics in 1989 and their major life transitions, such as retirement or death of a spouse, between 1989 and 1999. Parents' characteristics comprise their sex; age; education; ethnicity; area of residence; marital, health, work, and coresidence status; ownership of property, pensions, and assets; and help with household chores and childcare. The choice of these variables derives from previous findings concerning intergenerational transfers (see Section 2). The distributions of these characteristics are shown in Table 1. Although females have longer life expectancy than males, the number of male respondents is slightly greater than that of female respondents, because more than one million male migrants moved from Mainland China to Taiwan with the Nationalist Government around 1950. The fact that most of the male migrants were soldiers in their late teens or early 20s leads to an imbalance in the sex ratio. Three-quarters of the older adults in the sample were in their 60s in 1989. The population in Taiwan consists of four ethnic groups: Aborigines, Fukienese, Hakka, and Mainlanders. Aborigines were the earliest migrants to Taiwan, followed by Fukienese (originating from Fukien Province in China), Hakka (originating from Kwangtung Province in China), and then Mainlanders (mainly the Nationalist party army and its supporters who left Mainland China during the 1949 Civil War). Fukienese constitute the major ethnic group in our sample (62.2 percent).

Approximately 80 percent are illiterate or received only a primary school education. Slightly less than three-quarters of the older adults in the sample are married; less than two percent are divorced or separated. Most older adults reported excellent, very good, or good health in 1989. Because sons traditionally bear the major responsibility of providing support to their ageing parents, we distinguish the number of living sons from the number of living daughters. The average numbers of living sons and of living daughters are both about two. Over three-quarters of the older adults lived with their children in 1989. More older adults lived in rural or urban townships than in big cities (71.6 percent versus 28.4 percent). One-third of the older adults were working in 1989. Slightly less than half of the older adults had not yet divided their property and about a quarter had no property in 1989. More than two-thirds of the older adults owned assets, such as a house, land, savings in a bank or farmer's association, mutual fund, stocks, bonds, jewelry, or other investments. About two in ten older adults relied on pensions as their major income source and one in ten assisted their children with household chores or childcare as of the baseline survey.

We expect major life transitions of ageing parents to be associated with the provision of support that parents receive from their children. Thus, we examine a series of changes occurring between 1989 and 1999: death of a spouse; changes in health status, coresidence, work status, property division, assets and pensions status; and parents' provision of support to their children. Approximately 16.2 percent of the older adults in the sample had a spouse who died over the 10-year period. A majority of the older adults (51.4 percent) reported health declines, less than one-third said their health remained the same, and 17.5 percent reported that their health improved over time. About 58.6 percent of the older adults lived with their children at both survey dates; 7.1 percent of the older adults who did not live with their children at baseline were coresiding in 1999. Two-thirds of the older adults were not working at the time of both surveys; 26.8 percent of the older adults were working at baseline but retired by 1999. One in ten older adults in the sample divided their property during the 10-year period. Approximately 43.1 percent of the older adults had assets in both years and about one-tenth had assets only at follow-up. Over 70 percent of the older adults did not have a pension at either survey date. Very few (less than 1 percent) of the parents provided help with household chores and childcare to their children in both years.

3.3. ANALYTIC STRATEGY

In the first part of the analysis, we consider parents surveyed in 1989 and 1999 as separate samples and examine whether the association between receipt of support and parents' characteristics varies by the survey year. The cross-sectional analysis consists of two stages: (1) estimation of two independent logistic regression models, one for each survey year; and (2) estimation of a single model based on a pooled sample of responses from both years, along with interaction terms between each of the explanatory variables and the survey year. The second step, which provides an equivalent model to those obtained in the first, permits us to test differences in coefficients between the two models. In the second part of the analysis, we look at the association between parents' characteristics, their major life transitions, and the patterns of receiving financial support from their children over the decade. This longitudinal analysis is conducted by estimating a multinomial logistic regression model with four outcomes—parents received support in both years (yes–yes), received no support in either year (no–no), received support at baseline but not at follow-up (yes–no), and received no support at baseline but had support at follow-up (no–yes).

Because there are more than two outcomes in the multinomial model, the exponential coefficients represent relative risk ratios (rather than odds ratios) and are difficult to interpret. To facilitate interpretation, we present predicted probabilities in addition to the relative risk ratios. The hypothetical probabilities of receiving financial support from children are calculated in the following manner. For the calculation of predicted probabilities by sex, we first assume that all respondents in the sample are female and calculate the predicted probability of each outcome by holding all characteristics other than sex constant at their observed values. We then assume that all respondents are male and recalculate the predicted probabilities. All of the explanatory variables are categorical except for the numbers of living sons and daughters. For these variables, we compute the predicted probabilities at three preselected values (0, 1, and 2). All estimates are obtained using the statistical package STATA (StataCorp 2001).

4. Results

Table 2 shows the extent to which the association between receipt of support and parents' characteristics varies by the survey year. The first model presents the likelihood of receiving financial support from children in 1989. All of the explanatory variables are measured in the 1989 survey. An asterisk is shown next to the odds ratio (OR) if the coefficient differs significantly ($p < 0.05$) from that for the reference category (OR = 1.00). The results suggest that parents' ethnic background, educational attainment, marital status, number of living children, and economic resources are related to whether they receive financial support from their children. Mainlanders are less likely than Fukienese to receive support from their children (OR = 0.60). A college degree is associated with a lower likelihood of having support than no education (OR = 0.60). The odds that widowed parents receive financial assistance from their children are 1.42 times the odds for married parents or parents with a companion. A larger number of children, regardless of the children's sex, is related to a higher likelihood of having support (OR = 1.23 and 1.13 for sons and daughters, respectively). Parents with greater economic resources—those who are working, have pensions, or have assets—are less likely than their counterparts to receive financial support from their children. In this multivariate model, the parents' sex, age, health status, coresidence status, area of residence, whether or not they retain property, and the provision of support related to household chores and childcare are not significantly associated with the propensity to receive support.

The second model in Table 2 is analogous to the first model but uses information obtained from the 1999 survey.[7] Similar to the results shown in the first model, Mainlanders are less likely than Fukiense to receive financial support from their children. Parents with fewer children or with greater economic resources are less likely than their counterparts to receive financial assistance. Having a college degree, however, is not significantly associated with the propensity to receive support from children, although the direction of the association remains the same as in 1989. Parents who divorced or separated and who have poorer health are less likely to receive support than their counterparts in 1999. Again, these relationships are consistent with those observed in 1989, but the coefficients are statistically significant only in 1999.

When we compare coefficients across the two models, we find four variables for which the relationship with the receipt of support differs significantly between the two survey years. First, widowed parents are more likely than married parents to receive support from their children in 1989 but not in 1999. Second, being single is negatively associated with receipt of support in 1989 but positively associated with receipt of support in 1999. Third,

[7] The respondents' sex, age, ethnicity, and educational attainment in the second model are based on the 1989 report because these variables do not change over time.

Table 2. Odds ratios (OR) of receiving financial support from children in 1989 and 1999 (estimates are based on logistic regression models).†

Characteristics‡	1989	1999	Difference between 1989 and 1999
Sex			
Male	1.00	1.00	
Female	1.17	1.17	n.s.
Age in 1989			
60–69 years old	1.00	1.00	
70–79 years old	1.15	0.84	n.s.
≥ 80 years old	0.56	0.65	n.s.
Ethnicity			
Fukienese	1.00	1.00	
Mainlander	0.60*	0.73*	n.s.
Hakka	1.31	1.02	n.s.
Other ethnicity (Aboriginal, Foreign)	2.59	3.39*	n.s.
Education completed by 1989			
Received no formal education	1.00	1.00	
Primary school	0.80	1.10	n.s.
Junior high school	0.89	1.35	n.s.
Senior high school	0.65	0.94	n.s.
College or higher	0.60*	0.79	n.s.
Marital status			
Married or cohabiting	1.00	1.00	
Separated or divorced	0.53	0.41*	n.s.
Widowed	1.42*	0.89	$p < 0.05$
Never married	0.15	2.70	$p < 0.05$
Self-reported health status			
Excellent or very good health	1.00	1.00	
Good health	0.95	0.77*	n.s.
Fair or poor health	0.74	0.71*	n.s.
Number of living sons	1.23*	1.30*	n.s.
Number of living daughters	1.13*	1.11*	n.s.
Coresidence status			
Does not coreside with a child	1.00	1.00	
Coresides with a child	1.20	0.80*	$p < 0.05$
Area of residence			
City	1.00	1.00	
Urban township	1.08	0.83	n.s.
Rural township	1.15	0.88	n.s.
Work status			
Not working	1.00	1.00	
Working, full time	0.25*	0.44*	$p < 0.05$
Working, part time	0.44*	0.56	n.s.

(*continued*)

Table 2. (*Continued*)

Characteristics‡	1989	1999	Difference between 1989 and 1999
Property division status			
Has divided all or part of property	1.00	1.00	
Has not divided property	1.25	0.88	n.s.
Has no property	0.83	1.01	n.s.
Assets status			
Does not have assets	1.00	1.00	
Has assets	0.59*	0.75*	n.s.
Pension status			
Does not have a pension	1.00	1.00	
Has a pension	0.44*	0.61*	n.s.
Help with household chores and childcare			
No	1.00	1.00	
Yes	1.37	1.31	n.s.
Number of cases§	2181	2184	
Log likelihood	−1094.12	−1353.77	

Note. An asterisk * is shown next to the odds ratio if the coefficient significantly differs from that for the reference category (OR = 1.00) at the 0.05 level. "n.s." indicates no significant difference between 1989 and 1999.
†A series of dummy variables indicating missing information on ethnicity, education, health status, employment status, property division, and ownership of assets and pensions are included in the model for estimation.
‡Unless specified, all characteristics are measured in the year of the survey.
§The number of cases is slightly smaller than 2,187 because some dummy variables for missing information are dropped from the regression due to no variation in the dependent variable.

coresidence with a child is associated with a low likelihood of receiving support in 1999 but the reverse in 1989. Finally, working full time is negatively related to the likelihood of receiving support in both years, but the relationship is stronger in 1989 than in 1999. In sum, these results suggest that the association between receipt of support and parents' characteristics varies only modestly between the two interviews.

Next, we examine the association between parents' characteristics, their major life transitions, and the patterns of receiving financial support from their children over time. The estimation is based on a multinomial regression model in which the dependent variable has four categories: receiving no support in 1989 and 1999, receiving no support in 1989 but having support in 1999, receiving support in 1989 but not in 1999, and receiving support in 1989 and 1999. We present the results in terms of predicted probabilities in Table 3. The corresponding relative risk ratios and significance levels are shown in Appendix A. Below we restrict the discussion to relative risk ratios that are significant at $p < 0.05$.

The predicted probabilities indicate that older men are as likely as older women to receive (no) support from their children. The results do not reveal a consistent association between age and receipt of support, however, the likelihood of gaining support over the 10-year

Table 3. Predicted probabilities of receiving financial support from children in 1989 and 1999 by selected characteristics in 1989 and 1999 (estimates are based on multinomial logistic regression model)*

Characteristics	No support in 1989 No support in 1999	No support in 1989 Support in 1999	Support in 1989 No support in 1999	Support in 1989 Support in 1999
Sex				
Male	0.17	0.13	0.22	0.47
Female	0.16	0.14	0.22	0.48
Age in 1989				
60– 69 years old	0.17	0.14	0.21	0.49
70–79 years old	0.16	0.13	0.25	0.46
≥ 80 years old	0.20	0.23	0.24	0.33
Ethnicity				
Fukienese	0.14	0.15	0.23	0.48
Mainlander	0.24	0.14	0.21	0.41
Hakka	0.15	0.10	0.22	0.53
Other ethnicity (Aboriginal, Foreign)	0.00	0.16	0.17	0.67
Education completed by 1989				
Received no formal education	0.16	0.12	0.24	0.48
Primary school	0.17	0.15	0.20	0.48
Junior high school	0.13	0.16	0.21	0.50
Senior high school	0.21	0.14	0.19	0.46
College or higher	0.21	0.15	0.23	0.40
Marital status in 1989				
Married or cohabiting	0.17	0.15	0.21	0.47
Separated or divorced	0.33	0.05	0.22	0.40
Widowed	0.15	0.10	0.25	0.50
Never married	0.26	0.43	0.00	0.31

(continued)

Table 3. (*Continued*)

Characteristics	No support in 1989 No support in 1999	No support in 1989 Support in 1999	Support in 1989 No support in 1999	Support in 1989 Support in 1999
Number of living sons in 1989[†]				
0 son	0.31	0.11	0.23	0.35
1 son	0.24	0.12	0.23	0.41
2 sons	0.18	0.14	0.23	0.46
Number of living daughters in 1989[†]				
0 daughter	0.22	0.13	0.22	0.43
1 daughter	0.20	0.14	0.22	0.45
2 daughters	0.17	0.14	0.22	0.47
Area of residence in 1989				
City	0.15	0.17	0.18	0.50
Urban township	0.17	0.13	0.21	0.48
Rural township	0.18	0.11	0.26	0.45
Property division status in 1989				
Has divided all or part of property	0.19	0.13	0.24	0.45
Has not divided property	0.15	0.13	0.22	0.49
Has no property	0.20	0.15	0.19	0.46
Death of a spouse, 1989–1999				
No death of a spouse	0.17	0.14	0.22	0.47
Death of a spouse	0.18	0.10	0.23	0.49
Change in health status, 1989–1999				
Same health status	0.17	0.14	0.21	0.48
Health declines	0.17	0.13	0.22	0.47
Health improves	0.15	0.14	0.22	0.49
Change in coresidence status, 1989–1999				
Does not live with a child in either year	0.19	0.12	0.18	0.52
Lives with a child in 1989 only	0.16	0.16	0.20	0.48

Lives with a child in 1999 only	0.15	0.22	0.22	0.42
Lives with a child in both years	0.17	0.12	0.24	0.47
Change in work status, 1989–1999				
Does not work in either year	0.12	0.10	0.24	0.54
Works in 1989 only	0.21	0.22	0.19	0.38
Works in 1999 only	0.15	0.18	0.26	0.41
Works in both years	0.37	0.15	0.20	0.28
Property divided between 1989 and 1999				
No	0.17	0.14	0.22	0.47
Yes	0.18	0.09	0.23	0.50
Change in assets status, 1989–1999				
Does not have assets in either year	0.12	0.11	0.25	0.52
Has assets in 1989 only	0.14	0.16	0.20	0.50
Has assets in 1999 only	0.12	0.14	0.27	0.47
Has assets in both years	0.22	0.13	0.21	0.44
Change in pension status, 1989–1999				
Does not have pensions in either year	0.15	0.12	0.21	0.51
Has a pension in 1989 only	0.18	0.21	0.21	0.40
Has a pension in 1999 only	0.16	0.12	0.32	0.40
Has a pension in both years	0.25	0.18	0.23	0.34
Help with household chores and childcare, 1989–1999				
Does not provide support in either year	0.17	0.14	0.22	0.47
Provides support in 1989 only	0.17	0.09	0.27	0.47
Provides support in 1999 only	0.14	0.15	0.20	0.51
Provides support in both years	0.12	0.09	0.21	0.58
Number of cases		2187		

*A series of dummy variables indicating missing information on ethnicity, education, health status, employment status, property division, and ownership of assets and pensions are included in the model for estimation.

†Numbers of living sons or daughters are modeled as continuous variables. The predicted probabilities are shown for three values (0, 1, and 2 sons or daughters).

period is significantly higher for persons over age 80 compared with 60–69 years old (0.23 versus 0.14). There is no consistent relationship between level of education and receipt of support. Hakka are less likely than Fukienese to gain support over time. Similar to the results shown in the cross-sectional analysis (see Table 2), the likelihood that Mainlanders have no support from their children in either year is 1.7 times the probability for Fukienese (0.24/0.14).

The difference between Mainlanders and Fukienese might relate to the fact that Mainlanders were more apt to postpone marriage because of their migration from Mainland China to Taiwan during the 1949 Civil War. Marriage postponement is associated with having younger children at the time of the interviews, i.e., children who may not be able to provide financial support to their parents. In order to test this hypothesis, we incorporated the age of the eldest child in preliminary models (data not shown). We found that the average age of the eldest child in 1989 was about 32 for Mainlanders and 44 for Fukienese. The differences between Mainlanders and Fukienese in terms of receipt of financial support from children disappeared after controlling for the age of the eldest child in the cross-sectional analysis but not in the longitudinal analysis.

Parents who are separated or divorced at baseline are almost twice as likely to receive no support from their children in either year compared with parents who are married or have a companion (0.33 versus 0.17). In addition, the likelihood that separated or divorced parents gain support between interviews is only one-third of the probability for married parents (0.05/0.15). These results suggest that children growing up in a family in which parents were less committed to the marriage are less likely to provide support to their parents later on. However, unlike previous findings based on Americans, the association between parents' marital disruption (divorce or separation) and provision of support from children to their parents does not vary with parents' sex (data not shown). Widowed parents are less likely than married parents to gain support over the decade.

The greater the number of children, regardless of their sex, the more likely the parents are to have support in both years. For example, for the numbers shown in Table 3, one additional son reduces the probability of receiving no support from children by about one-fourth ((0.24–0.31)/0.31 or (0.18–0.24)/0.24). The association is weaker for daughters. Parents living in urban or rural townships are more likely than parents living in cities to lose support between 1989 and 1999 (0.21 for urban townships and 0.26 for rural townships versus 0.18 for cities). Compared with parents who do not live with their children in either year, parents who live with their children only at follow-up are almost twice as likely to gain support between 1989 and 1999 (0.22/0.12); but parents who live with children in both years are more likely to lose support over time (0.24 versus 0.18).

Ageing parents who have greater financial resources—those who are working or own assets or pensions—are less likely to receive financial support from their children. In particular, the likelihood that parents working at both survey dates receive no financial support from their children is three times the probability for parents who do not work in either year (0.37/0.12). The direction of causation is unclear. For example, it is possible that parents who work do so because their children do not provide sufficient financial support, instead of the other way around. Compared with those who do not work in either year, parents who worked in 1989 but not in 1999 were twice as likely to gain support from their children (0.22 versus 0.10). The probabilities that parents who have assets or pensions at baseline and follow-up receive no support from their children in either year are 1.8 (0.22/0.12) and 1.7 (0.25/0.15) times the probabilities for parents who have no assets and who have no pensions in either year, respectively. The likelihood that parents who first receive pensions during the follow-up period lose support from their children is 1.5 (0.32/0.21) times the probability for parents who have no pensions in either year.

We find no significant association between receipt of support and the following variables: death of a spouse, property division, changes in health status, and provision of support to children. There are several possible explanations for these results. First, our previous research (Lin et al. 2003) shows that more than one child in the family typically provides financial support to parents, and that children's provision of financial support depends on the support behaviour of their siblings. Thus, it is plausible that parents' receipt of support depends more on changes in their children's resources than on changes in their own needs. Second, because the provision of financial support to parents has been perceived as a filial obligation in Chinese culture, such support may be independent of assistance received from ageing parents. Finally, in view of the fact that the recent implementation of national health insurance protects many older, sick parents from experiencing economic hardship, financial support from children may be only weakly related to parental declines in health.

5. Conclusion

In this paper, we address two questions. First, do patterns of transfers in the same family change over time? Second, how are older parents' demographic, social, and economic characteristics associated with changes in the pattern of transfers? With regard to the first question, we find that approximately two-thirds of the older adults have a stable pattern of financial transfers, with 47.6 percent receiving support and 16.8 percent receiving no support at both survey dates. We also find that older adults are more likely to lose rather than gain support (22.0 percent versus 13.5 percent).

As to the second question, we find some evidence in support of all three explanations of why children provide support to their ageing parents. Consistent with the altruistic explanation, we find that parents who are not working and those without pensions or assets are more likely than their counterparts to receive financial assistance from their children. In addition, parents' receipt of pensions only at follow-up is associated with a loss of financial support from children. Whereas moving to a coresidential arrangement is associated with a gain in support, other changes in parents' needs are not consistently associated with the loss or gain of financial support from their children. For example, death of a spouse and deterioration in health status are not significantly related to parents' receipt of financial support from their children at follow-up.

Regarding the reciprocal explanation, we find that parents' separation or divorce is negatively associated with the likelihood that they obtain financial support from their children, as compared with married parents. Nevertheless, parents' provision of help with household chores and childcare is not related to the likelihood of receiving financial support from their children. Finally, consistent with previous studies in Taiwan regarding the structural explanation of provision of support from children to their parents, we find that parents who have more children (particularly sons) are more likely to receive support from them.

In conclusion, despite rapid demographic, economic, and social changes over the last few decades, adult children today continue to assume the important role of supporting their parents in Taiwan. In the face of an ageing society with smaller younger generations to support older adults (because of low fertility and high out-migration), the Taiwanese government has designed various social policies to enhance the well-being of older adults. For example, the government launched a pension program in 2002 to provide all citizens who are over age 65 with 3,000 NT dollars per month for living expenses. In 2004, the total number of adults aged 65 and older was slightly more than two million (2,150,475, Republic of China National Statistics 2006). Without taking into account inflation, the size of pension is about 60 percent of the average amount of financial support that ageing parents received from their children and relatives as reported in the 1999 survey (5,032 NT dollars per month, data not shown). Although this program was well received by the public, the extent to which the implementation of such a universal, monetary assistance program will affect the national budget and alter patterns of financial transfer from children to their parents merits future research.

Acknowledgement

An earlier version of the paper was presented at the International Meeting on Age-Structure Transitions and Policy Dynamics: The Allocation of Public and Private Resources across Generations, Taipei, Taiwan, December 6–8, 2001. We gratefully acknowledge the support of the Behavioural and Social Research Program of the National Institute on Ageing under grant numbers R01-AG166661 and R01-AG16790, the Graduate School of Arts and Sciences, Georgetown University, and the Chiang Ching-Kuo Foundation. We thank C. Y. Cyrus Chu, Diane Macunovich, and participants at the International Meeting on Age-Structure Transitions and Policy Dynamics for their helpful suggestions.

Appendix A. Relative risk ratios of receiving financial support from children in 1989 and 1999 by selected characteristics in 1989 and 1999 (the reference category is receiving support in both years).[†]

Characteristics	No support in 1989 No support in 1999	No support in 1989 Support in 1999	Support in 1989 No support in 1999
Sex			
Male	1.00	1.00	1.00
Female	0.89	1.05	0.95
Age in 1989			
60–69 years old	1.00	1.00	1.00
70–79 years old	0.99	0.95	1.25
≥ 80 years old	2.18	2.86*	1.76

Appendix A. (*Continued*)

Characteristics	No support in 1989 No support in 1999	No support in 1989 Support in 1999	Support in 1989 No support in 1999
Ethnicity			
Fukienese	1.00	1.00	1.00
Mainlander	2.41*	1.16	1.08
Hakka	0.91	0.58*	0.85
Other ethnicity (Aboriginal, Foreign)	0.00	0.65	0.48
Education completed by 1989			
Received no formal education	1.00	1.00	1.00
Primary school	1.12	1.30	0.86
Junior high school	0.76	1.36	0.85
Senior high school	1.56	1.40	0.83
College or higher	1.81	1.73	1.20
Marital status in 1989			
Married or cohabiting	1.00	1.00	1.00
Separated or divorced	2.96*	0.45	1.24
Widowed	0.77	0.60*	1.10
Never married	3.52	6.05	0.00
Number of living sons in 1989	0.60*	0.93	0.86*
Number of living daughters in 1989	0.80*	0.95	0.95
Area of residence in 1989			
City	1.00	1.00	1.00
Urban township	1.24	0.78	1.18
Rural township	1.39	0.71	1.57*
Property division status in 1989			
Has divided all or part of property	1.00	1.00	1.00
Has not divided property	0.67	0.91	0.84
Has no property	1.11	1.15	0.78
Death of a spouse, 1989–1999			
No death of a spouse	1.00	1.00	1.00
Death of a spouse	1.06	0.68	1.04

(*continued*)

Appendix A. (*Continued*)

Characteristics	No support in 1989 No support in 1999	No support in 1989 Support in 1999	Support in 1989 No support in 1999
Change in health status, 1989–1999			
Same health status	1.00	1.00	1.00
Health declines	1.08	0.93	1.08
Health improves	0.83	0.94	1.02
Change in coresidence status, 1989–1999			
Does not live with a child in either year	1.00	1.00	1.00
Lives with a child in 1989 only	0.90	1.45	1.24
Lives with a child in 1999 only	1.05	2.50*	1.56
Lives with a child in both years	1.00	1.17	1.53*
Change in work status, 1989–1999			
Does not work in either year	1.00	1.00	1.00
Works in 1989 only	3.09*	3.67*	1.21
Works in 1999 only	1.73	2.62*	1.49
Works in both years	8.92*	3.54*	1.69
Property divided between 1989 and 1999			
No	1.00	1.00	1.00
Yes	0.99	0.60	0.98
Change in assets status, 1989–1999			
Does not have assets in either year	1.00	1.00	1.00
Has assets in 1989 only	1.29	1.61	0.83
Has assets in 1999 only	1.11	1.50	1.21
Has assets in both years	2.63*	1.64	1.00
Change in pension status, 1989–1999			
Does not have pensions in either year	1.00	1.00	1.00

Appendix A. (*Continued*)

Characteristics	No support in 1989 No support in 1999	No support in 1989 Support in 1999	Support in 1989 No support in 1999
Has a pension in 1989 only	1.75*	2.37*	1.29
Has a pension in 1999 only	1.50	1.31	1.97*
Has a pension in both years	3.30*	2.59*	1.65
Help with household chores and childcare, 1989–1999			
Does not provide support in either year	1.00	1.00	1.00
Provides support in 1989 only	0.92	0.65	1.24
Provides support in 1999 only	0.67	1.02	0.86
Provides support in both years	0.46	0.46	0.77
Number of cases		2187	
Log Likelihood		–2376.43	

Note. An asterisk * is shown next to the relative risk ratio (RRR) if the coefficient significantly differs from that for the reference category (RRR=1.0) at the 0.05 level.

†A series of dummy variables indicating missing information on ethnicity, education, health status, employment status, property division, and ownership of assets and pensions are included in the model for estimation.

References

Agree, E.M., Biddlecom, A.E., and Valente, T.W. (2005), "Intergenerational transfers of resources between older persons and extended kin in Taiwan and the Philippines." *Population Studies*, 59(2):181–195.

Becker, G. (1991), *A Treatise on the Family*. Harvard University Press: Cambridge, MA.

Beckett, M., Goldman, N., Weinstein, M., Lin, I.-F., and Chuang, Y.-L. (2002), "Social environment, life challenge, and health among the elderly in Taiwan." *Social Science and Medicine*, 55(2):191–209.

Chang, M.-C. (1999), "A longitudinal study on living arrangements of the elderly in Taiwan." In: *Emerging Social Economic Welfare Programs for Aging in Taiwan in a World Context*, Chaonan Chen, Albert I. Hermalin, Sheng-Cheng Hu and Hames P. Smith, eds. pp. 43–64. Institute of Economics, Academia Sinica: Taipei, Taiwan.

Chattopadhyay, A. and Marsh, R. (1999), "Changes in living arrangement and familial support for the elderly in Taiwan: 1963–1991." *Journal of Comparative Family Studies*, 30(3):523–537.

Cox, D. and Rank, M.R. (1992), "Inter-vivos transfers and intergenerational exchange." *Review of Economics and Statistics,* 74:305–314.

Eggebeen, D.J. (1992), "Family structure and intergenerational exchanges." *Research on Aging,* 14(4):427–447.

Eggebeen, D.J. and Dennis P.H. (1990), "Giving between generations in American families." *Human Nature,* 1(3):211–232.

Furstenberg, F.F. Jr., Hoffman, S.D., and Shrestha, L. (1995), "The effect of divorce on intergenerational transfers: New evidence." *Demography,* 32(3):319–333.

Gouldner, A.W. (1960), "The norm of reciprocity: A preliminary statement." *American Sociological Review,* 25:161–178.

Greenberg, M.S. (1980), "A theory of indebtedness." In: *Social Exchange: Advances in Theory and Research,* Kenneth J. Gergen, Martin S. Greenberg and Richard H. Willis, eds. pp. 3–26. Plenum Press: New York.

Hashimoto, A. (1996), *The Gift of Generations: Japanese and American Perspectives on Aging and the Social Contract.* Cambridge University Press: New York.

Henretta, J.C., Hill M.S., Li W., Soldo, B.J., and Wolf, D.A. (1997), "Selection of children to provide care: The effect of earlier parental transfers." *Journal of Gerontology: Social Sciences,* 52B:S110–S119.

Hermalin, A.I., Ofstedal, M.B., and Chi, L. (1992), "Kin availability of the elderly in Taiwan: Who is available and where are they?" Population Studies Center Report No. 92-18. Ann Arbor: University of Michigan.

Hermalin, A.I., Ofstedal, M.B., and Lee, M.-L. (1992), "Characteristics of children and intergenerational transfers." Comparative Study of the Elderly in Asia Research Report No. 92-21, University of Michigan, MI: Ann Arbor.

Hermalin, A.I., Ofstedal, M.B., and Chang, M.-C. (1996), "Types of supports for the aged and their providers in Taiwan." In: *Aging and Generational Relations over the Life Course: A Historical and Cross-Cultural Perspective.* Tamara K. Hareven, ed. pp. 400–437. Walter de Gruyter: New York.

Hsu, H.-C., Lew-Ting, C.-Y., and Wu, S.-C. (2001), "Age, period, and cohort effects on the attitude toward supporting parents in Taiwan." *The Gerontologist,* 41(6):742–750.

Lee, M.-L., Lin, H.-S., and Chang, M.-C. (1995), "Living arrangements of the elderly in Taiwan: Qualitative evidence." *Journal of Cross-Cultural Gerontology,* 10:53–78.

Lee, Y.-J., Parish, W.L., and Willis, R.J. (1994), "Sons, daughters, and intergenerational support in Taiwan." *American Journal of Sociology,* 4:1010–1041.

Lee, Y.-J. and Xiao, Z. (1998), "Children's support for elderly parents in urban and rural China: Results from a national survey." *Journal of Cross-Cultural Gerontology,* 13:39–62.

Lillard, L. A. and Willis, R.J. (1997), "Motives for intergenerational transfers: Evidence from Malaysia." *Demography,* 34(1):115–134.

Lin, I.-F., Goldman, N., Weinstein, M., Lin, Y.-H., Gorrindo, T., and Seeman T. (2003), "Gender differences in adult children's support of their parents in Taiwan." *Journal of Marriage and Family,* 65(1):184–200.

Litwak, E. (1985), *Helping the Elderly: The Complementary Roles of Informal Networks and Formal System.* The Guilford Press: New York.

Logan, J. R. and Spitz, G.D. (1996), *Family Ties: Enduring Relations between Parents and Their Grown Children.* Temple University Press: Philadelphia.

McGarry, K. and Schoeni, R.F. (1995), "Transfer behavior in the health and retirement study: Measurement and the redistribution of resources within the family." *Journal of Human Resources,* 30:S184–S226.

Pezzin, L.E. and Schone, B.S. (1999), "Parental marital disruption and intergenerational transfers: An analysis of lone elderly parents and their children." *Demography,* 36(3):287–297.

Republic of China National Statistics. (2006), Available at http://www.stat.gov.tw.

Riley, M.W. (1983), "The family in an aging society: A matrix of latent relationships." *Journal of Family Issues,* 4:439–454.

Roan, C.L., Hermalin, A.I., and Ofstedal, M.B. (1996), "Intergenerational contact and support in Taiwan: A comparison of elderly parents' and children's reports." Population Studies Center Report No. 96-36. Ann Arbor: University of Michigan.

Silverstein, M. (1995), "Stability and change in temporal distance between the elderly and their children." *Demography,* 32(1):29–45.

Silverstein, M., Conroy, S.J. Wang, H., Giarrusso R., and Bengtson, V.L. (2002), "Reciprocity in parent-child relations over the adult life course." *Journal of Gerontology: Social Sciences,* 57B(1):S3–S13.

Soldo, B.J. and Hill, M.S. (1995), "Family structure and transfer measures in the health and retirement study." *The Journal of Human Resources,* 30:S108–S137.

StataCorp. (2001), *Stata Statistical Software: Release 7.0.* College Station, TX: Stata Corporation.

Sun, T.-H. and Liu, Y.-H. (1994), "Changes in intergenerational relations in the chinese family: Taiwan's experience." In: *Tradition and Change in the Asian Family*, Lee-Jay Cho and Moto Yada, eds. pp. 319–361. The East-West Center: Hawaii.

Taiwan Provincial Institute of Family Planning and Institute of Gerontology University of Michigan. (1989), *1989 Survey of Health and Living Status of the Elderly in Taiwan: Questionnaire and Survey Design.* Comparative Study of the Elderly in Asia Research Report No. 1, University of Michigan. MI: Ann Arbor.

Thibaut, J.W. and Kelley, H.H. (1959), *The Social Psychology of Groups.* Wiley: New York.

Weinstein, M., Sun, T.-H., Chang, M.-C., and Freedman, R. (1990), "Household composition, extended kinship, and reproduction in Taiwan: 1965–1985." *Population Studies,* 44:217–239.

Whitbeck, L.B., Simmons, R.L. and Conger, R.D. (1991). "The effects of early family relationships on contemporary relationships and assistance patterns between adult children and their parents." *Journal of Gerontology: Social Sciences,* 46(6):S330–S337.

Whitbeck, L.B., Hoyt, D. R., and Huck, S.M. (1994), "Early family relationships, intergenerational solidarity, and support provided to parents by their adult children." *Journal of Gerontology: Social Sciences,* 49(2):S85–S94.

Yang, H. (1996), "The distributive norm of monetary support to older parents: A look at a township in China." *Journal of Marriage and the Family,* 58:404–415.

CHAPTER 4. DEMOGRAPHIC EVENTS AND THE TIMING OF MONETARY TRANSFERS: SOME EVIDENCE FROM GERMANY

SUMON KUMAR BHAUMIK
Centre for Economic Development and Institutions
Brunel University

1. Background

The genesis for the exploration of motives underlying *inter vivos* transfers, of which intergenerational transfers are a special case, lay in the seminal papers of Barro (1974) and Becker (1974). The issue has survived the test of time, and has once again emerged as an important part of the economists' research agenda.[1] There are primarily two reasons behind this surge of interest. First, an important implication of Barro's analysis is that any increase in social security benefits accruing to a member of the "dynastic" household would crowd out, perhaps even dollar for dollar, private intrahousehold transfers to that individual. Since the efficacy of publicly provided social security benefits, including nonmonetary transfers like health care, is the focus of a major public policy debate, a better understanding of the motives underlying intergenerational transfers has become imperative.

Second, it has been argued that social security provisions that can provide a cushion against a drop in consumption during old age may reduce "life-cycle" savings, and hence reduce the volume of aggregate domestic savings (Lee, Mason, and Miller 2000). It is fairly obvious that, in the presence of appropriate social norms, private intergenerational or *inter vivos* transfers may also provide the aforementioned cushion, and thereby affect household and aggregate domestic savings. Many developing countries do not have publicly funded social security but have social norms that encourage private transfers. Further, industrialized countries are finding it increasingly difficult to sustain well funded social security systems, and hence private transfers may become increasingly important in these countries in the foreseeable future. Once again, therefore, it is important to understand the factors determining the probability of occurrence and the magnitude of private transfers.

The series of studies that were motivated by Cox (1987) largely focus on whether, and to what extent, *inter vivos* transfers are motivated by altruism. Cox himself found that the amount of *inter vivos* transfers were positively related to the incomes of the recipients, an observation that contradicts the logic of altruism. Indeed, if parents and children are bound

[1] See, for example, Hayashi (1995); Altonji, Hayashi, and Kotlikoff (1995); Sloan and Zhang (1995); McGarry (1995, 1997, 2000); Rangel (1999); Berman and Rzakhanov (2000); and Lee, Mason, and Miller (2000).

Anne H. Gauthier et al. (eds.), Allocating Public and Private Resources across Generations, 89–115.

together by altruistic motives, then each of these two generations would be interested in preserving the consumption-welfare level of the other generation. This would imply that a transfer from one generation to the other would be induced by a *fall* in the income of the recipient generation. This logic was further extended by Altonji, Hayashi, and Kotlikoff (henceforth AHK) (1992). They argued that, in the presence of altruism on the part of (say) a parent, if the parent's income goes down by one dollar, and simultaneously the child's income goes up by one dollar, then there would be one dollar less of transfers from the parent to the child.[2]

The basic Cox–AHK line of argument has been extended, and tested in many different ways (Cox 1990; Cox and Jimenez 1992, 1993; Bhaumik and Nugent 2000; McGarry 2000). Further, alternative motives for *inter vivos* transfers like strategic motives have been brought into light (Bernheim, Shleifer, and Summers 1985; Cox and Stark 1994; Pezzin and Schone 2000). However, that altruism drives *inter vivos* transfers remains the null hypothesis for much of the empirical literature on intergenerational transfers, and, more generally, private transfers among members of dynastic households.

As such, empirical analyses aimed at testing the aforementioned null hypothesis have involved the use of a simple probit or Tobit model in which a binary indicator of the occurrence of transfer or the amount of transfer itself is the dependent variable.[3] In recent years, however, such empirical analyses have typically included the use of a two-step Heckman procedure to correct for sample selection bias, the rationale for which would be explained later in the paper. Other things remaining the same, if the coefficient of income and/or the variable representing publicly provided benefits is negative and significant, it is argued that the data manifests the type of crowding out of private transfers, by other sources of income and consumables, that is consistent with the logic of altruism.

This paper argues that the stylized analysis is seriously flawed on two different counts. First, income or publicly provided monetary and nonmonetary benefits per se may not be the determinant of the private transfers. Such transfers are more likely to depend on events in the recipients' lives, viz., marriage, illness, and childbirth. As we shall see later, if a specification does not include such events as explanatory variables for private transfers, the conclusion drawn about the existence of altruistic relations between donors and recipients of private transfers may be erroneous. Second, if the recall period of the data is as large as

---

[2] McGarry (2000) calls this the "derivatives" hypothesis, and argues that it would not hold "under reasonable assumptions about the formation of expectations about future income." Indeed, the linear relationship between income and private transfers implied by the "derivatives" hypothesis can be brought into question with a simple construction. If a fall in the income of (say) a child can be viewed as not only a decline in the current income, but also as a signal of the possibility that her future incomes might also be low, altruistic parents would have to transfer more to the child not only to counteract the latter's low income during the current period, but would also have to compensate her for lower income in the future periods. Hence, the strict dollar-for-dollar crowding-out implied by the "derivatives" hypothesis will not hold.

[3] The rationale for using a probit model with transfer (1) and no transfer (0) as the binary dependent variable is obvious. The case for use of a Tobit model is more involved. It can be argued, for example, that a transfer is made by a donor if the act of making the transfer pushes her utility beyond some unobserved threshold. If u denotes utility, and u^* denotes the threshold level of utility, we can argue that a transfer is observed if $u > u^*$, and not otherwise. In other words, the distribution of transfers is censored, and hence the appropriate econometric formulation for its analysis is the Tobit model.

one year, it is reasonable to assume that transfers may affect income even as income may affect transfers. For example, if the parents of a young couple provide financial support to the latter such that the grandchildren can be sent to a day care centre, the female in the younger household may be able to join the labour force more effectively than otherwise, and this would certainly have a positive impact on the latter household's income. If such simultaneity exists, and yet is not taken into consideration, the resultant empirical estimates may be biased.

The analysis presented in the subsequent sections of the paper goes on to show that "events" in the recipients' lives do matter, and have a significant impact on the probability of receiving a private transfer as well as the magnitude of the transfer. It also shows that income and transfers have a significant impact on each other, and that therefore the appropriate method of estimation would be one using a simultaneous equation system. Section 2 of the paper expands on the rationale underlying the argument that the existing empirical methodology is flawed. Section 3 discusses the data used for the subsequent empirical analysis, namely, the German Socio-Economic Panel (GSOEP). It also provides a framework for the empirical analysis. The empirical results are presented and analysed in Section 4. Section 5 gives the conclusion, and suggests possible extensions of the analysis.

2. The Lacunae in the Empirical Literature

The literature on *inter vivos* (and hence intergenerational) transfers suffers from two shortcomings. First, it does not take into account the fact that households and individuals neither give nor receive transfers continually.[4] Indeed, it is reasonable to assume that if households and individuals are altruistic, they would provide monetary support to others within the extended family or "dynastic" household in times of need,[5] such that the consumption level of the recipient individual or household is sustained in the face of some crisis. For example, incidence of monetary transfers within a parent–child cohort may be high both before and after marriages of the children, during periods in which the children suffer prolonged and serious illnesses, and when the children are out of work, in the absence of adequate unemployment compensation. Similarly, monetary transfers from the children to the parents are more likely to be observed during illnesses of the latter, and if the social security benefits of the aged parents are inadequate.[6] In other words, it might be possible to observe transfers from a parent to a child during the latter's year of marriage, year of

---

4 The transfers from children to elderly people in developing countries are often an exception to this "rule." In the absence of publicly funded social security provisions, and in the absence of life-cycle savings, a large section of the elderly population in such countries may require continual transfers from their children for survival. In those contexts, the stylized empirical analysis may be more appropriate (Cox and Jimenez 1992, 1993; Bhaumik and Nugent 2000).

5 A household is in need of financial assistance whenever its expenditure exceeds its income. Hence, need is consistent with both increase and decrease in the income of the potential recipient.

6 Note that a regression model's control for the health of an elderly parent is not necessarily adequately specified in so far as "events" driven transfers are concerned. For example, an elderly parent may be in good health in general and, *ceteris paribus*, may not attract any transfer from her children. However, during a particular year, this elderly parent may suffer a heart attack, thereby attracting transfers only during that year of her golden years. In order to capture the rationale underlying this flow of transfers, we would have to control for the "event" of hospitalization, rather than the overall health of the parent, which may be fine.

major illness, or year of temporary unemployment, and not during the other years of the adult child's lifetime.[7]

The implications of this proposition are significant. As mentioned above, the usual "test" for altruism is to observe whether the probability of receiving a transfer and the magnitude of transfer vary inversely with the income of the recipient individual or household, and/or with the existence and extent of public support in the form of social security benefits.[8] Let us assume that there are two households with the same income x, and that each household has a potential donor within its extended family, each donor's household income being y. Let us assume further that each of the potential recipients have the same nonincome characteristics. Does this imply that during any time period either both or neither of these households would receive a transfer? It is obvious that this is very unlikely. For example, if one of the potential recipients fall prey to some illness, she would be more likely to receive transfers from her extended family. In other words, the (lack of) variation in x, y, and other explanatory variables would not be able to explain the fact that one of the potential recipients would actually receive a transfer while the other household would not. If, however, the specification includes an indicator of the health of the potential recipients, the asymmetry in the occurrence of transfer can be explained. Clearly, any examination of the motives underlying private transfers without controlling for income-reducing and expenditure-augmenting events in the lives of the recipient would not be meaningful.[9]

Indeed, if the aforementioned x is regressed on y, the coefficient of y may not be significantly different from zero, thereby indicating, *à la* AHK view of altruism, that the donor and the recipient are not linked by way of altruism. However, if altruism induces donors to make transfers to recipients in the extended household if and when the latter's current incomes are insufficient to cover their current expenses, a transfer made during an event like illness or childbirth would tantamount to altruism. However, if a variable controlling for the "event" is introduced into the specification, the coefficient of that variable is likely to be positive and significant. In other words, the empirical test for altruism should involve not only a negative and significant coefficient for the income explanatory variable, but also positive and significant coefficients for the "events" variables on the right-hand side of the regression equation.

---

[7] Note that the rationale for providing support during or immediately after such life course events may differ significantly across events (Henretta et al. 1997). For example, parents may provide substantial help during a normative life course event like marriage partly because of altruism, but also partly because of intangible factors like "status" and "prestige" of the household. On the other hand, in the case of non-normative events like sickness, transfers may be induced not only by altruism but also by an urge on the part of the potential donors to insure themselves against adverse events in the future by strengthening pillars of social institutions like social obligation and reciprocation. However, a detailed discussion of the motives associated with the different forms of transfer lies outside the scope of the paper.

[8] If information is available about the donor's individual or household income, it may be used as a control variable because a low-income household, for example, may not receive a transfer from another low income household within the "dynasty," even if the former household is in need for such a transfer per se. However, few surveys provide the relevant information about the donor, and hence studies in general are unable to control for this factor.

[9] The only important piece of research acknowledging the limitation of the stylized methods of testing can be attributed to McGarry (1997).

There is one other aspect of *inter vivos* (and hence intergenerational) transfers that has not been adequately addressed in the existing literature, namely, the impact of the transfers on the earning potential of the recipients. It is obvious, for example, that people incur search costs when looking for jobs, and it would be reasonable to assume that the reservation wage of an individual depends on her ability to bear these costs. Hence, if, in the event of unemployment, an individual receives support in the form of private transfers, she is likely to be able to search for an appropriate job for a longer period of time, thereby increasing the probability of getting a higher paid job than she otherwise would. Similarly, as highlighted by the aforementioned example about grandparents paying for a grandchild's day care, transfers may enable a member of the household to effectively join the labour force, and thereby augment both her own and the household's income. In other words, transfers can have an impact on an individual's or a household's income.[10] Arguably, transfers are unlikely to have an impact on household income in the short run. But if the recall period for a data set is large, say, one year, as with the GSOEP data, the "reverse" link between transfers and income cannot be ignored.

One should note, however, that it is by no means obvious that the impact of private transfers on household income would necessarily be positive, as suggested by the above example. For example, if a parent supports a child during a spell of unemployment, the child may spend much of the recall period searching for an appropriate job. In such an event, the child's income registered for the recall period might be lower than what it would have been in the absence of a transfer, which may have forced the child to reduce the search period. Similarly, following a childbirth, an event which would increase financial pressure on a household, if the parents of either of the spouses provide support in the form of private transfers, the female of the couple may stay back at home during the recall period to take care of the child. In that case, the impact of the private transfers on the household's income would be negative. In sum, unlike in the case of the impact of household income on private transfers, the impact of private transfers on personal and household income is indeterminate. The importance of this possible reverse causality lies in the fact that if such causality exists, the appropriate estimation technique would involve the use of a simultaneous equation system with household income and private transfers as the two endogenous variables, rather than the single equation regression models that are stylized in the literature.

An interesting deviation from this trend is the research of Pezzin and Schone (1998), which attempted to "predict jointly the combination of household membership (i.e., coresidence of parents and children), labour-force participation, informal care, and financial transfer provision resulting from parent and daughter bargaining." However, the explanatory power of the Pezzin–Schone model has been limited by the absence of controls for transfer-inducing events and by the selectivity problem that arises because of the study's focus on female children.

[10] In light of the argument put forward by McGarry (2000), it can be argued that, therefore, if a household or individual receives a transfer in period t, such that this transfer has a positive impact on its/his income, then the conditional probability that the household or individual would receive transfers in the future would decline. But in order to capture such dynamics one would have to use panel data models and such an analysis lies outside the scope of this paper.

This paper addresses the lacunae in the literature in two stages. First, data is used to justify the sample selection process, which aims to simulate intergenerational transfers in the absence of information that can help match donors with recipients. The data indicates that transfers to children and children-in-law by far dominate transfers made by the older generation which, for reasons explained below, comprises the potential donors in the subsequent analysis. Next, it uses a stylized single-equation regression model to verify whether "events" like marriage, hospitalization, and childbirth matter, i.e., whether such events determine the probability and amount of transfers received to a significant extent.[11] The analysis unambiguously indicates that "events" determine both the probability and the amounts of transfers significantly, and that they should be included in the specification to sharpen the null hypothesis for altruistic relationship between donors and recipients, and the test for the same. Finally, it introduces a simultaneous equation system with household income and private transfers received by households as the two endogenous variables, thereby extending the empirical analysis to address the second lacuna in the literature. This exercise too provides insights into the modelling of the income-transfers relationship. However, it has caveats that are discussed later in the paper.

3. Data and Structure of Empirical Analysis

The data used for empirical analysis are taken from the 1996 and 1997 rounds of the GSOEP.[12] The first round of survey for the GSOEP data was conducted in 1984 and it is a longitudinal survey of private households and persons in the Federal Republic of Germany (Haisken-Denew and Frick 2000). Since then, each year the respondents are asked a core set of questions which include queries about demographic features of the households, income and social security benefits of the household members, their education, health, and labour market performance, and expectations about the future. In 1984, the GSOEP survey covered 5,921 West German households. In 1990, households in the former German Democratic Republic were included in the survey, providing retrospective information about income, transfers, and events. The approximate number of households included in the survey during a year is about 6,800, the number of individuals in each survey being about 13,000.

The GSOEP data provides a wealth of information about the possible determinants of both transfers and income. Specifically, it provides information about the education, extent of labour market participation, and household wealth of individual respondents. It also provides demographic information like age, gender, and the number of children, as well as information about the extended families of individuals, namely, parents and siblings. Importantly, it provides information about events like marriage, childbirth, divorce and illness in the lives of the respondents, information that is crucial for the

---

[11] As we shall see later, the GSOEP data facilitates analyses of transfers received by individuals and households, rather than analyses of transfers made to members of the extended household.

[12] Certain information like age and level of education attained were available in the 1996 data set. On the other hand, the data on income, transfers etc. were available from the 1997 data, the recall period being the previous year. Hence, the relevant information from the 1996 and 1997 data sets had to be combined to obtain information about all relevant variables.

empirical exercise suggested above. Finally, the data provides information about monetary transfers.[13]

However, while the surveys provide extensive information about several aspects of an individual's (and household's) "economic" life, it does not provide as much detail about the (monetary) transfers given and received by households. Specifically, the survey asks the respondents whether they transferred money to certain categories of people during the recall period, namely, parents/parents-in-law, children/children-in-law, separated/divorced spouse, other relatives, and unrelated persons. On the other hand, while each respondent is asked whether or not she received private transfers during the recall period, she is not required to identify the sources of such transfers. As a consequence, it is extremely difficult to match the donors with the corresponding recipients.[14]

Ideally, an analysis of *inter vivos* transfers should bring together the information about both the donors and the recipients. But since it is not always possible to obtain information about both donors and recipients, it is not unusual to focus either on the recipients alone (Bhaumik and Nugent 2000), or on donors and recipients separately. As has already been mentioned, the GSOEP data does not allow one to match donors and recipients of specific transfers. Hence, this paper would have to opt for separate analyses of both transfers made and received. However, since any test of altruism requires, at the very least, an examination of the relationship between transfers received by an individual or household, on the one hand, and the individual or household's income and "events" in the individual or household's life, on the other hand, the analysis will focus more on the transfers received by individuals or households.[15]

It has already been mentioned that the analysis in this paper would use the stylized empirical methodology until the simultaneous relationship between income and transfers (received) is brought under scrutiny. The stylized practice in the existing literature is to examine individually the determinants of the probability of receiving (or giving) a transfer, and those for the magnitude of the transfer. If the focus of the analysis is the determinants of the probability of receiving transfers, for example, the stylized approach involves the use of a probit or a logit model, with the dependent variable having a value of zero or one,

---

[13] The information on income, labour market participation, life-cycle events like marriage, childbirth, sickness, and monetary transfers is collected on the basis of a one-year recall period. Specifically, the GSOEP round for year t provides information on these variables for year $t-1$.

[14] For example, after an attempt was made to match donors with the corresponding recipients for the survey year 1997, only about 30 matches could be made unambiguously. One way to overcome this problem would be to pool matched observations over a number of years, so as to create a sample of a reasonable size. However, in order to undertake meaningful econometric exercises for such a sample, one would have to assume that the observations are independent of each other. However, if indeed transfers are precipitated by "events" in a recipient's life, the probability of receiving a transfer in period t would be strongly correlated with the probability of receiving a transfer in an earlier period. For example, if a household receives a private transfer in period t on account of childbirth, the absence of childbirth in the household in the previous j periods would explain why the household did not receive such transfers between the $(t-j)^{\text{th}}$ period and the tth period.

[15] As we shall see later, the analysis initially focuses on transfers received by individuals. However, once simultaneity between net income and transfers received is introduced as a possibility, the unit of analysis shifts from an individual to a household.

for transfers received and not received, respectively. Hence, if *TRANSDUM* represents the dummy dependent variable, the specification would be given by

$$\text{TRANSDUM} = \gamma_0 + \gamma_1 \text{INCOME} + \Gamma' X + \text{e} \tag{1}$$

where *INCOME* refers to the household income of the (potential) recipient, and *X* corresponds to the values of the other variables determining the probability of a transfer. After taking into account the correlations between the explanatory variables, the specification for the GSOEP data included information on income and wealth of the recipients (and the households to whom they belonged), the geographical location of their home and workplace,[16] demographic information,[17] and the various demographic and lifecycle events (see Table 1).

The stylized approach to exploring the relationship between the magnitude of transfer and its possible determinants involves the use of a Tobit model. If the amount of transfers received by the households is given by *TRANS96*, the specification for the Tobit equation would be

$$\text{TRANS96} = \beta_0 + \beta_1 \text{INCOME} + \text{B}' X + \text{u} \tag{2}$$

As has been mentioned earlier in the paper, the data on transfers is vulnerable to a selection process, namely, if a household has a high probability of receiving a transfer then the magnitude of the transfer it receives is also likely to be high (Jurges 1999). Hence, the Tobit model given by equation (2) is often estimated by the two-stage Heckman process, which corrects the bias associated with the aforementioned problem of selection (Jurges 1999; Bhaumik and Nugent 2000).

4. Results and Inference

4.1. SAMPLE SELECTION AND DESCRIPTIVE STATISTICS

Since this paper intends to focus on intergenerational transfers, it is necessary to construct the samples of potential donors and recipients such that the empirical results obtained from analyses of these samples have implications for private transfers across generations. In other words, the paper can focus on either upstream transfers, from the younger to the older generation, or downstream transfers, from the older to the younger generation, or

[16] This distinction is necessitated by the fact that the economic disparities of West Germany (former FDR) and East Germany (former GDR) have persisted over time. In other words, there is reason to believe that the economic opportunities which determine the need for support from "dynastic" households differ between residents of and workers in these two parts of Germany.

[17] Two things should be noted in this context: first, it has been assumed that many of the transfers one likely to originate from "dynastic" households of parents and siblings. Second, the marital status of individuals was significantly correlated with some of the other explanatory variables, and hence a dummy variable to indicate the marital status of an individual was omitted from the specification. However, coefficient estimates were also obtained with a dummy variable indicating the marital status of an individual in the specification, and its introduction into the specification did not alter the qualitative analysis to any extent.

Table 1. Description of variables explaining incidence and amount of transfers received (for probit and Tobit analyses).

Income and wealth	
NETINC96	Net nontransfer income of the household to which the potential recipient belongs
NOYEDU96	Number of years of education attained by the potential recipient as of 1996, a measure of employability and "permanent" income
HOWNER96	Dummy variable which takes the value unity if the household to which the potential recipient belongs owns the house in which it lives, and is zero otherwise
LIQAST96	Dummy variable which takes the value unity if the household to which the potential recipient belongs has liquid assets in the form of savings with banks or building societies, stocks etc., and is zero otherwise
Geographical location	
LIVFDR96	Dummy variable which takes the value unity if the potential recipient lives in former West Germany (FDR), including West Berlin, as of 1996, and is zero otherwise
WRKGDR96	Dummy variable which takes the value unity if the potential recipient works in former East Germany (GDR), including East Berlin, as of 1996, and is zero otherwise
Demographic characteristics	
AGE96	Age (in years) of the potential recipient in 1996
MALE96	Dummy variable which takes the value unity if the potential recipient is male, and is zero otherwise
NOCHLD96	Number of dependent children in the potential recipient's household
PALIVE96	Dummy variable which takes the value unity if at least one parent of the potential recipient is alive, and is zero otherwise
SALIVE96	Dummy variable which takes the value unity if at least one sibling of the potential recipient is alive, and is zero otherwise
Demographic "events" in 1996	
NOHVST96	Number of times the potential recipient had to spend at least one night at a hospital on account of significant health problems
MARAGE96	Dummy variable which takes the value unity if the potential recipient got married in 1996, and is zero otherwise
MOVEIN96	Dummy variable which takes the value unity if the potential recipient moved in with a partner in 1996, and is zero otherwise
DIVORC96	Dummy variable which takes the value unity if the potential recipient got divorced in 1996, and is zero otherwise
CBIRTH96	Dummy variable which takes the value unity if the potential recipient became a parent in 1996, and is zero otherwise

both. However, younger people typically experience a greater number (and greater variety) of "events" in their lives—for example, moving in with a partner, marriage, divorce, and childbirth—relative to older people; people in both age groups can experience health related "events." Therefore, given that the paper seeks to establish a link between transfers received and "events" in the lives of the recipients, it would be prudent to focus on downstream transfers. Hence, people who are between 25 and 40 years of age are defined as potential recipients, while people who are more than 50 years of age are defined as potential donors.[18]

The data indicates that the incidence of private transfers in the sample of all individuals in the 25–40 age group is very low, namely, 2.6 percent. This poses a significant problem because meaningful estimation of probit, logit, and Tobit models typically require that at least 8–10 percent of the observations belong to either the transfer (one) or the no transfer (zero) category. In order to overcome this problem, the relevant probit model for transfer recipients is estimated with two different samples: all the individuals within the 25–40 age group, and people in the 25–40 age group with a household income of less than 35,000 marks, the incidence of transfers for the latter group being in excess of 8 percent. The regression estimates indicate that qualitatively, and from the point of view of significance, the coefficients of explanatory variables are similar for the two samples; both regressions indicate that events play a significant role in determining the probability of transfers received by an individual. In the light of this finding, and given the desire to have meaningful econometric estimates, the sample for the subsequent analysis of transfers (received) is restricted to people in the 25–40 age group whose household income is less than 35,000 marks.[19] The resultant samples for potential recipients include 818 observations.[20]

The descriptive statistics for the potential donors (Table 2) highlight certain interesting patterns. It can be seen that people who make transfers to parents and parents-in-law

---

[18] The relative numbers of upstream and downstream transfers suggest that the latter is clearly the dominant form of transfers in Germany, at least as indicated by the 1996–1997 GSOEP data. For example, of the 4,169 individuals in the full data set of potential donors, only 67 reported transfers to parents or parents-in-law, while 390 reported transfers to children or children-in-law. It can be argued that people who are more than 50 years of age are more likely to make downstream rather than upstream transfers, often simply because they do not have living parents. However, even among the potential "recipients" in the 25–40 age group, only about 3 percent of the people reported making transfers to parents or parents-in-law. Hence, it would be reasonable to argue that, given the data, an empirical analysis of upstream transfers would not be very meaningful.

[19] This *ad hoc* restriction on the income of the individuals, which results in the use of the people in the bottom 20 percent of the income distribution for the analysis, is likely to sharpen the relationship between transfers and events because private inter-household transfers that aim to alleviate financial (or liquidity) pressures are more likely to occur among lower income households; to that extent the estimation might be vulnerable to selection bias. Indeed, as will be seen later, the events that influence transfers for the lower income households are quite different from the events that matter for all the households, manifesting the difference between the lower income and upper income individuals or households. However, since the central hypothesis of the analysis—incidence and amount of transfers are significantly influenced by the occurrence of demographic events—is validated by both sets of regressions, the decision to proceed with the sample of lower income individuals or households is largely vindicated.

[20] The sample for potential donors includes 1765 observations.

Table 2. Descriptive statistics (older generation).

Variable	PTRANSD = 1	CTRANSD = 1	RTRANSD = 1	UTRANSD = 1	TRANSD = 0
Income and wealth:					
NETINC96	18932.10	9055.57	6751.28	9550.94	8592.74
	(11715.09)	(11355.19)	(9771.35)	(10516.62)	(11076.82)
PVTTRN96	0	38.53	67.60	165.51	156.50
	(0.00)	(405.47)	(569.65)	(891.33)	(1851.79)
NOYEDU96	11.09	11.92	11.11	12.53	10.77
	(2.54)	(1.63)	(2.23)	(2.44)	(2.10)
SIZOHH96	81.93	80.62	72.76	75.72	78.39
	(32.85)	(34.39)	(35.90)	(30.49)	(31.44)
HOWNER96	0.37 (0.50)	0.41 (0.49)	0.35 (0.48)	0.41 (0.50)	0.39 (0.48)
LIQAST96	0.68 (0.47)	0.90 (0.29)	0.94 (0.23)	0.96 (0.18)	0.85 (0.35)
Geographical location:					
WRKGDR96	0.06 (0.25)	0.03 (0.18)	0 (0.00)	0.03 (0.18)	0.03 (0.17)
LIVFDR96	0.37 (0.50)	0.67 (0.46)	0.67 (0.47)	0.48 (0.50)	0.73 (0.44)
Individual and household characteristics:					
AGE96	57.43 (4.83)	66.97 (8.44)	70.23 (9.08)	65.72 (8.14)	66.26 (8.59)
MALE96	0.43 (0.51)	0.51 (0.50)	0.33 (0.47)	0.44 (0.50)	0.38 (0.48)
ESTAT96	0.50 (0.51)	0.19 (0.39)	0.09 (0.30)	0.17 (0.38)	0.13 (0.34)
NOCHLD96	1.56 (1.15)	2.20 (1.32)	1.45 (1.43)	1.00 (1.06)	1.79 (1.44)
AEQUIV96	2.06 (0.87)	1.77 (0.59)	1.61 (0.56)	1.72 (0.59)	1.90 (0.70)
PALIVE96	0.68 (0.47)	0.11 (0.31)	0.08 (0.28)	0.10 (0.30)	0.13 (0.33)
SALIVE96	0.68 (0.47)	0.66 (0.47)	0.60 (0.49)	0.62 (0.49)	0.60 (0.48)
Events:					
NOHVST96	0 (0.00)	0.27 (0.97)	0.21 (0.50)	0.31 (0.76)	0.91 (1.57)
MARAGE96	0 (0.00)	0 (0.00)	0 (0.00)	0 (0.00)	0.002 (0.04)
MOVEIN96	0 (0.00)	0.004 (0.06)	0 (0.00)	0 (0.00)	0.002 (0.05)
DIVORC96	0 (0.00)	0 (0.00)	0 (0.00)	0 (0.00)	0 (0.00)
CBIRTH96	0.06 (0.25)	0 (0.00)	0.01 (0.11)	0 (0.00)	0 (0.00)
Incidence of transfer:					
PTRANSD		0.009 (0.09)	0.01 (0.11)	0 (0.00)	0 (0.00)
CTRANSD	0.12 (0.34)		0.22 (0.42)	0.31 (0.47)	0 (0.00)
STRANSD	0 (0.00)	0.004 (0.06)	0.01 (0.11)	0 (0.00)	0 (0.00)
RTRANSD	0.06 (0.25)	0.07 (0.26)		0.31 (0.47)	0 (0.00)
UTRANSD	0 (0.00)	0.04 (0.19)	0.14 (0.35)		0 (0.00)
Amount of transfer:					
PTRANS96	2179.38	39.44	28.16	0	0
	(1812.44)	(416.37)	(237.35)	(0.00)	(0.00)
CTRANS96	537.50	7110.54	560.56	834.46	0
	(1490.81)	(13383.97)	(1581.27)	(1630.57)	(0.00)
STRANS96	0	1.83	5.63	0	0
	(0.00)	(27.09)	(47.47)	(0.00)	(0.00)
RTRANS96	93.75	132.56	3090.14	1779.31	0
	(375.00)	(685.12)	(4652.48)	(4288.89)	(0.00)
UTRANS96	0	17.56	157.74	817.24	0
	(0.00)	(110.54)	(491.83)	(946.36)	(0.00)
Number of individuals:					
N	16	218	71	29	1465

Note: standard deviations in parentheses.

($PTRANSD = 1$) are younger than those who provide transfers to children and children-in-law ($CTRANSD = 1$), other relatives ($RTRANSD = 1$), and nonrelatives ($UTRANSD = 1$); and that a greater proportion of them are employed.[21] On average, they have higher household income, but a smaller proportion of them has liquid assets. In other words, it can be hypothesized that while transfers to elderly parents are made out of current income, transfers to children and others are made out of the stock of savings.[22] This implies that altruistic parents with foresight are likely to save in order to insure their children (and perhaps other members of "dynastic" households) against negative income or positive expenditure shocks. There is, at the very least, *prima facie* evidence about the possibility that the savings behaviour of rational individuals is likely to be influenced by their expectations about income and expenditure related shocks in the lives of their children and other people belonging to the "dynastic" households—the shocks themselves being precipitated by demographic and other events in the lives of the potential recipients.

Further, the individuals in the sample of potential donors experienced practically no demographic events—marriage, moving in with a partner, divorce, and childbirth—in their lives during the recall period. However, the nondonors reported significantly greater incidence of illness, on average, than the donors. Hence, illness as an "event" possibly influences not only the probability of receiving private transfers, but also the probability of making transfers.

It is not immediately obvious from the descriptive statistics whether making transfers to one category of potential recipients (for example, parents) crowds out the transfers made to other categories of potential recipients.[23] Indeed, it can be seen that a significantly large proportion of those who make transfers to other relatives and nonrelatives also make transfers to children and children-in-law. A possible explanation for this apparent positive relationship is that those who make transfers to individuals who are not within the core of the "dynastic" families have significant liquid assets and are, therefore, able to accommodate the needs of both children and those further away from the core of the "dynasty."

The descriptive statistics for the potential recipients (Table 3) indicate that, on average, the household income of the recipients, net of taxes and private transfers, are about 25 percent lower than that of nonrecipients. In other words, it can be hypothesized that the coefficient of *NETINC96* in the probit and Tobit estimations would be negative. More importantly, however, it is evident that recipients of private transfers ($TRANSDUM = 1$) during the recall period experienced higher incidence of marriage, moving in with partners, and childbirth

[21] A very small proportion of the respondents seemed to have made transfers to spouses, and hence the characteristics of those individuals ($STRANSD = 1$) are not reported in Table 1.

[22] Interestingly, the people who do not make transfers to anyone during the recall period ($TRANSD = 0$) have incomes that are comparable to incomes of those who make transfers to children, other relatives and nonrelatives; but fewer individuals in the nondonor category have liquid assets compared to the aforementioned donors.

[23] Since people in Eastern Germany were, on average, worse off than their Western Germany counterparts, they were likely to have been in greater need of transfers. This, in turn, implies that people living and/or working in Eastern Germany, who were part of East German dynastic households, could a priori be expected to make more transfers than their West German counterparts.

Table 3. Descriptive statistics (younger generation).

	Transdum = 0		Transdum = 1	
Variable	**Mean**	**Std. dev.**	**Mean**	**Std. dev.**
Dependent variable:				
TRNS96	0	0	8534.08	5722.22
Income and wealth:				
NETINC96	24032.41	8704.22	17949.98	9953.57
NOYEDU96	11.87	2.53	12.21	2.63
SIZOHH96	74.90	33.86	72.75	30.74
HOWNER96	0.19	0.39	0.10	0.31
LIQAST96	0.74	0.43	0.71	0.45
Geographical location:				
WRKGDR96	0.17	0.37	0.06	0.24
LIVFRG96	0.27	0.44	0.15	0.36
Individual and household characteristics:				
AGE96	31.83	4.18	31.13	3.92
MALE96	0.49	0.50	0.33	0.47
ESTAT96	0.67	0.46	0.31	0.46
NOCHLD96	0.27	0.66	0.36	0.79
AEQUIV96	2.13	0.95	2.04	0.84
PALIVE96	0.84	0.35	0.98	0.12
SALIVE96	0.81	0.39	0.83	0.37
Events:				
NOHVST96	0.15	0.45	0.12	0.37
MARAGE96	0.01	0.12	0.07	0.26
MOVEIN96	0.03	0.18	0.09	0.28
DIVORC96	0.01	0.10	0.01	0.12
CBIRTH96	0.04	0.20	0.09	0.28
Number of individuals:				
N	752		66	

than the nonrecipients (*TRANSDUM* = 0). This is consistent with our a priori hypothesis that "events" play a significant role in determining the probability of receiving transfers, and perhaps also the magnitude of the transfers.

Finally, while almost all of those who received transfers have at least one parent alive, the probability of having at least a live parent is 14 percentage points lower among nonrecipients. This suggests that parents are perhaps the single most important source of transfers to those in the 25–40 age group. This is also corroborated by the data on the potential donors. While 218 of these individuals reported making transfers to children and children-in-law during the recall period, many fewer individuals made transfers to the other categories of people.

4.2. REGRESSION RESULTS: IMPACT OF EVENTS ON TRANSFERS

A probit analysis[24] (Table 4) indicates that the probability of making an intergenerational transfer to children and children-in-law is positively and significantly related to the educational attainment of a potential donor (NOYEDU96), whether or not she has liquid assets like savings accounts and insurance policies (LIQAST96), and the number of children she has (NOCHLD96).[25] The positive relationship between the aforementioned probability and the number of children is easily explained. However, the significance of *NOYEDU96* and *LIQAST96* in determining this probability possibly implies that savings for old age are meant not only to support old age consumption but also to provide support to "dynastic" households if such a need arises. Indeed, if the educational attainment of an individual is viewed as an indicator of her permanent income, which, in turn, determines the quantum of savings, the relationship between savings for old age and precautionary motives for supporting "dynastic" households during a donor's old age comes sharply into focus. The policy implication of this finding would be discussed in the concluding section.

The importance of this precautionary motive for savings, by way of savings for old age as indicated by the potential donors' permanent incomes, becomes even more evident when one takes into consideration the results of the Tobit analyses (Table 4). It can be seen that the magnitude of the transfers depend only on the educational attainment of the potential donors, education being a well accepted proxy for permanent income and hence lifetime savings.

The analyses involving the sample of potential recipients were undertaken using two different specifications. In the first specification, household income (*NETINC96*) was used as an explanatory variable, and hence the probit and Tobit models are as given by equations (2) and (3). However, since the magnitude of the household income itself does not indicate the financial condition of the household, the second specification replaced *NETINC96* with per adult equivalent income (*PAEINC96*).[26] Probit analyses involving these two specifications, and the two samples—one including all individuals in the 25–40 age group (full sample) and the other including people in the 25–40 age group whose household incomes were less than 35,000 marks in 1996 (subsample)—(Table 5) indicate that, as expected,

[24] Note that the probability of allocation of funds, in the form of transfers, to the different types of recipients might, arguably, be related. In that case, variables PTRANSD, STRANSD, UTRANSD, and RTRANSD in the probit analysis may be endogenous. However, an attempt to endogenize these variables would lead to a five-equation system of probit models whose statistical properties are uncertain, and whose computational cost is significant. Since the focus of the paper is the role of events in determining the probability and quantum of transfers received, and the analysis of the behaviour of the donors is used simply to motivate the rationale for examining intergenerational (as opposed to lateral) transfers, the paper retracts from this costly yet uncertain exercise. However, any discussion about the relationship between transfers to children/children-in-law and "permanent" income should be tempered in the light of the possible endogeneity, and its impact on the regression estimates.

[25] Since the paper seeks to explore the determinants of intergenerational transfers, and given that more than 70 percent of the donors made transfers to children and children-in-law, it seemed reasonable to restrict the analysis to transfers to the aforementioned younger generation.

[26] The determination of adult equivalence for consumers of different age and gender is a complicated exercise in itself. Since the purpose of this exercise is to test the robustness of the qualitative results obtained from the Tobit analyses, it was simply assumed that a child's adult equivalence is 0.5.

Table 4. Determinants of probability and magnitude of transfers given to children and children-in-law.

Variables	Probit	Tobit	Tobit (corrected for selectivity bias)
Constant	−2.77***	−3185.78	−3232.33
	(0.00)	(0.21)	(0.18)
INVMILLS			7110.14 ***
			(0.00)
Income and wealth:			
NETINC96	1.31*E*-06	−0.001	0.02
	(0.76)	(0.93)	(0.35)
NOYEDU96	0.08***	180.71*	186.82**
	(0.00)	(0.07)	(0.05)
HOWNER96	0.09	516.95	516.16
	(0.26)	(0.28)	(0.25)
LIQAST96	0.25**	508.28	503.10
	(0.04)	(0.44)	(0.41)
Geographical location:			
WRKFDR96	0.03	280.66	298.94
	(0.87)	(0.83)	(0.81)
LIVGDR96	−0.08	−419.06	−410.11
	(0.36)	(0.43)	(0.41)
Individual characteristics:			
AGE96	0.001	8.08	17.15
	(0.83)	(0.79)	(0.55)
MALE96	−0.009	71.27	81.92
	(0.91)	(0.87)	(0.85)
NOCHLD96	0.14	190.98	196.72
	(0.00)	(0.23)	(0.19)
Events:			
NOHVST96	−0.01	−0.51	−2.47
	(0.65)	(0.99)	(0.98)
Crowding out:			
PTRANSD	0.20	−276.20	−287.12
	(0.64)	(0.91)	(0.90)
STRANSD	−0.71	−1335.13	−1432.62
	(0.25)	(0.64)	(0.59)
UTRANSD	0.53	−227.39	−241.81
	(0.04)	(0.86)	(0.83)
RTRANSD	0.49***	−294.12	−310.28
	(0.00)	(0.86)	(0.85)
Tobit parameter:			
σ		7835.37	7437.69
Statistic:			
Log likelihood	−622.12	−3620.55	−3566.74

Note: p-values in parentheses. *Significant at the 10 percent level. **Significant at the 5 percent level. ***Significant at the 1 percent level.

Table 5. Determinants of probability of receiving transfers (*probit analysis*).

	Specification 1		Specification 2	
Variable	Subsample	Full sample	Subsample	Full sample
Constant	−1.46*	−1.12**	−1.67**	−1.02**
	(0.07)	(0.01)	(0.04)	(0.02)
Income and wealth:				
NETINC96	−4.16E-05***	−1.99E-05***		
	(0.00)	(0.00)		
PAEINC96			−6.82E-05***	−3.53E-05***
			(0.00)	(0.00)
NOYEDU96	0.04*	0.03**	0.05**	0.03**
	(0.10)	(0.05)	(0.04)	(0.01)
HOWNER96	−0.30	−0.22*	−0.45**	−0.35***
	(0.18)	(0.07)	(0.05)	(0.00)
LIQAST96	0.03	0.01	0.05	−0.04
	(0.84)	(0.92)	(0.72)	(0.72)
Geographical location:				
WRKGDR96	−0.43*	−0.28**	−0.45*	−0.30**
	(0.08)	(0.04)	(0.07)	(0.03)
LIVFDR96	−0.32*	−0.31***	−0.36**	−0.33***
	(0.08)	(0.00)	(0.05)	(0.00)
Individual and household characteristics:				
AGE96	−0.01	−0.007	−0.01	−0.01*
	(0.51)	(0.52)	(0.29)	(0.07)
MALE96	−0.33**	−0.56***	−0.28*	−0.52***
	(0.02)	(0.00)	(0.06)	(0.00)
NOCHLD96	0.02	0.03	0.03	−0.01
	(0.36)	(0.48)	(0.70)	(0.79)
PALIVE96	1.08***	0.27	1.21***	0.36**
	(0.00)	(0.16)	(0.00)	(0.05)
SALIVE96	−0.004	−0.05	0.06	−0.05
	(0.98)	(0.63)	(0.75)	(0.64)
Events:				
NOHVST96	−0.42**	0.15**	−0.30	0.15***
	(0.05)	(0.00)	(0.14)	(0.00)
MARAGE96	0.85**	0.60***	0.95***	0.70***
	(0.01)	(0.00)	(0.00)	(0.00)
MOVEIN96	0.57**	0.19	0.78**	0.37
	(0.04)	(0.41)	(0.01)	(0.12)
DIVORC96	−0.15	0.62**	0.17	0.80***
	(0.81)	(0.03)	(0.79)	(0.00)
CBIRTH96	0.63**	−0.03	0.34	−0.11
	(0.03)	(0.83)	(0.25)	(0.50)
Statistics:				
Log likelihood	−190.43	−426.06	−186.03	−437.84
Percentage of sample receiving transfers	8.06	2.66	8.06	2.66

Note: p-values in parentheses. *Significant at the 10 percent level. **Significant at the 5 percent level. ***Significant at the 1 percent level.

the probability of receiving a transfer varies inversely with both the absolute and per adult equivalent household incomes, as well as with home ownership, and positively with the existence of at least one live parent. It is not obvious as to why this probability varies positively with the educational attainment of the potential recipients. One possible explanation is that parents who bear the cost associated with higher educational attainment are benevolent in general, and hence are also willing to provide private transfers more readily.

Importantly, however, the probit analyses indicate that events play an important role in determining the probability of transfers received by individuals or households. Indeed, while marriage, moving in with partners, and perhaps childbirth significantly increase the probability of receiving transfers among the lower income people, health problems and divorce have a positive impact on this probability for all 25–40-year-old people.[27] Further, the Wald test rejects the null hypothesis that the coefficients of *all* the events included in the specifications are simultaneously equal to zero, at the 1 percent level of significance. The first criticism of the stylized empirical literature, explored in some detail in Sections 1 and 2, is, therefore, vindicated. The importance of such events in determining the probability and magnitude of private transfers is further highlighted by the Tobit analyses (Tables 6 and 7). Indeed, the results of the Tobit estimations indicate that marriage, moving in with partners, and childbirth have a fairly significant impact on the magnitude of transfers received.[28]

The Tobit analyses also suggest that, as with the probability of receiving transfers, the magnitude of transfers varies negatively with absolute and per adult equivalent income, and positively with the existence of at least one live parent. Both the probit and Tobit estimates also indicate that the probability of receiving transfers, as also the magnitude of the transfers, is lower for men. Further, the probability and the magnitude decline if a potential recipient lives in West Germany or works in the former GDR. Since residence in West Germany implies access to more occupational (and perhaps financial) opportunities, thereby increasing the probability of obtaining and retaining stable and well paid jobs, the negative sign of the *LIVFDR96* variable is not surprising. The similar sign of the *WRKGDR96* variable can perhaps be explained by the possibility that if a person has a job in former GDR, he can expect to have greater financial stability over his lifetime compared

---

[27] In other words, events matter irrespective of the income class to which the people in the sample belong. Lower income people presumably require support from the extended family in the event of the formation of a union, i.e., a marriage, and in the event of childbirth. Presumably, higher income people are likely to be better prepared for (by and large) planned events like marriage and childbirth, and more vulnerable to events like health problems and divorce which often take people by surprise. Not unsurprisingly, therefore, once a large number of people with higher incomes are included in the sample, health problems and divorce are seen to be more important determinants of transfers than marriage and childbirth. But, while the selection of the sample on the basis of household income may have biased the results in favour of events like marriage and childbirth, the importance of events per se cannot be brought into question.

[28] It is somewhat difficult to explain why the magnitude of transfers would vary inversely with the number of hospital visits. A possible explanation is that in the presence of publicly provided medical coverage and disability insurance the financial burden of illness is perhaps not as severe in Germany as in some other countries, and hence the coefficient of *NOHVST96* is picking up effects of correlation between transfers and the number of hospital visits that cannot be explained by any causal relationship between hospital visits, financial burden, and transfers.

Table 6. Determinants of magnitude of transfer received (*Tobit analysis*).

	Specification 1		Specification 2	
Variable	Coefficient	p-value	Coefficient	p-value
Constant	−021549.39*	0.06	−24739.17**	0.03
Income and wealth:				
NETINC96	−0.63***	0.00		
PAEINC96			−1.01***	0.00
NOYEDU96	586.09	0.13	758.83**	0.05
HOWNER96	−5154.64	0.11	−7074.76**	0.03
LIQAST96	943.67	0.67	1286.91	0.57
Geographical location:				
WRKGDR96	−7245.59**	0.05	−7404.90**	0.04
LIVFDR96	−4560.22*	0.08	−5068.72**	0.05
Individual and household characteristics:				
AGE96	−129.78	0.62	−243.07	0.35
MALE96	−4480.18**	0.03	−3677.43*	0.08
NOCHLD96	1460.98	0.29	709.51	0.60
PALIVE96	16020.27*	0.00	17736.78***	0.00
SALIVE96	2.39	0.99	1097.29	0.68
Events:				
NOHVST96	−6358.31**	0.03	−4484.77	0.12
MARAGE96	13073.06***	0.00	14583.18***	0.00
MOVEIN96	8447.992**	0.03	11096.20***	0.00
DIVORC96	−1798.48	0.83	3040.47	0.73
CBIRTH96	8942.52**	0.03	4434.28	0.28
Tobit parameter:				
σ	14357.82		14308.69	
Statistic:				
Log likelihood		−845.99		−842.47

*Significant at the 10 percent level. **Significant at the 5 percent level. ***Significant at the 1 percent level.

to a vast majority of his peers and his "dynastic" households, and that hence he is less likely to receive transfers.[29]

4.3. SIMULTANEITY BETWEEN INCOME AND TRANSFER

The stylized probit and Tobit models indicate that there is a strong case in favour of using events as explanatory variables for transfers received. However, as explained in an earlier

[29] In other words, the implicit assumption underlying the use of *WRKGDR* and *LIVFDR* as control variables is that the economic environment in East and West Germany differ significantly, and that the specification should control for this difference. Numerically, these variables are not highly correlated and hence using them together in a specification would not lead to multicollinearity problems.

Table 7. Determinants of magnitude of transfer (*Tobit analysis with correction for selection bias*).

	Specification 1		Specification 2	
Variable	Coefficient	p-value	Coefficient	p-value
Constant	−17593.31**	0.04	−21240.03**	0.04
INVMILLS	26910.02***	0.00	27401.08***	0.00
Income and wealth:				
NETINC96	−0.54***	0.00		
PAEINC96			−0.80***	0.00
NOYEDU96	290.95	0.25	337.87	0.21
HOWNER96	−7800.36***	0.00	−8401.69***	0.00
LIQAST96	1960.40	0.23	2570.55	0.13
Geographical location:				
WRKGDR96	−6985.99***	0.00	−6198.58**	0.02
LIVFDR96	−1485.36	0.44	−1299.71	0.52
Individual and household characteristics:				
AGE96	179.62	0.36	34.28	0.86
MALE96	−1720.81	0.31	−879.32	0.61
NOCHLD96	1062.68	0.21	788.27	0.37
PALIVE96	3371.45	0.43	6174.44	0.24
SALIVE96	360.13	0.83	1954.51	0.28
Events:				
NOHVST96	−5739.72***	0.00	−3460.74*	0.08
MARAGE96	11369.89***	0.00	12606.02***	0.00
MOVEIN96	6278.45***	0.00	8258.91***	0.00
DIVORC96	561.29	0.91	5979.21	0.26
CBIRTH96	5839.04**	0.01	1874.45	0.42
Tobit parameter:				
σ	4788.16		4933.76	
Statistic:				
Log likelihood		−654.13		−655.72

*Significant at the 10 percent level. **Significant at the 5 percent level. ***Significant at the 1 percent level.

section, single equation models may not be appropriate for the analysis of transfers if the recall period is significantly large, namely, one year. Indeed, results obtained from the Hausman test suggests that *NETINC96* (net income) is endogenously determined and hence cannot be treated as an exogenous variable in the Tobit equation that is used to estimate the relationship between the magnitude of transfers and its possible determinants.[30] Hence,

[30] The variable *NETINC96* was regressed on all exogenous variables, and the residuals were saved as variable *NETINC96R*. Next, in the Tobit specification, *NETINC96R* was introduced as an additional explanatory variable. The coefficient of *NETINC96R* was significantly different from 0 at the 1 percent level of significance. This suggested that the estimates of the Tobit model, which treated *NETINC96* as an exogenous variable, are not consistent.

in order to account for the endogenous nature of *NETINC96*, a system of equations with *NETINC96* and private transfers as endogenous variables has to be jointly estimated.

In the earlier analyses, the unit of analysis was the individual. Since two people in the same household may have different probabilities of receiving transfers, depending on their gender, whether or not their parents and other "dynastic" individuals are alive, the (unaccounted for) differences in the financial abilities of the potential donors and so on, this approach was meaningful. However, in order to analyse the determinants of transfers within a simultaneous equation framework, it is necessary to have similar units of measure for income and transfers. Since *NETINC96* is a measure of household income, the measure of transfer used for the analysis should be gross private transfers received by the household, i.e., it is the sum of private transfers received by all individuals within the household (*PVTTRN96*).

Given that the household became the new unit of analysis, and given that GSOEP data on age, education and a number of other variables are available for individual household members, some other changes had to be introduced. Specifically, individual-specific variables like age and educational attainment were reconstructed as average of the age and numbers of years of education of the adults in the household. Further, the variable *NOHVST96* was reconstituted to reflect the total number of hospital visits made by adult individuals in the households. At the same time, the dummy variables *PALIVE96* and *SALIVE96* were accorded value unity if any member of the household had a live parent and sibling, respectively. Similarly, the dummy variable *WRKGDR96* was given the value unity if any one of the adult household members worked in West Germany. Finally, the variable *MALE96* was replaced by *SPOUSE96*, a dummy variable, which indicated whether or not at least one pair of household adults were married.

The aforementioned system of equations, therefore, were given by:

$$\mathrm{PVTTRN96} = \phi_0 + \phi_1 \mathrm{NETINC96} + \Phi' X_1 + \nu_1 \tag{3}$$

and

$$\mathrm{NETINC96} = \sigma_0 + \sigma_1 \mathrm{PVTTRN96} + \Omega' X_2 + \nu_2 \tag{4}$$

where X_1 and X_2 comprise information about the exogenous variables. The explanatory variables for transfers in equation (3) are the same as those in equation (2) with one exception, namely, *MALE96* was replaced by *SPOUSE96* (marital status). The explanatory variables for *NETINC96* in equation (4) included *PVTTRN96*, average educational attainment of household members, the geographical location of the households' home and workplace, and dummy variables to capture existence of wealth and "dynastic" individuals like parents and siblings. In addition, it included a dummy variable (*ESTAT96*) to indicate whether or not at least one adult member of the household was employed during the recall period, and an interaction between *PALIVE96* and *SALIVE96*. This interaction term was introduced to capture the dynamics whereby the existence of an elderly parent may adversely affect effective labour-force participation, but the coexistence of a sibling may ease the demand on the time of a household. Finally, the "events" were included.

While marriage and moving in with a partner were expected to augment household income additively, and divorce impact household income negatively through the same arithmetic algorithm, childbirth was expected to have an impact on household income by way of a negative impact on the "time constraint" of the households.

Joint estimation of this system of equations, one of which involves Tobit while the other involves OLS, is computationally costly. Hence, the estimation of the coefficients of equations (3) and (4) were undertaken by way of the two-stage procedure suggested in Maddala (1983).[31] The estimates of the regression coefficients are presented in Table 8.

Since the focus of this analysis is on transfers, it would be reasonable to focus only on the signs and significance levels of the coefficients of the explanatory variables in the Tobit equation. Importantly, income continues to have a negative and significant coefficient, and existence of live parents and "events"—specifically, divorce and childbirth—continue to have the expected impact on the magnitude of private transfers.[32] At the same time, control variables like *NOCHLD96* (number of children) and *WRKGDR96* (working in East Germany) have experienced a reversal of their signs. However, the most striking difference between the results of the single and simultaneous equation frameworks is that in the former *NOYEDU96* (education) had a positive sign while in the latter the proxy for educational attainment has a negative sign. As mentioned above, it is possible to accord some plausible explanation to a positive sign of the *NOYEDU96* variable. However, if education is a proxy for permanent income, it is reasonable to assume that higher levels of educational attainment would have a negative impact on the magnitude of private transfers. Indeed, irrespective of the magnitude of his current income, if a potential recipient is perceived to have a high permanent income, it is likely to be assumed that he has greater capacity to borrow against his future (higher) income stream, and hence he is likely to receive less support from "dynastic" households.

At the very least, the reversal of signs of coefficients of key explanatory variables in the transfer/Tobit equation suggests that it is imperative to take into consideration the simultaneity between private transfers and income. Indeed, while it is reassuring to note that the signs of important determinants of private transfers like income and events, that are rooted in economic theory, have expected signs, and are hence "qualitatively" robust, the

[31] Given the nature of the GSOEP data, identification of the system of equations pose a major challenge; all the relevant variables that determine the magnitude of transfers received also influence the incomes of the recipients. In a panel estimation the time element could perhaps be used for the purpose of identification, something that cannot be done with cross-section data. Hence, identification has been attempted by the introduction of *ESTAT96* in the specification for the OLS model with *NETINC96* as the dependent variable, and the introduction of an interaction of *PALIVE96* and *SALIVE96* in the specification for the Tobit model with *PVTTRN96* as the dependent variable. This is hardly the best possible way to identify the system of equations, but the best attainable, given the nature of the data and the questionnaire.

[32] Note that although "events" continue to matter, the relative importance of the events differ between the single equation Tobit model and the simultaneous equation framework. In much of the earlier analysis, marriage and moving in with a partner were the triggers for a flow of private transfers. However, in the simultaneous equation framework, childbirth and divorce have gained importance while marriage and moving in with a partner no longer have a significant impact on the magnitude of private transfers.

Table 8. Simultaneous determination of magnitude of transfer and income.

	NETINC96		PVTTRN96	
Variable	**Coefficient**	**p-value**	**Coefficient**	**p-value**
Constant	21683.82***	0.00	44726.12***	0.00
Income and wealth:				
NETINC96(F)			−2.37***	0.00
PVTTRN96(F)	−4.91***	0.00		
NOYEDU96	264.34***	0.00	−337.39***	0.00
HOWNER96	−2689.62***	0.00	5.59	0.29
LIQAST96	370.97	0.52	4829.79***	0.00
Geographical location:				
WRKGDR96	−3252.66***	0.00	5112.39***	0.00
LIVFDR96	−2974.12***	0.00	−3004.98***	0.00
Individual and household characteristics:				
AGE96	−64.80	0.28	193.25***	0.00
SPOUSE96	−3264.38***	0.00	−914.10***	0.00
ESTAT96	5339.02***	0.00		
NOCHLD96	663.35*	0.07	−757.39***	0.00
PALIVE96	3219.93**	0.01	8166.89***	0.00
SALIVE96	−1783.65	0.19	−202.72***	0.00
PALIVE96* SALIVE96	3519.08**	0.02		
Events:				
NOHVST96	−3816.17***	0.00	−5165.61***	0.00
MARAGE96	15806.23***	0.00	−4.63	0.64
MOVEIN96	4585.11**	0.02	3.14	0.66
DIVORC96	−4228.61**	0.04	89.74***	0.00
CBIRTH96	7201.89***	0.00	5285.98***	0.00
Tobit parameter:				
σ			23.92	
Statistic:				
Log likelihood				−429.45
Adjusted R^2		0.46		

Note: PVTTRN96(F) predicted values for PVTTRN96, after the variable was regressed on all exogenous variables and instruments; NETINC96(F) predicted values for NETINC96, after the variable was regressed on all exogenous variables and instruments. *Significant at the 10 percent level. **Significant at the 5 percent level. ***Significant at the 1 percent level.

importance of the exercise involving the simultaneous equation system is not diminished. Indeed, while events matter as much in the simultaneous equation framework as in the single equation framework, thereby vindicating once again the first criticism leveled against the existing literature; a policy discussion about transfers cannot ignore the impact of key variables like educational attainments on the quantum of transfers; and the analysis suggests that coefficients of such variables may be crucially dependent on the choice of the appropriate regression paradigm. Since, in empirical analyses of economic phenomena, robustness of estimates, at least with respect to their signs, is important for meaningful policy discussions, ignoring the possibility that household income is endogenously determined might be a costly oversight.

5. Concluding Remarks

The literature on *inter vivos* and intergenerational transfers that has developed significantly since Cox's (1987) seminal contribution has two lacunae. First, it has ignored the role of specific events like marriage and childbirth in triggering private transfers. Second, it has ignored the possibility that household or individual income, an important explanatory variable for private transfers, may be endogenous, and that therefore a simultaneous equation framework may be more appropriate for analysing the determinants of private transfers. This paper has addressed both of these issues.

However, an empirical analysis that takes into consideration the relationship between the timing of transfers and demographic or life-cycle events in the lives of the recipients cannot overlook the possibilities that the "events" themselves might be endogenous—people, for example, might give birth because they anticipate transfers that would ease the financial burden of the "event"—and that transfers may be delayed and hence not observed during the years of the "events" themselves. While these considerations have merit, they are perhaps not so significant as to have an impact on the efficacy of the analysis itself. To begin with, some events like sickness and divorce are not undertaken in anticipation of transfers. Other events like marriage and childbirth are certainly planned, at least in a postmodern society, but it is unlikely that the possibility of receiving transfers is an important determinant of such life-changing events. Indeed, it is factors like age, stage of career to which the potential recipients belong, and intangible aspects of human behaviour that are likely to determine the timing of such events. There is a likelihood of a transfer occurring with a lag, but this poses a significant problem: the window of time within which a transfer might occur is unknown, and there is no theoretical justification to choose one window size over another. Therefore, given that there seems to exist a statistically significant relation between "events" and transfers that co-occur within a one-year window, it would be reasonable to accept the significance of this relationship.

The analysis embodied in the paper has taken into consideration different samples, specifications as well as different modelling paradigms—single equation probit and Tobit models, as well as a Tobit–OLS simultaneous equation system—and the signs and significance levels of important determinants of private transfers are fairly robust. Specifically, current household income unambiguously has a negative impact on receiving private transfers, while the aforementioned events, except (in general) for hospital visits, unambiguously have a positive impact on such transfers. At the same time, however, reversal in the sign of an important control variable like the educational attainment of the potential recipients, a proxy for their permanent incomes, indicates the necessity to check for endogeneity of household or personal income.

It should be noted that, apart from highlighting the lacunae in the econometric methodology used for empirical analyses of the determinants of private transfers, the paper also throws light on an issue that may have substantial relevance for macroeconomic analysis. The fact that events like marriage, divorce, and childbirth in the life of an individual or household increases both the probability and quantum of private transfers received suggests that savings of individuals (i.e., potential donors) are aimed not only to support old age consumption but also to insure children and other members of "dynastic" families against negative income shocks and positive expenditure shocks. This view is also supported by the fact that the probability and the quantum of transfers made by potential donors—transfers which are presumably aimed at alleviating the financial constraints of "dynastic" households which face "events"—vary positively with their education status which, in turn, determines their "permanent" income and hence savings behaviour. At the same time, the marginal effect of transfers is fairly significant; an event like childbirth can induce transfers equal to about 30 percent of the average annual income of the recipient's household (Tables 3 and 8). In other words, as financial engineering adds to completeness of the financial market, and allows people to insure against all foreseeable events, such as childbirth and sickness, it is possible to envisage circumstances under which aggregate household savings would be negatively and significantly affected.

The study has been limited by the fact that, given the nature of the questionnaire, it is not possible to match sufficient number of donors with recipients during a survey year. Hence, one simple yet important extension of this research endeavour is to examine the impact of "events" and the possible endogeneity of income with data that allows matching of donors and recipients. Further, this study has abstracted from an important form of *inter vivos*, and especially intergenerational, transfers, namely, nonmonetary transfers. Therefore, a second possible extension involves a fusion of this study with that by Pezzin and Schone (1998). Finally, it would be interesting to determine whether the intuition embodied in this paper holds true for private transfers in countries with public institutions and social norms that are significantly different from those in Western developed countries.

Acknowledgement

This paper was prepared at the Max Planck Institute for Demographic Research for presentation at the Annual Meeting of the Population Association of America, Washington, D.C., March 2001. The author would like to thank Donald Cox, Alexia Prskawetz, Ulf-Christian Ewert, Arnstein Aassave, and Hans Peter Kohler for their helpful remarks; Michaela Kreyenfeld for advice about the GSOEP data; and Barbara Schone, Arjun Bedi, Manisha Chakrabarty, Anne Gauthier, and participants at the IUSSP meeting on Age Structure in Taipei (December 2001) for their detailed comments. The views expressed in this paper do not necessarily reflect the views of MPIDR, or Brunel University and the author remains responsible for all remaining errors.

Appendix

List of Variables

Private Transfers Made	
PTRANSD	Dummy variable with value unity if a transfer is made by an individual to parents or parents-in-law.
CTRANSD	Dummy variable with value unity if a transfer is made by an individual to children or children-in-law.
STRANSD	Dummy variable with value unity if a transfer is made by an individual to spouse.
RTRANSD	Dummy variable with value unity if a transfer is made by an individual to relatives.
UTRANSD	Dummy variable with value unity if a transfer is made by an individual to unrelated people.
TRANSD	Dummy variable with value zero if no transfer is made by an individual to anyone.
PTRANS96	Amount of transfers made to parents or parents-in-law.
CTRANS96	Amount of transfers made to children or children-in-law.
STRANS96	Amount of transfers made to spouse.
RTRANS96	Amount of transfers made to relatives.
UTRANS96	Amount of transfers made to unrelated people.
Private Transfers Received	
TRANSDUM	Dummy variable with value unity if an individual receives a private transfer from any source.
TRNS96	Amount of private transfers received by an individual.
PVTTRN96	Amount of private transfers received by a household.
Income and Wealth	
NETINC96	Household income for individual or household receiving private transfers.
HOWNER96	Dummy variable with value unity if the individual or household owns the house they live in.
SIZOHH96	Size of the house that an individual or household lives in, in square meters.
LIQAST96	Dummy variable with value unity if the individual or household owns a liquid financial asset.
Individual's Characteristics	
AGE96	Age in 1996
MALE96	Dummy variable with value unity if the individual is a male.
NOYEDU96	Number of years of education that an individual has.
ESTAT96	Dummy variable with value unity if an individual is employed.
Household Characteristics	
SPOUSE96	Dummy variable with value unity if at least one couple in the household is married.
NOCHLD96	Number of children in the household.

AEQUIV96	Adult equivalence of the household.
PALIVE96	Dummy variable with value unity if either the father or the mother of an individual is alive.
SALIVE96	Dummy variable with value unity if either the brother or the sister of an individual is alive.
Events	
NOHVST96	Number of hospital visits during 1996.
MARAGE96	Dummy variable with value unity if an individual was married during 1996.
MOVEIN96	Dummy variable with value unity if an individual moved in with a partner, aside from marriage, during 1996.
DIVORC96	Dummy variable with value unity if an individual was divorced during 1996.
CBIRTH96	Dummy variable with value unity if an individual had a child during 1996.

Note: The interpretation for some of the variables change in the context of the simultaneous equation model, and the changes have been indicated in the text.

References

Altonji, J.G., Hayashi, F. and Kotlikoff, L. (1992), "Is the extended family altruistically linked? Direct evidence using Micro data." *American Economic Review*, 82:1177–1198.

Altonji, J.G., Hayashi, F. and Kotlikoff, F. (1995), "Parental altruism and inter vivos transfers: Theory and evidence," *Working Paper 1995*, National Bureau of Economic Research: Cambridge, Massachusetts.

Barro, R.J. (1974), "Are government bonds net wealth?" *Journal of Political Economy*, 82:1095–1117.

Becker, G.S. (1974), "A theory of social interactions." *Journal of Political Economy*, 82:1063–1093.

Berman, E. and Rzakhanov, Z. (2000), "Fertility, migration and altruism," *Working Paper 7545*, National Bureau of Economic Research: Cambridge, Massachusetts.

Bernheim, B.D., Shleifer, A. and Summers, L. (1985), "The strategic bequest motive." *Journal of Political Economy*, 93:1045–1076.

Bhaumik, S.K. and Nugent, J.B. (2000), "Wealth accumulation, fertility and transfers to elderly heads in Peru." In: *Sharing the Wealth,* A. Mason and G. Tapinos, eds. Oxford University Press: New York.

Cox, D. (1987), "Motives for private income transfers." *Journal of Political Economy*, 95:508–546.

Cox, D. (1990), "Intergenerational transfers and liquidity constraints." *Quarterly Journal of Economics*, 105: 187–218.

Cox, D. and Jimenez, E. (1992). "Social security and private transfers in developing countries: The case of Peru." *World Bank Economic Review*, 6:155–169.

Cox, D. and Jimenez, E. (1993), *Private Transfers and Effectiveness of Public Income Redistribution in the Philippines*, Mimeo, The World Bank: Washington, DC.

Cox, D. and Stark, O. (1994), *Intergenerational Transfers and the Demonstration Effect*, Mimeo, Boston College: Boston, MA.

Hayashi, F. (1995), "Is the Japanese extended family altruistically linked? A test based on Engel curves," *Working Paper 5033*, National Bureau of Economic Research: Cambridge, MA.

Henretta, J.C., Hill, M.S., Li, W., Soldo, B.J. and Wolf, D.A. (1997), "Selection of children to provide care: The effect of earlier parental transfers," *Journal of Gerontology*, 52(B):110–119.

Haisken-DeNew, J.P. and Frick, J.R. (2000), *Desktop Companion to the German Socio-Economic Panel*, version 3, DIW: Berlin.

Jurges, H. (1999), "Parent-child-transfers in Germany: A study of magnitudes and motivations." *Zeitschrift fur Wirtschaft*, 119(3):422–453.

Lee, R., Mason, A. and Miller, T. (2000), *From Transfers to Individual Responsibility: Implications for Savings and Capital Accumulation in Taiwan and the United States*, Mimeo, University of California: Berkeley.

Maddala, G.S. (1983), *Limited-Dependent and Qualitative Variables in Econometrics*, Econometric Society Monographs, Cambridge University Press.

McGarry, K. and Schoeni, R.F. (1995), "Transfer behavior within the family: Results from the asset and health dynamics survey," *Working Paper 5099*, National Bureau of Economic Research: Cambridge, Massachusetts.

McGarry, K. (1997), "Inter vivos transfers and intended bequests," *Working Paper 6345*, National Bureau of Economic Research: Cambridge, Massachusetts.

McGarry, K. (2000), Testing parental altruism: Implications of a dynamic model, *Working Paper 7593*, National Bureau of Economic Research: Cambridge, Massachusetts.

Pezzin, L.E. and Schone, B.S. (1998), "Intergenerational household formation, female labor supply and informal care giving." *Journal of Human Resources*, 34(3):475–503.

Pezzin, L.E. and Schone, B.S. (2000), *Examining the Motives for Intergenerational Transfers: A New Test Based on Siblings' Behavior*, Mimeo, Johns Hopkins University School of Medicine: Baltimore, Maryland.

Rangel, A. (1999), *Forward and Backward Intergenerational Goods: A Theory of Intergenerational Exchange*, Mimeo, Stanford University, Stanford: California.

Sloan, F.A. and Zhang, H.H. (1995), *Upstream Intergenerational Transfers*, Mimeo, Duke University: Durham, North Carolina.

CHAPTER 5. MATERNAL CORESIDENCE AND CONTACT: EVIDENCE FROM CROSS-NATIONAL SURVEYS

JUDITH TREAS
Department of Sociology, University of California, Irvine

PHILIP N. COHEN
Department of Sociology, University of North California

1. Introduction

Since families provide a safety net that compensates for the limitations of public support systems, family exchanges between the generations remain an integral component of well-being, even in the mature welfare states of the developed world. Finding the right balance of state and family transfers constitutes a central issue for public policy. As they confront ageing populations, nations come under increased pressure to reconcile contradictory goals. They are urged to do more to help beleaguered families care for their dependents, to hold the line against rising welfare and social service costs, and to guard against permitting public transfers to undermine private assistance. Although most developed countries confront this public policy challenge, they come to the intergenerational transfer debate constrained by their unique cultural traditions, by their distinctive histories of public welfare, and by their different demographic age structures. Comparative and cross-national studies enrich our understanding of these demographic and welfare contexts (Hantrais and Letablier 1996). Although cross-national estimates and comparisons of public intergenerational transfers are easy to come by, more knowledge is needed about the private support that flows between younger and older family members in different societies.

The opportunities are ripe for cross-national investigation due to a proliferation of cross-national surveys (Smith 1992) and new methodological approaches for analysing these data (Bryk and Raudenbush 1992; Treas and Widmer 2000). Using a larger, more diverse set of developed countries than was available to earlier studies of family intergenerational integration and exchange, we investigate the likelihood of maternal coresidence as well as the frequency of maternal contact for adults with surviving mothers. While maternal coresidence and maternal contact fail to capture all facets of intergenerational assistance within families (Hashimoto, Kendig, and Coppard 1992), they are a useful point of departure. Significant intergenerational support between family members occurs face-to-face. These intimate exchanges often take place within the household where kin share resources, provide services to one another, and enjoy the social interaction that gives meaning and

Anne H. Gauthier et al. (eds.), Allocating Public and Private Resources across Generations, 117–137.

importance to individual lives. With the growth of an aged population in need of personal assistance in daily life, living together and getting together take on greater significance for adult family members. Since both coresidence and contact facilitate support, we examine how these two mechanisms of intergenerational transfers are related across developed nations with different cultural traditions and social welfare histories.

2. Theoretical Background

We define family support as the give and take of valued goods and services, that is, the social, emotional, instrumental, and economic exchanges engaged in by related persons over the life course. Shared housing offers an efficient context for private resource transfers, particularly to dependent family members. Living together, family members get access to shelter, personal care, domestic services, companionship, and valued goods. Research has focused on the composition of households and the living arrangements of individuals, particularly more dependent persons such as children, youth, and old people (Kobrin 1976; Kiernan 1986; White 1994; Wolf 1994; Hogan and Lichter 1995; Gierveld and Van Tilburg 1999). Households, however, have limitations when it comes to understanding family support patterns (Day 1989).

Even when family members share a household, the direction and balance of intergenerational resource flows is often unclear (Cohen and Casper 2002). Although most people assume that ageing parents are the net beneficiaries of coresidence, the needs of adult offspring dictate this living arrangement in places as diverse as the U.S., England, and urban China (Acquilino 1990; Grundy and Harrop 1992; Logan and Spitze 1996; Treas and Chen 2000). Although family members who live together are generally assumed to pool resources, some members do not (Treas and Chen 2000). And, however important intrahousehold exchanges may be, they cannot tell us about interhousehold transfers (e.g., cash remittances or delivered meals). It is important to study the exchanges between parents and grown children who live together and those who do not, but few studies model both family transfers and multigenerational living simultaneously. Palloni (2000) points out that coresidence models that ignore transfers and transfer models that omit coresidence are both misspecified. At the very least, it is important to complement information on living arrangements by documenting the flows of other types of support and assistance between adult generations (Rossi and Rossi 1992; Farkas and Hogan 1995; Knipscheer et al. 1995; Logan and Spitze 1996).

2.1. INDIVIDUAL-LEVEL CHARACTERISTICS

Intergenerational exchanges must consider the age, gender, and marital status of individuals and populations, because coresidence and contact reflect the gendered life course of family members as well as their socioeconomic resources for achieving independence. Coresidence is most common among the young and the old. Therefore, the age structure of the population will affect coresidence rates. Because the young have greater kin contact than their seniors, according to data from a multi-country study (Farkas and Hogan 1995), the age structure can also impact frequency of visits. "Nest-leaving," the process of moving away from the parental home, has been the focus of much research (Kiernan 1989; White 1994; Dey and Morris 1999; Mayer and Schwarz 1989; White 1994; Dey and Morris 1999; Corijn and Klijzing 2001). The likelihood of living with parents falls off sharply after age 18, although the cross-national variation is substantial (Kiernan 1986). Depending on their school status, adolescents and young adults make a gradual and often unsteady transition from dependence on their parents to economic self-sufficiency and autonomous living. Having a job and a good income markedly increases the likelihood of leaving home in both Europe and the U.S. (Short and Garner 1990), while unemployment or other financial problems can prompt a return. In the U.S. (Goldscheider and DaVanzo 1986), Britain (Kerckhoff and Macrae 1992), and Australia (Young 1989), about half of the young people return home after their first spell of independent living. Personal development ideologies are associated with age norms for leaving home, but most people agree that individual circumstances justify return (Settersten 1998). For a minority of unmarried adults, coresidence with parents continues well into middle age.

Because their offspring grow up and leave home, the likelihood of living in the same household as a grown-up child decreases for persons in their 50s, 60s, and 70s (Gierveld, De Valk, and Blommesteija 2000). At advanced ages when adult offspring are in their middle years, however, the likelihood of coresidence increases, as parents' need for intimate support grows (Cohen and Casper 2002). The well-being of older people is particularly dependent on intergenerational family supports (Kendig, Hashimoto, and Coppard 1992). Besides mitigating loneliness in old age (Knipscheer et al. 1995), shared housing is a strategy to cope with late-life poverty. In the U.S., Australia, Poland, Finland, Germany, Canada, and Taiwan, poverty rates are higher for aged persons who live alone as opposed to living with others besides a spouse (Smeeding and Saunders 1998). Where well-developed pension systems protect from poverty in later life, older people are less likely to need the family economic support provided by coresidence, but frail and disabled older people still need the emotional support of kin as well as family assistance with the activities of daily living (e.g., managing money, keeping house, getting dressed) (Treas 1995). Although studies in the U.S. find relatively low levels of routine exchange and intergenerational support in families (Cherlin and Furstenberg 1986), those with higher needs—parents in poor health and grown offspring with young children—do receive more assistance (Hogan, Eggebeen, and Clogg 1993; Logan and Spitze 1996).

Gender and marital status also affect coresidence and contact. A study of seven developed countries found that women have greater contact with kin than do men (Farkas and Hogan 1995). As for coresidence, women leave home earlier than do men, in part because they marry at younger ages (Kiernan 1986; Goldscheider and Goldscheider 1993). Sons may also gain more from coresidence; according to U.S. data, they receive more domestic services from their mothers and do less household work than do daughters who remain at home (Logan and Spitze 1996). Divorce, single parenthood, or the end of a cohabiting relationship can prompt a return to the parental household. At the other end of the life course, elderly women are more likely than elderly men to share a home with an adult child. Given longer life expectancies, women are more likely to live to experience the disabilities of advanced old age. Women are also more likely to outlive the companionship of a spouse. Widows are more likely than older married people to live with their children, just as unmarried offspring are more likely to remain in the parental home than their married counterparts. Perhaps because they do not have other family obligations and supports, the never-married have been found to have more kin contact than those who are (or who have been previously) married (Farkas and Hogan 1995).

Demographic factors like age and gender predict coresidence, but these factors cannot fully explain the marked differences in the household status of young people from country to country (Kiernan 1986; Lesthaeghe 2000). Women, aged 20–24, in 20 largely European countries in the 1990s demonstrate the diversity. The percent living with parents ranges from 8 percent in Sweden to 87 percent in Italy (Lesthaeghe 2000). Patterned by region, these differences suggest disparities in economic opportunities and welfare state provisions (Dey and Morris 1999). In Southern European countries like Italy, where young people are highly dependent on parents' financial support, women usually remain at home until they marry (Lesthaeghe 2000). In East-Central European states like Poland, Slovenia, and Hungary, about half of the young women coreside with parents, but others leave home for early marriage and motherhood. In Northern Europe's generous social democratic welfare states (e.g., Sweden and Norway), women leave home early not to marry, but rather to live independently or to cohabit. In Western countries (e.g., Netherlands, Canada, Austria, Germany) where jobs and student fellowships are a bigger part of young people's budgets, fewer than half of women, aged 20–24, coreside with parents; most of the remainder either live alone or cohabit (with or without having children). Nest-leaving peaks in the late teens in the U.S., where college students are apt to live apart from parents, and in Denmark where housing has been relatively cheap (Kiernan 1989). Nest-leaving in Great Britain has been more drawn out, apparently because young people wait to leave home until they are ready to live as couples.

The living arrangements of older adults also show cross-national variation. Pampel (1992), for example, finds that there are national differences in the likelihood of older people living alone in ten countries of the European Community. Rates of solitary living are lowest among the largely Catholic populations of Ireland, Northern Ireland, and Italy, and they are highest in Denmark. Living alone increased in all countries between 1975 and 1989, but country-to-country differences were maintained. These cross-national differentials remain even after controls for individual-level variables (age, gender, marital status, and socioeconomic status) and for aggregate-level variables (GNP, social spending, housing stock, postmaterial values). As a determinant of intercountry differences in late-life living arrangements, Pampel points to cultural values, namely postmaterialism, which emphasizes personal fulfillment over the dictates of restrictive social institutions like church and family (Inglehart 1977).

2.2. REGIONS AND REGIMES

Although socio-demographic characteristics of populations affect intergenerational exchange, cultural, social, and political factors are also at play. To account for differences in family life, one argument points to long-standing cultural contrasts in family organization between the regions of Europe (Macfarlane 1978; Hajnal 1982; Reher 1998). The individualism of Northwestern Europe and the English-heritage countries of the globe may be contrasted with the lingering Eastern and Southern European tradition of familism. Looking backward, these regional distinctions map to religious (i.e., Catholic and Protestant) differences in family traditions and gender values. As we might expect, these cultural differences are reflected not only in behaviour, but also in attitudes toward intergenerational obligations. Southern Europeans are more likely than Northwestern Europeans to agree that children owe unconditional love and respect to their parents and that parents must do their best for their children (van den Akker, Halman, and de Moor 1994).

While regional differences are consistent with cultural traditions, contemporary observers argue that cultural differences have long since been incorporated into distinctive state approaches to social welfare. These welfare regimes influence both the material circumstance of family life and its ideological underpinnings. Perhaps the most influential formulation of state policy distinctions is found in Esping-Anderson's (1990) typology of capitalist welfare state regimes, which asks whether the social rights of citizenship guarantee a livelihood regardless of labour market attachment. His recent work emphasizes the welfare state's "de-familialization," i.e., the extent to which the state assumes the family's responsibility of caring for dependents (Esping-Anderson 1999). Whether ranked by state support of family services, daycare, or home-help for the aged, the Nordic social democratic regimes demonstrate far and away the greatest de-familialization. Conservative European states (e.g., Germany, Austria, the Netherlands) are a distant second followed by Japan and *laissez faire*, liberal regimes like the U.K., U.S., and Canada. Late to develop public services, Southern European countries show the least state support for the caregiving functions that traditionally fall to families.

Intergenerational coresidence of old people—as well as unemployed youth—is inversely related to state de-familialization (Esping-Anderson 1999). Rates of coresidence are very low in the Nordic social democratic countries; there rich services and generous benefits not only reduce economic need for intergenerational coresidence, but also presumably diminish normative expectations that parents and grown children must rely on one another. Coresidence is higher in the "service-passive" liberal and conservative states, and highest in "service-poor" Southern Europe and Japan. It remains to be seen how capitalist social welfare regime types relate to intergenerational contact as opposed to intergenerational coresidence. Nor do we know how formerly socialist states, facing economic dislocations and eroded public services in their transition to capitalism, rank in terms of coresidence and contact. Compared to Western Europe, Eastern Europe has high rates of multigenerational living among older women (Koropeckyj-Cox, Agree, and Botev 2000), perhaps reflecting perennial housing shortages in these countries. Certainly, in attitudes toward women's gender roles, these formerly socialist states have the most in common with Southern European countries, which are characterized by late economic development, traditional gender beliefs, and non-Protestant heritage (Treas and Widmer 2000). As further evidence of regional distinctions, people in Southern Europe and formerly socialist Eastern Europe are less likely than their Northern European counterparts to agree that children are responsible for taking care of their ageing parents (Van Peer 1998).

2.3. FAMILISM OR "INTIMACY AT A DISTANCE?"

Given that parents and grown children can exchange assistance either by living together or by getting together, the relationship between coresidence and contact is—from a comparative perspective—an empirical question of theoretical interest. We might hypothesize that coresidence and contact will be positively associated across nations. This expectation is based on the argument that some cultures adhere to collective values of familism, while others embrace individualistic orientations (Triandis 1995). Close family ties characterize cultures based on familism. Individuals owe their allegiance to kin on whom they rely for advice, companionship, assistance, and support. In individualistic cultures, family ties are weaker. Individuals make their own way in the world, relying more on impersonal institutions and on persons who need not be kin. Thus, we would expect high rates of both coresidence and contact in family-oriented societies and low rates in individualistic ones. In other words, coresidence and contact complement one another.

In contrast to the familism hypothesis, the "intimacy at a distance" hypothesis leads us to expect no association—or even a negative one—between coresidence and contact. According to Rosenmayr (1977), intergenerational coresidence reflects economic or occupational requirements, rather than emotional closeness between the generations. Compelled by economic necessity, coresidence may be fraught with tensions, leading to mutual isolation. On the other hand, separate residence (frequently with grown children living near ageing parents) may foster a high level of contact and assistance between the generations. In societies where the residential independence of generations is valued and feasible, family members may prefer frequent contact (the so-called "intimacy at a distance"), rather than a shared household arrangement (Wenger 1992; Knipscheer et al. 1995). Since the preference for family assistance over formal care is documented even where coresidence is rare (Knipscheer 1992; Tornstam 1992), contact and coresidence may be cultural substitutes for one another.

The empirical relation between coresidence and contact remains an open question. Commentators on English-speaking countries with low rates of coresidence emphasize that separate residence can coexist with frequent contact and support between the generations (Wenger 1992). Cross-national studies to validate this assertion are largely lacking. A pioneering comparative study reported the share of unmarried persons, 65 and older, who lived alone in Britain, the U.S., and Denmark (Shanas et al. 1968). It also reported the share of elderly people in each country who had seen a child in the past two days. The relationship of coresidence and contact was negative. The Danes, while least likely to live with others, were most likely to have seen a child. While most likely to coreside, the British were least likely to report child contact. More recent data, however, show a positive relation between coresidence and contact, even controlling for residential proximity of the generations (Hollinger and Haller 1990; Van Peer 1998). In the late 1980s, fewer than 40 percent of never-married persons in the U.S. lived with their mothers compared to 90 percent or more of their counterparts in Hungary and Italy. Countries like the U.S., however, did not compensate for low rates of maternal coresidence with high levels of daily contact.

3. Data and Method

This paper makes use of data from the 1994 International Social Survey Program (ISSP). The data is made available by the *Zentralarchiv für empirische Sozialforschung* in Cologne, Germany, which bears no responsibility for this analysis nor our interpretation. The ISSP is an established program of cross-national collaboration that has coordinated annual surveys on various topics since 1985 (Smith 1992). The 1994 survey focused on gender and family issues. The study was carried out by independent research institutions in 24 largely Western and industrialized countries, usually as a supplement to national probability surveys. Twenty countries report data suitable for our analysis. The countries represent different cultural regions and types of social welfare regimes: the Nordic social democracies of Sweden and Norway; the conservative welfare states of Austria, the Netherlands, and West Germany; the liberal welfare states of Australia, Canada, Great Britain, Ireland, Japan, New

Zealand, Northern Ireland, and the U.S.; the formerly socialist states of the Czech Republic, East Germany, Hungary, Poland, Russia, and Slovenia; and the Southern European state of Italy.[1] (For a variety of reasons regarding data availability and consistency, we did not include Bulgaria, Israel, the Philippines, or Spain.) The total sample size is 16,296 adults age 18 or older. For men, country sample sizes range from 173 in Northern Ireland to 798 in Norway. For women, they range from 179 in Northern Ireland to 690 in West Germany. Since the surveys are carried out by different organizations in each country, response rates vary (see Appendix A).

We focus on two ISSP questions. The first asked, "Is your mother still alive?" Respondents answering "yes," were then asked, "How often do you see or visit your mother?" Valid responses include: lives in the same household, daily, at least several times a week, at least once a week, at least once a month, several times a year, and less often. We limit our study to respondents, 18 and older, whose mothers are still alive, i.e., the population "at risk" of living with or visiting their mothers. One model for maternal coresidence analyses a dummy dependent variable, whether the respondent *lives with* the mother (yes = 1, no = 0). For those who do not live with their mothers, a model for maternal contact focuses on an ordinal dependent variable; the frequency of face-to-face *visits* ranges from 2 ("less than several times a year") to 7 ("daily"). Responses to the contact variable are approximately normally distributed, with a mean of 3.5, a median of 4 ("once a week"), and a skewness of 0.001. Given this distribution, we treat the contact variable as continuous in linear models. We do not interpret the scores as literally representing the frequency of visits, but rather we assume only that higher scores represent more frequent visits. This seems appropriate given the arbitrary nature of the scale and the likely inaccuracies in reporting.[2]

As independent variables, we consider individual-level social and demographic characteristics, previously found to be associated with coresidence and kin contact. Respondent's age is measured in 10-year categories with ages 18–24 as the omitted group. Marital status, also measured by dummy variables, contrasts the *formerly married* and the *never married* with the omitted *married* category. To achieve a relatively comparable measure for various educational systems, we create a dummy variable for *college or higher* educational attainment. Employment status is also a marker of socioeconomic circumstances. Dummy variables distinguish *employed full-time*, *employed part-time*, and *student* from the omitted *not employed* (*nonstudent*) category. Since socioeconomic resources may also depend on the spouse's employment status, we use a dummy variable for *employed spouse* (employed spouse = 1, no employed spouse = 0 for married as well as unmarried respondents). Because there are relatively few missing values for most variables, we simply drop cases with missing data. The exception is spouse's employment, where higher levels of missing

---

[1] Data for the two parts of the newly unified Germany were collected from separate samples in 1994, and we consider them separately here. For this early date, we consider the two parts of Germany to be separate welfare regimes, because in the few years after reunification policies and practices were slow to change; we would not expect intergenerational relations formed over many years to respond so immediately to new (or incipient) changes in policy. The differences apparent in our results, especially for maternal visiting, provide support for this decision.

[2] If we take the variable at its literal meaning, and transform it into the number of times per year the respondent visits, it becomes much more skewed, with a mean of 83 visits per year and a median of 52 (skewness = 1.64).

data require that we substitute the gender- and country-specific proportion employed, as derived from the employment of married respondents in the data set. In Sweden, for example, where 71 percent of the married men are employed, we impute a value of 0.71 for the employed spouse variable of a married woman who is missing this data. We add a dummy variable to the models indicating whether spouse's employment has been imputed in this manner. (Descriptive statistics are presented in Appendix B, Tables B.1 and B. 2.)

We first examine the individual-level determinants of maternal coresidence across all 20 countries, and the differences across countries in the rates of coresidence, net of these individual-level independent variables. We repeat this exercise for frequency of maternal contact. Then, we test whether maternal contact varies across countries as a function of coresidence rates that are adjusted for key life-course variables. In other words, we construct gender-specific, country-level coresidence rates, reflecting the proportion of adults predicted to coreside at the mean of age, age-squared, and marital status variables. This adjusted coresidence rate is utilized as a country-level independent variable, enabling us to test whether societal coresidence practices are positively associated with contact (the familism hypothesis) or not (the intimacy-at-a-distance hypothesis). For this multivariate test of the hypotheses, we use the HLM software package to estimate hierarchical linear models incorporating the country-level coresidence variable (Bryk and Raubenbush 1992).

The basic equation for the individual-level is

$$Y_{ij} = \beta_0 + \beta_1(\text{Male}) + \Sigma\beta_{kj}X_{ikj} + R_{ij}$$

where Y_{ij} equals the odds of respondent i living with (or the frequency of visiting) the respondent's mother, in country j. Visiting frequency can be analysed with linear models. For analyses of maternal coresidence, Y_{ij} is formulated as the log odds, and the model is logistic. β_0 is the individual-level intercept and β_1 is the difference between male and female respondents. X_{ikj} is the set of individual-level variables and β_{kj} is the vector of coefficients associated with those variables. Finally, R_{ij}, the individual-level error term, is assumed to be normally distributed with a zero mean and a constant variance. Individual-level control variables are centred at their grand means, so the intercept is interpreted as the average odds of living with (or frequency of visiting) the mother for a woman with average characteristics.

The first country-level equation takes the form:

$$\begin{aligned}\beta_{0j} &= \gamma_{00} + U_{0j}\\ \beta_{1j} &= \gamma_{10} + \mathrm{U}_{1j}\\ X_{kj} &= \gamma_k\end{aligned}$$

where γ_{00} and γ_{10} are the intercepts for the country-level models; U_{0j} and U_{1j} are the error terms at the country level; and γ_k are the constant coefficients across all countries (i.e., the individual-level control variables are constrained to have fixed effects across countries). With the error terms included, HLM produces predicted values for each country of both the intercept and the effect of being male, which yield predicted coresidence rates and visit frequencies, net of the controls. For females, the intercepts give the predicted coresidence rate or the predicted visit frequency. Adding the coefficient for being male to the intercept yields the male predicted coresidence rate or visiting frequency. These predicted values are similar to those that would be derived from an OLS model that contained a set of dummy control variables, i.e., one for each country (see Bryk and Raudenbush 1992:40).

In the second set of models, we estimate the effect of country coresidence rate on the frequency of visiting. This modifies the country-level equation as follows:

$$\beta_{0j} = \gamma_{00} + \gamma_{01}(\text{coresidence rate}_j) + U_{0j}$$

where γ_{01} is the effect of the gender-specific, country-level coresidence rate on β_0. We estimate these models separately by sex instead of modelling β_1 across countries.

4. Empirical Results

4.1. HOW MUCH DO THE TWENTY COUNTRIES DIFFER IN MATERNAL CORESIDENCE AND CONTACT?

Although sharing the same household is an efficient way to transfer resources between generations, there is a remarkable degree of variation in the extent to which this coresidence occurs. For women whose mothers are still alive, the percent living with mother ranges from a scant 4 percent in Sweden and 7 percent in Great Britain to highs of 32 percent in Japan and 38 percent in Italy. For men, the percent living with mother ranges from 11 percent in the U.S. and Sweden to 47 percent in Japan (see Appendix B). Figure 1 arrays the 20 countries in terms of the percentages of male and female respondents coresiding with their mother. Consistent with previous research, Northwestern European countries and former English colonies are characterized by low rates of maternal coresidence. Eastern and Southern European nations are distinguished by high rates of coresidence, as is Japan. Male respondents are more likely than their female counterparts to report maternal coresidence. Although Northern Ireland shows gender parity, its estimate is based on a small sample. The pattern of disproportionately high rates of maternal coresidence for men is especially apparent for Ireland, Slovenia, and Japan. Although the nature and direction of private intergenerational resource transfers involved in coresidence is unclear, the data show that the coresidence between sons and mothers is more common than between daughters and mothers.

If grown-up children do not coreside, visits offer an opportunity to exchange services and resources with their mothers. Figure 2 shows the variation in the visiting patterns of those adults who do not coreside with their mothers. On average, Italians say that they see their mothers between once a week and once a month. At the other extreme are the Japanese who, on average, visit their mothers several times annually. The extreme cases of the Japanese and Italians are instructive, because these two countries had the highest rates of maternal coresidence among the 20 cases considered. Thus, high rates of intergenerational living can go along with either high or low frequency of face-to-face interaction for those grown offspring who do not live with their mothers. Whether the pattern of positive association predominates, as implied by the arguments of familism, remains to be investigated below. Although male respondents are much more likely than their female counterparts to live

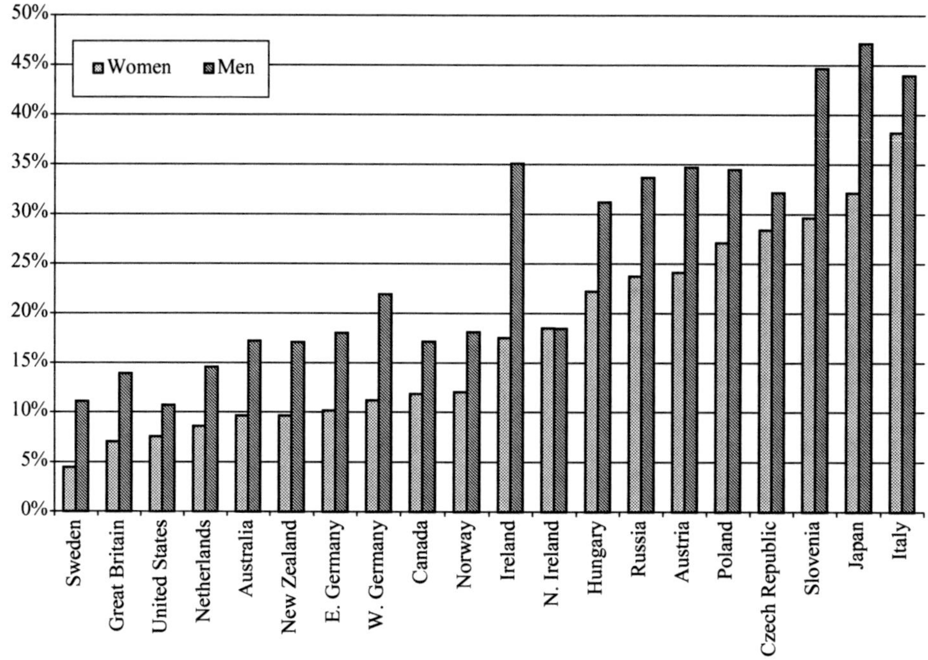

Figure 1. Percent living with mother.

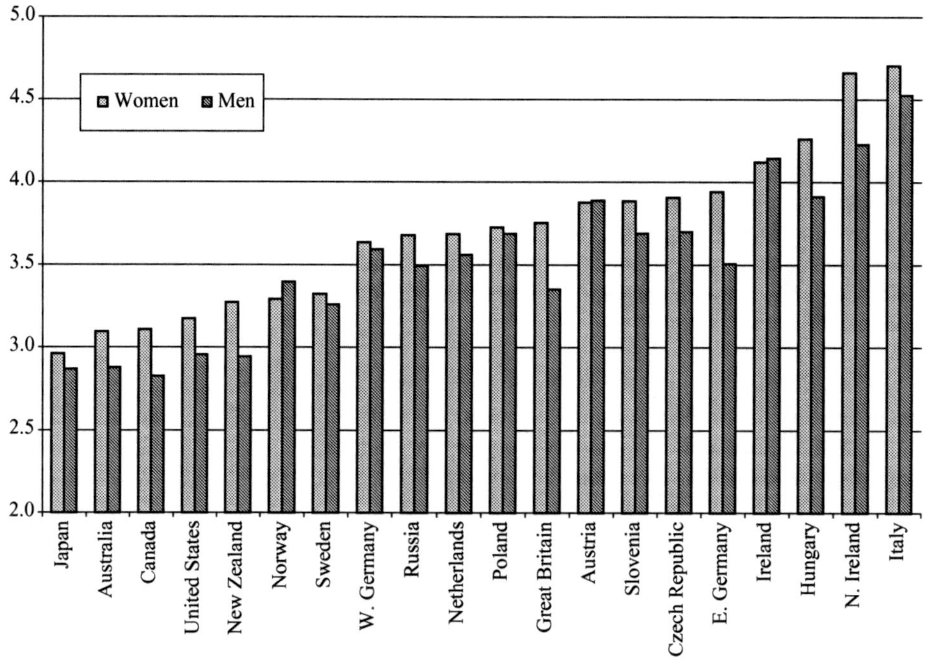

Figure 2. Frequency of visits with mother.

with their mothers, they are not more likely to visit them. With rare exceptions like Austria and Ireland, women visit more frequently than do men. Certainly, gender differences in contact are less pronounced than those for coresidence.

4.2. HOW DO INDIVIDUAL-LEVEL CHARACTERISTICS AFFECT CORESIDENCE AND CONTACT?

To understand which individual-level characteristics determine these patterns of family solidarity, we examine cross-national differences in maternal coresidence and contact, controlling for respondent's age, marital status, education, employment status, and spouse's employment status. The HLM results for models of coresidence and of visiting frequency are presented in Table 1.

The likelihood of coresidence is highest for the 18–24-year-olds; it declines until middle age (35–44) and then rises, presumably as maternal needs increase. Men are more likely to coreside than are women. Compared to married people, respondents who are never-married are substantially more likely to live with their mothers and those who were previously married are somewhat more likely to do so. A college education decreases the likelihood of coresidence. Employment status has surprisingly little effect. Respondents who are full-time or part-time workers are no less likely to coreside than the omitted category of not employed respondents. Students, however, do display a greater likelihood of living with their mothers. While respondent's employment has little effect, having an employed spouse does reduce the likelihood of maternal coresidence ($p < 0.001$).

Maternal contact declines with age only to increase for respondents aged 55 and older. Men, of course, visit less frequently than do women. Although marital status has marked effects on living arrangements, it does not influence the frequency of visits. There are no significant differences in maternal contact between the married respondents and the never-married or previously married ones. Higher education is negatively associated with contact. Respondent's employment status is not statistically significant, except that students who do not live with their mothers see them less often than do not employed children. In contrast to coresidence, spouse's employment has no effect on maternal contact.

4.3. DO INDIVIDUAL-LEVEL DETERMINANTS ACCOUNT FOR COUNTRY DIFFERENCES IN MATERNAL CORESIDENCE AND CONTACT?

We use the models in Table 1 to derive predicted values for each country at the mean of the independent variables. These gender-specific predicted probabilities of coresidence and frequency of visiting appear in the bottom panel of Table 1. For the 20 countries, the rank–order correlations between the observed and adjusted sets of measures are quite high. For the probability of coresidence, the correlations are 0.86 for men and 0.89 for women, suggesting very little change in the ordering of countries as a result of controlling for the social and demographic variables. For the frequency of visiting, the rank–order correlations are even higher—0.98 for both men and women. Although many of the individual-level, independent variables in the model have significant effects on coresidence and on frequency of contact, these factors do not account for the rank ordering of behaviours at the aggregate level of the countries.

Table 1. Hierarchical linear models for coresiding with and visiting mother.

	Coresiding	Visiting
Intercept	−1.884***	3.671***
Age 25–34	−0.919***	−0.278***
Age 35–44	−1.950***	−0.514***
Age 45–54	−1.041***	−0.632***
Age 55+	−0.722***	−0.495***
Male	0.347***	−0.135**
Formerly married	0.732***	0.094
Never married	1.955***	0.008
College	−0.508***	−0.282***
Employed full-time	−0.020	−0.033
Employed part-time	0.070	0.009
Student	0.273**	−0.287***
Spouse employed	−0.474***	−0.026
Spouse employment imputed	0.551***	−0.147*
Individual-level R^2	†	0.316

Predicted rates by country and gender

	Coresiding		Visiting	
Country	Women	Men	Women	Men
Australia	0.18	0.22	3.16	2.93
W. Germany	0.11	0.18	3.53	3.52
E. Germany	0.14	0.18	3.85	3.45
Great Britain	0.11	0.15	3.67	3.27
Northern Ireland	0.16	0.20	4.53	4.18
United States	0.10	0.13	3.16	2.97
Austria	0.25	0.28	3.81	3.82
Hungary	0.26	0.30	4.18	3.84
Italy	0.36	0.38	4.63	4.51
Ireland	0.19	0.29	4.10	4.11
Netherlands	0.07	0.08	3.68	3.62
Norway	0.06	0.11	3.27	3.43
Sweden	0.07	0.11	3.34	3.27
Czech Republic	0.26	0.30	3.82	3.70
Slovenia	0.32	0.42	3.82	3.66
Poland	0.29	0.34	3.66	3.65
Russia	0.30	0.37	3.77	3.56
New Zealand	0.17	0.19	3.32	3.04
Canada	0.15	0.16	3.26	3.00
Japan	0.33	0.43	2.97	2.93

*$p < 0.05$; **$p < 0.01$; ***$p < 0.001$.

†HLM does not produce a standard measure of explained variance for nonlinear models.

4.4. DO NATIONAL DIFFERENCES IN MATERNAL CORESIDENCE AND CONTACT CORRESPOND TO REGIONAL AND WELFARE REGIME TYPOLOGIES?

Thus far, we have related coresidence and contact to individual-level variables of the gendered life course, and we have demonstrated that differences in these variables provide an insufficient explanation of country-to-country differences. Now, we ask whether country differences, net of individual-level variables, correspond to broader regional or welfare regime patterns of family solidarity. Table 2 presents women's mean predicted values on coresidence and visiting, controlling for individual-level variables, for the countries grouped by capitalist welfare-state regime type (social democratic, liberal, and conservative). We also distinguish formerly socialist states. Italy and Japan are shown separately, because they are outliers in contact and coresidence and have had distinctive experiences as welfare states (Esping-Anderson 1999).

The results confirm the high levels of intergenerational family solidarity in the formerly socialist states, namely, the Czech Republic, East Germany, Hungary, Poland, Russia, and Slovenia. Controlling for individual-level characteristics, the mean predicted proportion of women coresiding is 0.26 while the mean predicted frequency for visits is 3.64—nearly once a month. Among the capitalist welfare states, the social democratic countries of Norway and Sweden stand out, because their levels of coresidence and contact (0.07 and 3.32) are lower than those of their European counterparts.

Falling between the social democratic and formerly socialist countries on both coresidence and contact, the liberal and conservative states are virtually indistinguishable from one another. Italy and Japan, of course, display the highest proportions coresiding, but Italian women visit their mothers very often while Japanese women visit relatively infrequently.

Table 2. Mean predicted maternal coresidence proportion and contact frequency for women, by welfare regime type.

	Coresidence	Visiting
Social democratic Norway, Sweden	0.07	3.31
Liberal Australia, Canada,Great Britain, New Zealand, Northern Ireland, United States, Ireland	0.15	3.60
Conservative Austria, West Germany, Netherlands	0.14	3.67
Formerly socialist Czech Republic, East Germany, Hungary, Poland, Russia, Slovenia	0.26	3.64
Italy	0.36	4.63
Japan	0.33	2.97

Table 3. Hierarchical linear model for frequency of visiting mother in 19 countries (respondents living with their mothers excluded).

	Women	Men
Intercept	3.138***	2.959***
Country maternal coresidence rate	2.510**	2.493**
Age 25–34	−0.245***	−0.308***
Age 35–44	−0.511***	−0.512***
Age 45–54	−0.552***	−0.732***
Age 55+	−0.426***	−0.544***
Formerly married	0.145*	0.062
Never married	0.059	−0.017
College	−0.252***	−0.321***
Employed full-time	−0.030	−0.051
Employed part-time	0.009	0.107
Student	−0.284**	−0.327**
Spouse employed	0.030	−0.043
Spouse employment imputed	−0.123	−0.171+

Note: Dependent variable is frequency of visits (2–7); Japanese respondents excluded.
$^{+}p < 0.10$; $^{*}p < 0.05$; $^{**}p < 0.01$; $^{***}p < 0.001$.

4.5. DO MATERNAL CORESIDENCE RATES IN A COUNTRY AFFECT THE FREQUENCY OF MATERNAL CONTACT?

Having considered the determinants of coresidence and contact separately, we now consider how maternal coresidence practices influence the frequency of visiting for adults who do not live with their mothers. Thus, in addition to the individual-level variables, our gender-specific models incorporate a country-level variable—the gender-specific proportion coresiding (adjusted by age and marital status). We exclude Japan from this analysis, because it is an extreme outlier with the lowest frequency of visits but the second-highest coresidence rate. The HLM results for the remaining 19 countries are presented in Table 3. The results show that the higher the proportion coresiding in a country, the more frequently do other offspring have maternal contact. The significant coefficients for the effect of coresidence rates imply that the average score for maternal contact would be 2.5 points higher on the 2–7 ordinal scale in a country where 100 percent of the adults live with their mothers than in a country where no adults live with their mothers.[3] However, the effect of coresidence does not differ significantly by gender.

5. Discussion and Conclusion

Many researchers have observed that there are cross-national differences in the popularity of multigenerational living. Our analysis offers a refinement and extension of earlier

[3] In the model with Japan included, the effect of coresidence was marginally significant in the same direction for both men and women.

studies. Marshalling data from 20 countries and focusing analysis on respondents who are "at risk" of coresidence by virtue of having a surviving mother, we confirm that there is substantial cross-national variation in the likelihood that an adult will live with his or her mother. Furthermore, we demonstrate that adults who do not live with their mothers differ from country to country in the frequency with which they visit their mothers. These intergenerational patterns of support and exchange are strongly gendered. Men are more likely than women to make their home with their mothers. Women, however, see their mothers somewhat more often than do men. Significant differences in the behaviour of men and women remain even after differences in age, marital status, education, employment, and spouse's employment are controlled. Furthermore, these controls for individual-level factors do not explain the cross-national differences observed in maternal coresidence and maternal contact.

Maternal coresidence and contact are behaviours that take place in national contexts with distinctive cultural traditions, unique religious heritages, and particular state policies. As noted, these histories have given rise to different welfare approaches and varying degrees of commitment to helping families provide for dependent kin. In the social democratic countries of Norway and Sweden, where public support for families is most fully developed, we find that there is less intergenerational contact and markedly less maternal coresidence. By contrast, maternal coresidence and contact is high where de-familialization by the state is limited: 1. the formerly socialist countries, where the public safety net has unraveled in the face of economic dislocation and 2. Italy, where public welfare programs were slow to develop. In the formerly socialist countries of Eastern Europe, intergenerational solidarity is not limited to coresidence but extends to contact as well. We infer that familism and/or the general economic deprivation of transition economies, rather than merely housing shortages, account for the patterns observed. Other European nations and English-heritage countries with liberal or conservative approaches to social welfare policies fall somewhere in the middle in terms of coresidence, contact, and de-familialization.

Rather than being substitutes for one another, maternal coresidence and contact seem to complement each other as mechanisms of intergenerational support. Countries where more adults live with their mothers are countries where those adult children who do not coreside visit more frequently. Japan, having high rates of coresidence and low frequency of visiting, are a singular exception to this pattern. Presumably, contact offers a way for grown-up children who do not share a household with their mothers to share other resources, exchange services, and interact with kin. Previous research has indicated that children crowd their siblings out of the parental home, thus discouraging their coresidence (Treas and Chen 2000). The presence of brothers and sisters in a household, however, may prompt more frequent visits—not only by modelling norms of family togetherness, but also by permitting people to enjoy siblings' and parents' company simultaneously. In any case, the results favour the familism argument relating coresidence to family solidarity over the "intimacy at a distance" argument that holds that coresidence is a poor indicator of the strength of intergenerational ties.

Although we find that coresidence and contact are associated with each other and with state welfare efforts, public and private transfers are not interchangeable. At the public level, intergenerational support such as public pensions typically consists of impersonal transfers taking place at arm's length. Family transfers, on the other hand, are highly personalized exchanges that benefit from long-standing associations, frequent interaction, and relationship-specific investments that permit parties to know one another's needs, preferences, capacities, and contributions. Even in countries characterized by low coresidence and less frequent visiting, people look to kin, rather than to formal systems, for help with various personal problems (Knipscheer 1992; Tornstam 1992). Ageing populations, however, present challenges to both systems.

The number of people in need of family support will increase in response to the growth of the older population as well as its ageing. Population ageing will place unprecedented demands on public systems for pensions, health care, and social services (Birg, in press; Golini, in press; Treas, in press). The low fertility contributing to population ageing also limits the availability of children—an important factor contributing to both coresidence and contact in later life (Wolf 1994; Farkas and Hogan 1995). The postponement of marriage and childbearing, however, increases the proportion of adults who will live for extended periods with their parents—a life-course development with poorly understood implications for the intergenerational support of ageing parents.

Acknowledgements

Support for this research comes from the Center for European and German Studies, University of California. We thank Jeanne Batalova for research assistance, and Anne Gauthier for her helpful comments.

Appendix A. Response rates by country.

Country	Response rate (%)
Australia	93.5
Austria	70.4
Canada	72.5
Czech Republic	82.2
E. Germany	54.7
W. Germany	52.8
Great Britain	55.7
Hungary	76.4
Italy	69.3
Japan	77.8
New Zealand	69.7
Norway	60.7
Poland	82.1
Slovenia	43.8
Sweden	64.0
United States	78.0

Note: Not available for Ireland, Northern Ireland, Netherlands, or Russia.

Appendix B.

Table B.1. Descriptive statistics by country: Adults with living mothers.

	Australia	W. Germany	E. Germany	Great Britain	Northern Ireland	United States	Austria	Hungary	Italy	Ireland
Women										
Live with mother	0.10	0.11	0.10	0.07	0.18	0.08	0.24	0.22	0.38	0.17
Visit frequency*	3.09	3.64	3.94	3.75	4.66	3.17	3.88	4.26	4.70	4.12
Age 25–34	0.26	0.39	0.31	0.38	0.32	0.33	0.27	0.27	0.29	0.33
Age 35–44	0.35	0.26	0.28	0.25	0.26	0.30	0.27	0.27	0.23	0.31
Age 45–54	0.25	0.13	0.15	0.14	0.12	0.16	0.19	0.20	0.16	0.17
Age 55+	0.09	0.05	0.13	0.08	0.08	0.08	0.08	0.11	0.10	0.02
Finished college	0.23	0.13	0.13	0.22	0.18	0.35	0.04	0.14	0.18	0.30
Employed FT	0.36	0.41	0.50	0.38	0.33	0.55	0.40	0.57	0.31	0.37
Employed PT	0.30	0.16	0.09	0.24	0.16	0.14	0.16	0.02	0.12	0.11
Student	0.06	0.06	0.05	0.04	0.06	0.05	0.09	0.04	0.09	0.06
Spouse employed	0.74	0.55	0.54	0.55	0.34	0.43	0.59	0.53	0.51	0.53
Was married	0.05	0.09	0.13	0.18	0.22	0.24	0.08	0.20	0.04	0.06
Never married	0.15	0.29	0.18	0.19	0.32	0.25	0.25	0.15	0.35	0.30
N	467	608	296	284	173	479	274	419	338	223
Men										
Live with mother	0.17	0.22	0.18	0.14	0.18	0.11	0.35	0.31	0.44	0.35
Visit frequency*	2.88	3.59	3.51	3.35	4.23	2.95	3.89	3.91	4.52	4.15
Age 25–34	0.21	0.35	0.28	0.42	0.34	0.32	0.28	0.28	0.30	0.33
Age 35–44	0.34	0.27	0.31	0.25	0.30	0.32	0.23	0.29	0.26	0.25
Age 45–54	0.25	0.14	0.13	0.15	0.13	0.18	0.14	0.15	0.13	0.14
Age 55+	0.14	0.06	0.09	0.05	0.06	0.06	0.10	0.10	0.07	0.04
Finished college	0.20	0.16	0.18	0.23	0.26	0.34	0.06	0.14	0.17	0.25
Employed FT	0.83	0.81	0.77	0.72	0.76	0.75	0.76	0.64	0.71	0.66
Employed PT	0.05	0.02	0.01	0.02	0.01	0.08	0.02	0.01	0.08	0.01
Student	0.04	0.08	0.04	0.04	0.07	0.03	0.12	0.06	0.10	0.07
Spouse employed	0.44	0.23	0.44	0.36	0.41	0.35	0.32	0.34	0.28	0.21
Was married	0.03	0.07	0.07	0.13	0.06	0.16	0.03	0.08	0.02	0.03
Never married	0.23	0.39	0.29	0.27	0.25	0.30	0.40	0.30	0.44	0.44
N	483	690	311	245	179	383	245	359	323	211

*For those not living with their mothers.

Table B.2. Descriptive statistics by country: Adults with living mothers.

	Netherlands	Norway	Sweden	Czech Republic	Slovenia	Poland	Russia	New Zealand	Canada	Japan
Women										
Live with mother	0.09	0.12	0.04	0.28	0.30	0.27	0.24	0.10	0.12	0.32
Visit frequency*	3.69	3.29	3.32	3.91	3.88	3.73	3.68	3.27	3.11	2.96
Age 25–34	0.32	0.29	0.28	0.28	0.29	0.28	0.32	0.31	0.28	0.26
Age 35–44	0.29	0.25	0.25	0.23	0.31	0.31	0.30	0.27	0.34	0.27
Age 45–54	0.15	0.13	0.21	0.14	0.13	0.15	0.13	0.18	0.14	0.20
Age 55+	0.04	0.04	0.08	0.02	0.06	0.06	0.04	0.09	0.03	0.06
Finished college	0.26	0.25	0.29	0.21	0.17	0.18	0.52	0.50	0.64	0.25
Employed FT	0.37	0.56	0.39	0.76	0.67	0.48	0.57	0.37	0.47	0.37
Employed PT	0.00	0.00	0.30	0.03	0.02	0.04	0.07	0.18	0.22	0.13
Student	0.13	0.20	0.12	0.14	0.08	0.07	0.10	0.03	0.14	0.12
Spouse employed	0.34	0.62	0.61	0.62	0.61	0.52	0.60	0.63	0.62	0.65
Was married	0.10	0.09	0.05	0.12	0.07	0.12	0.12	0.11	0.07	0.04
Never married	0.34	0.42	0.22	0.26	0.21	0.21	0.16	0.18	0.29	0.28
N	665	798	426	310	321	480	755	342	676	436
Men										
Live with mother	0.15	0.18	0.11	0.32	0.45	0.34	0.34	0.17	0.17	0.47
Visit frequency*	3.56	3.40	3.26	3.70	3.69	3.69	3.49	2.94	2.82	2.87
Age 25–34	0.28	0.28	0.29	0.29	0.31	0.28	0.32	0.29	0.34	0.22
Age 35–44	0.28	0.25	0.27	0.23	0.30	0.33	0.28	0.26	0.23	0.24
Age 45–54	0.16	0.16	0.20	0.17	0.11	0.15	0.13	0.14	0.17	0.20
Age 55+	0.06	0.05	0.06	0.03	0.03	0.06	0.01	0.08	0.04	0.09
Finished college	0.34	0.32	0.26	0.27	0.16	0.13	0.48	0.63	0.72	0.29
Employed FT	0.65	0.72	0.73	0.79	0.76	0.63	0.77	0.79	0.61	0.78
Employed PT	0.00	0.00	0.03	0.01	0.01	0.02	0.06	0.03	0.10	0.01
Student	0.17	0.17	0.11	0.14	0.07	0.08	0.10	0.07	0.16	0.15
Spouse employed	0.13	0.44	0.48	0.56	0.50	0.40	0.49	0.43	0.44	0.27
Was married	0.04	0.07	0.05	0.04	0.01	0.04	0.07	0.06	0.07	0.01
Never married	0.46	0.48	0.32	0.34	0.36	0.28	0.21	0.30	0.40	0.38
N	523	679	361	314	291	435	505	223	397	369

*For those not living with their mothers.

References

Acquilino, W.S. (1990), "The likelihood of parent-adult child coresidence: Effects of family structure and parental characteristics." *Journal of Marriage and the Family*, 52:405–419.

Birg, H. (2002), "Demographic ageing and population decline in 21st century Germany: Consequences for systems of social insurance." *Policy Responses Population Ageing and Population Decline. U.N. Population Bulletin*, 44/45:103–134.

Bryk, A.S. and Raudenbush, S.W. (1992), *Hierarchical Linear Models: Applications and Data Analysis Methods*. Sage: Newbury Park, CA.

Cherlin, A.J. and Furstenberg, F.F., Jr. (1986), *The New American Grandparent: A Place in the Family, A Life Apart*. Basic Books: New York.

Cohen, P.N. and Casper, L.M. (2002), "In whose home? Multigenerational families in the United States, 1998–2000." *Sociological Perspectives*, 45:1–20.

Corijn, M. and Klijzing, E. (2001), *Transitions to Adulthood in Europe*. Kluwer Academic Publishers: Boston, MA.

Day, A.T. (1989), "Kinship networks and informal support in the later years." In: *Later Phases of the Family Life Cycle: Demographic Aspects*. E. Grebenik, C. Hohn, and R. Mackensen, eds. Clarendon Press: Oxford.

Dey, I. and Morris, S. (1999), "Parental support for young adults in Europe." *Children and Youth Services Review*, 21:915–935.

Esping-Anderson, G. (1990), *The Three Worlds of Welfare Capitalism*. Polity Press: Cambridge.

Esping-Anderson, G. (1999), *Social Foundations of Postindustrial Economies*. Oxford University Press: Oxford.

Farkas, J.I. and Hogan, D.P. (1995), "The demography of changing intergenerational relations." In: *Adult Intergenerational Relations: Effects of Societal Change*, V.L. Bengtson, K.W. Schaie, and L.M. Burton, eds. pp. 1–25. Springer: New York.

Gierveld, J. De Jong, de Valk, H. and Blommesteija, M. (2000), "Living arrangements of older persons and family support in more developed countries." In: *Technical Meeting on Population Ageing and Living Arrangements of Older Persons: Critical Issues and Policy Responses*. pp. 1–21. U.N. Population Division, Department of Economic and Social Affairs: New York.

Gierveld, J. De Jong and Van Tilburg, T. (1999), "Living arrangements of older adults in the Netherlands and Italy: Coresidence values and behaviour and their consequences for loneliness." *Journal of Cross-Cultural Gerontology*, 14:1–24.

Goldscheider, F.K. and DaVanzo, J. (1986), "Semiautonomy and leaving home in early adulthood." *Social Forces*, 65:187–201.

Goldscheider, F.K. and Goldscheider, C. (1993), *Leaving Home Before Marriage: Ethnicity, Familism, and Generational Relationships*. University of Madison Press: Madison, WI.

Golini, A. (2002), "Possible policy responses to population ageing and population decline: The case of Italy." *Policy Responses to Population Ageing and Population Decline. U.N. Population Bulletin*, 44/45:150–170.

Grundy, E. and Harrop, A. (1992), "Co-residence between adult children and their elderly parents in England and Wales." *Journal of Social Policy*, 21:325–348.

Hajnal, J. (1982), "Two kinds of preindustrial household formation systems." *Population and Development Review*, 8:449–494.

Hantrais, L. and Letablier, M.-T. (1996), *Families and Family Policies in Europe*. Longman: London.

Hashimoto, A. Kendig, H.L. and Coppard, L.C. (1992), "Family support to the elderly in international perspective." In: *Family Support for the Elderly: The International Experience*, H.L. Kendig, A. Hashimoto, and L.C. Coppard, eds. pp. 293–308. Oxford University Press: Oxford.

Hogan, D.P., Eggebeen, D.J. and Clogg, C.C. (1993), "The structure of intergenerational exchanges in American families." *American Journal of Sociology*, 98:1428–1458.

Hogan, D.P. and Lichter, D.T. (1995), "Children and youth: Living arrangements and welfare." In: *State of the Union: America in the 1990s*, vol. 2: Social Trends, R. Farley, ed. pp. 1–46. Russell Sage: New York.

Hollinger, F. and Haller, M. (1990), "Kinship and social network in modern societies: A cross-cultural comparison of among seven nations." *European Sociological Review*, 6:103–124.

Inglehart, R. (1977), *The Silent Revolution: Changing Values and Political Styles of Western Publics*. Princeton University Press: Princeton, NJ.

Kendig, H.L., Hashimoto, A. and Coppard, L.C. (1992), *Family Support for the Elderly: The International Experience*. Oxford University Press: Oxford.

Kerckhoff, A.C. and Macrae, J. (1992), "Leaving the parental home in Great Britain: A comparative perspective." *Sociological Quarterly*, 33:281–301.

Kiernan, K. (1986), "Leaving home: Living arrangements of young people in six West-European countries." *European Journal of Population*, 2:177–184.

Kiernan, K. (1989), "The departure of children." In: *Later Phases of the Family Cycle*, E. Grebenik, C. Hohn, and R. Mackensen, eds. pp. 120–144. Clarendon Press: Oxford.

Knipscheer, C.P.M., Gierveld, J. De Jong, Van Tilburg, T.G., and Dykstra, P.A. (1995), *Living Arrangements and Social Networks of Older Adults*. VU University Press: Amsterdam.

Knipscheer, K.C.P.M. (1992), "The Netherlands in European perspective." In: *Family Support for the Elderly: The International Experience*, H.L. Kendig, A. Hashimoto, and L.C. Coppard, eds. pp. 147–159. Oxford University Press: Oxford.

Kobrin, F.E. (1976), "The fall in household size and the rise of the primary individual in the United States." *Demography*, 13:127–138.

Lesthaeghe, R. (2002), "Europe's demographic issues: Fertility, household formation and replacement migration." *Policy Responses to Population Ageing and Population Decline. U.N. Population Bulletin*, 44/45:385–423.

Logan, J.R. and Spitze, G.D. (1996), *Family Ties: Enduring Relations Between Parents and their Grown Children*. Temple University Press: Philadelphia.

Macfarlane, A. (1978), *The Origins of English Individualism*. Blackwell: Oxford.

Mayer, K.U. and Schwarz, K. (1989), "The process of leaving the parental home: some German data." In: *Later Phases of the Family Life Cycle: Demographic Aspects*, E. Grebenik, C. Hohn, and R. Mackensen, eds. pp. 145–163. Clarendon Press: Oxford.

Palloni, A. (2000), "Programatic and policy aspects of population ageing and living arrangements." In: *Technical Meeting on Population Ageing and Living Arrangements of Older Persons: Critical Issues and Policy Responses*. pp. 1–46. U.N. Population Division, Department of Economic and Social Affairs: New York.

Pampel, F.C. (1992), "Trends in living alone among the elderly in Europe." In: *Elderly Migration and Population Redistribution*, A. Rogers, W. H. Frey, A. Speare, Jr., P. Rees, and A. Warnes, eds. Belhaven Press: London.

Reher, D. (1998), "Family ties in Western Europe: Persistent contrasts." *Population and Development Review*, 24:203–234.

Rosenmayr, L. (1977), "The family–A source of hope for the elderly." In: *Family, Bureaucracy, and the Elderly*, E. Shanas and M.B. Sussman, eds. pp. 132–157. Duke University Press: Durham, NC.

Rossi, A.S. and Rossi, P.H. (1992), *Of Human Bonding: Parent-Child Relations Across the Life Course*. A. de Gruyter: New York.

Settersten, R.A. (1998), "A time to leave home and a time never to return? Age constraints on the living arrangements of young adults." *Social Forces*, 76:1373–1400.

Shanas, E., Townsend, P., Wedderburn, D., Friis, H., Milhoj, P. and Stehuower, J. (1968), *Old People in Three Industrial Societies*. Atherton Press: New York.

Short, K. and Garner, T. (1990), "Living arrangements of young adults living independently: Evidence from the Luxembourg Income Study." U.S. Government Printing Office: Washington, DC.

Smeeding, T. and Saunders, P. (1998), "How do the elderly in Taiwan fare cross-nationally? Evidence from the Luxembourg Income Study (LIS) project." In: *Luxembourg Income Study Working Paper Series, Syracuse University*, vol. 183. Syracuse, NY.

Smith, T.W. (1992), "The international social survey program." *International Journal of Public Opinion*, 4:275–278.

Tornstam, L. (1992), "Formal and informal support to the elderly in Sweden." In: *Family Support for the Elderly: The International Experience*, H.L. Kendig, A. Hashimoto and L.C. Coppard, eds. pp. 138–146. Oxford University Press: Oxford.

Treas, J. (1995), "Older Americans in the 1990s and beyond." *Population Bulletin*, 50:1–48.

Treas, J. (2002), "Population aging in the United States: Trends and prospects." *Policy Responses to Population Ageing and Population Decline. U.N. Population Bulletin*, 44/45:358–370.

Treas, J. and Chen, J. (2000), "Living arrangements, income pooling, and the life course in urban Chinese families." *Research on Aging*, 22:238–261.

Treas, J. and Widmer, E. (2000), "Married women's employment over the life course: Attitudes in cross-national perspective." *Social Forces*, 79:1409–1436.

Triandis, H.C. (1995), *Individualism and Collectivism*. Westview: Boulder, CO. van den Akker, P., Halman, L., and de Moor, R. (1994), "Primary relations in western societies." In: *The Individualizing Society: Value Change in Europe and North America*, P. Ester, L. Halman, and R. de Moor, eds. pp. 97–127. Tilburg Press: Tilburg, Netherlands.

Van Peer, C. (1998), "Welfare and solidarity toward the family." In: *Population, Family, and Welfare: A Comparative Survey of European Attitudes*, vol. 2. R. Palomba and H. Moors, eds. pp. 179–217. Clarendon Press: Oxford.

Wenger, C.G. (1992), "The major English-speaking countries." In: *Family Support for the Elderly*, H.L. Kendig, A. Hashimoto, and L.C. Coppard, eds. pp. 117–137. Oxford University Press: Oxford.

White, L. (1994), "Coresidence and leaving home: Young adults and their parents." *Annual Review of Sociology*, 20:81–102.

Wolf, D.A. (1994), "The elderly and their kin: Patterns of availability and access." In: *Demography of Aging*, L.G. Martin and S.H. Preston, eds. National Academy Press: Washington, DC.

Young, C.M. (1989), "The effect of children returning home on the precision of the timing of the leaving-home state." In: *Later Phases of the Family Life Cycle: Demographic Aspects*, E. Grebenik, C. Holm, and R. Mackensen, eds. pp. 164–181. Clarendon Press: Oxford.

CHAPTER 6. SON PREFERENCE, MARRIAGE, AND INTERGENERATIONAL TRANSFER IN RURAL CHINA

MARCUS W. FELDMAN AND SHRIPAD TULJAPURKAR
Morrison Institute for Population and Resource Studies, Stanford University, California, USA

SHUZHUO LI AND XIAOYI JIN
Population Research Institute, Xi'an Jiaotong University, Xi'an, People's Republic of China

NAN LI
Department of Sociology, University of Victoria, Victoria, British Columbia, Canada

1. Introduction

Rural Chinese society has historically been dominated by a rigid male-centered patrilineal family system and virilocal marriage remains overwhelmingly dominant. Under this system, parents call in a daughter-in-law for each of their sons, and all sons are entitled to stay home after marriage and coreside with their parents for a period of time until family division occurs or until both of their parents pass away. A son's offspring use his father's surname to continue the family lineage. Family property is usually inherited equally among all sons, and sons are obliged to take care of their parents in their old age. By contrast, parents marry all of their daughters to other families, and these daughters as well as their future descendants are no longer regarded as members of their natal families. All rights, productivity, and services of daughters are transferred to their husbands' families at the time of marriage, and they can neither inherit their natal family's property nor do they have any formal obligation to take care of their natal parents in old age (Das Gupta and Li 1999).

The patrilineal family system produces strong preference for sons and various forms of discrimination against daughters (Skinner 1997; Lavely, Li, and Li 2001). Son preference has been strong throughout the histories of various Asian countries, such as China, Korea, and India (Dyson and Moore 1983; Das Gupta and Li 1999). In China, the phenomenon of "missing females," which existed more or less throughout the first half of the 20th century, was exacerbated in periods of war and famine but became less serious in the 1960s–1970s (Coale and Banister 1994; Das Gupta and Li 1999).

Anne H. Gauthier et al. (eds.), Allocating Public and Private Resources across Generations, 139–162.

The Chinese government has adopted a national-government-guided population-control policy, which has been changing through the years. In the early 1970s, a couple was encouraged to have fewer children. However, since the early 1980s, in urban areas a couple has only been allowed to have one child, while those in rural areas were *encouraged* to have only one child. Those couples who could demonstrate hardships because of having only one child (e.g., not enough male labourers for agricultural fieldwork or not having a son to support them in their old age) were allowed to have an additional child 3–5 years after the first birth. For minorities, the policy is less strict. Since the late 1970s, a dramatic socioeconomic transition driven by economic reform and the opening-up of the country, as well as a strict population-control policy, have resulted in rapid fertility decline, with the total fertility rate declining from 5.81 in 1970 to 1.85 in 1995. However, sustained low fertility has also intensified the manifestation of son preference, as evidenced both in a rising sex ratio at birth (SRB) and in deteriorating female infant survival. For instance, SRB in China increased from 107 in 1980 to 118 in 1995, and the ratio of male to female infant mortality declined from 1.06 in 1981 to 0.75 in 1995 (Li and Zhu 2001).

Many studies have examined the temporal trends and regional variation in the degree to which the SRB in China is biased (Hull 1990; Johansson and Nygren 1991; Wen 1993; Zeng et al. 1993; Gu and Roy 1995; Tuljapurkar, Li, and Feldman 1995; Poston et al. 1997). In general, high SRB promises to be a growing problem for Chinese society, in terms of the low status of women and girls, imbalance in the future marriage market, and potential social conflicts (Hull 1990; Zeng et al. 1993; Tuljapurkar, Li, and Feldman 1995; Das Gupta and Li 1999). According to the 5th National Population Census of China (taken in the year 2000), about 18.76 percent of the men aged 15 and over are never married; while for women this proportion is only about 13.75 percent. Chinese governments at various levels have been cognizant of this problem and since the early 1990s have adopted various measures to control the increase of SRB. However, the continuous increase in SRB since the 1990s suggests that these measures have yet to be effective. In this low fertility regime, son preference plays an important role in the increase of China's SRB. It is, therefore, unlikely that government controls such as those adopted in China, or rapid economic development such as occurred in South Korea, will change the strong son preference in the short term (Park and Cho 1995; Das Gupta and Li 1999).

Sustained low fertility has also resulted in accelerated population ageing in China. From the 5th National Population Census of China in 2000, the population of China is 1,295,330,000. The proportion of people aged 0–14 is 22.89 percent; that of people aged 15–64 is 70.15 percent; and that of people aged 65 and above is 6.96 percent. Compared with the 4th population census in 1990, the proportion of people aged 0–14 has declined 4.80 percent, while that of people aged 65 and above has increased 1.39 percent. Although the proportion of the population that lives in rural areas has declined from 73.77 percent in 1990 to 63.91 percent in 2000, the proportion of older people in rural areas is even higher than that in urban areas. The 2000 census shows that, in rural areas, the proportion of people aged 0–14 is 25.49 percent; that of people aged 15–64 is 67.16 percent; that of people aged 65 and above is 7.35 percent; while these relative proportions in urban areas are 18.43, 75.27, and 6.30 percent, respectively. Thus, the rural population has a greater burden in raising children and supporting the elderly. Related to this is old age security, an issue that is rapidly becoming important in rural China. However, a reliable pension system with broad coverage has not been established in rural China, while income is not high enough for most rural couples to enable them to save for their old age (Yu 1996). That children should show filial piety to, and be responsible for, the support of their elderly parents is a widely accepted personal value and community norm in traditional Chinese culture (Ganschow 1978). In addition, relevant laws in China protect old people's legal interests and rights, and define the norm of family responsibility for elder care (Davis-Friedmann 1991).

The 1994 Annual Population Change Survey conducted by China's State Statistical Bureau shows that for 57 percent of the elderly, children or relatives provide the main source of income, 25 percent have their own income, and 16 percent live off pensions. For rural areas, however, the corresponding values are 64, 29, and 4 percent, respectively (Du and Wu 1998). Further, the China 1990 census showed that 72.5 percent of the elderly coresided with their children in extended families (Du 1998). As a result, family support for the elderly, primarily provided by adult children, is and will continue to be the dominant form of care for the elderly in rural areas in the foreseeable future (Shi 1993; Gu, Chen, and Liang 1995; Xu and Yuan 1997). However, the general nature of the patrilineal family system and the characteristics of virilocal marriage shape the gender pattern of old age support. It is usually sons in a family, but not daughters, who provide the necessary support for their older parents (Xu 1996; Yang 1996). This inevitably intensifies the existing son preference.

On the other hand, sustained low fertility has resulted in a rapidly rising proportion of couples without a son, whose old age support in the future faces a serious challenge (Li et al. 1998; Li, Feldman, and Jin 2002c). Calling in a son-in-law for one of their daughters (uxorilocal marriage) could be a realistic choice for these couples in order to continue the family lineage and secure their old age. There may be a tremendous potential demand for adoption of uxorilocal marriage in rural China. Wider practice of uxorilocal marriage may not only alleviate son preference—since uxorilocal marriages provide a secure old age for the daughter's parents as well as guaranteeing the survival of the family name—but may also change power structures and improve women's status within the family (Li and Zhu 1999; Jin and Li 2003; Jin et al. 2002). However, the strength of the patrilineal family system and the high psychological cost related to uxorilocal marriage, together with the low proportion of no-son families in populations with high fertility, have all resulted in uxorilocal marriage being historically rare throughout rural China (Chuang and Wolf 1995; Wolf 1989).

Most studies of son preference in China have focused on the manifestations of son preference, not on son preference itself (Arnold and Liu 1986; Xie 1989; Li and Cooney 1993; Wen 1993; Poston et al. 1997; Graham, Larsen, and Xu 1998). Although researchers have realized the important role played by culture in shaping fertility patterns and behaviours (Hammel 1990; Pollak and Watkins 1993), and have suggested that it would be fruitful to investigate communication of attitudes toward and behaviours in childbearing among members of a population (Bongaarts and Watkins 1996), less attention has been paid to the dynamics of son preference in China, especially its formation, transmission, and evolution, and its quantitative impact on SRB as well as its relationship to diversity of marriage forms and old age security.

Over the last twenty years, theoretical research on cultural transmission and evolution has received considerable attention in academic circles. Several scholars have developed quantitative models for complex transmission and evolution of people's behaviours, attitudes, beliefs, and habits (Cavalli-Sforza and Feldman 1981; Cavalli-Sforza et al. 1982; Chen Cavalli-Sforza, and Feldman 1982; Kumm, Laland, and Feldman 1994; Laland 1994; Laland, Kumm, and Feldman 1995). These models provide a useful framework in which to address human behaviours related to son preference and other social issues. Thus, cultural transmission theory may provide an appropriate framework in which to address the transmission and evolution of attitudes and behaviours surrounding son preference, as well as their effects on actual childbearing behaviours and SRB.

Other papers in this volume have focused on the transfer of wealth and status between generations, primarily in the United States. These statistical analyses have often used techniques similar to those in behaviour genetics, with regression playing an important role. Familial aggregation of wealth and status is reported in terms of estimated heritability, which can be regarded as a measure of intergenerational transmission, but may also confound genetic and cultural effects. In the present study, we are interested in the attitudes toward sons and daughters, what the antecedents of these attitudes are, and how attitudinal bias is translated into behaviours that discriminate against female children. These attitudes are transmitted not only within nuclear families, but within whole clans and perhaps even larger population units. The perceived inequality in the value to families (and clans) of sons and daughters translates into an important and growing sex ratio imbalance in the whole country. Intergenerational cultural transmission of this bias varies across China but for the most part results in a continuing deficit in the number of girls born. In this basic demographic sense, sons and daughters certainly have unequal chances.

Over the last five years, the Population Research Institute of Xi'an Jiaotong University in China and the Morrison Institute for Population and Resource Studies of Stanford University in the United States have jointly conducted a project on son-preference culture, marriage, and intergenerational transfer in rural China. The project uses survey data combined with theories of cultural transmission and evolution to analyse the dynamic relationship among son preference, SRB, marriage form, and intergenerational transfer in rural China. The study includes three parts: measurement of son preference and the analysis of its transmission and evolution, especially its quantitative effects on SRB; transmission and determinants of uxorilocal marriage and its relationship with son preference; marriage form and intergenerational transfer. Here we review some of the theory we have developed and the results of our surveys.

2. Models and Estimates of Son Preference

In exploring the interaction of son preference and fertility decline on the increase in SRB, we identify two aspects of women's childbearing situations. The first, largely determined by fertility, is the *sex-selection pressure (y)*, which is the probability of the random event Y, that a woman is in the *sex-selection situation*, namely, that she has no son and is able to have only one more child. Second, we define *son-preference potency (p)*, as the probability that a woman has a sex-selected son, given that she is in the sex-selection situation. Obviously, son preference here is defined in terms of a woman's childbearing behaviour rather than attitudes, and depends on the extent of son preference of her family including her husband and herself.

Sex-selection pressure driven by fertility decline is a demographic effect, while son-preference potency reveals the relationship between cultural factors and individual behaviours. Thus, the factors leading to an increase in SRB are of two types: rising sex-selection pressure caused by fertility decline (y), and son-preference potency (p) determined by the economic and cultural background of a population. In order to formulate a relationship between son preference and SRB in the context of rural Chinese society, we make the following simplifying assumptions (Li, Feldman, and Tuljapurkar 1998, 1999): (1) sex selection is done only by women who have no previous son and are able (or permitted) to have only one more child; (2) fertility is constant over a long time; and (3) there is no mortality in the reproductive ages.

Then we have the following relation:

$$\text{SRB} = (\text{TFR}^* s_0 + p^* y)/(\text{TFR} - p^* y),$$

where TFR represents women's total fertility rate, s_0 represents normal SRB (usually about 1.05), and SRB is the observed sex ratio in the population.

Thus, the probability p of having a sex-selected son for a couple in the sex-selection situation depends on their social and cultural background. The complementary probability, $p_\text{N} = 1 - p$, is the probability that a couple have the trait of no-son preference, denoted by π_0. A couple who have trait π_0 will not have a sex-selected son when they are in the sex-selection situation. We focus on trait π_0 and its probability p_N.

The decision to have a sex-selected son for a couple in the sex-selection situation may be regarded as the result of the transmission of ideas from their parents or neighbours, or acquisition of information from outside the community about why a son is necessary. Differences in decisions made by couples in different generations may be interpreted in terms of the extent of transmission of cultural trait π_0 across generations. Although it is difficult to characterize this process of transmission in detail, its effect on changes in the value of p_N can be clarified using the general theory of cultural transmission (Cavalli-Sforza and Feldman 1981). According to this theory, individuals from a younger generation acquire the cultural trait π_0 from their parents at a vertical transmission rate C_v, given that their parents have this trait. They may also acquire π_0 from the members of an older generation at an oblique transmission rate C_o, or horizontally from the mass media at a rate M_0, if their parents do not have trait π_0. For technical reasons, we assumed $C_\text{v} = C_\text{o}$.

We have constructed a dynamic demographic model including cultural transmission of son preference (Li, N., Feldman, and Li, S. 2000) and applied the model to data collected in surveys of two counties in Shaanxi province—Lueyang and Sanyuan. These surveys were carried out in 1997. In 2000 we extended the survey to include questions about intergenerational transfer of financial assistance between children and parents as well as assistance with housework and field labour. The 2000 survey was carried out in Songzi county, in Hubei province.

Sanyuan county is located in the central part of the Guanzhong plain in central Shaanxi province, close to the Wei River and only 30 kilometres away from Xi'an city. The county's population was close to 390,000 in 1997. Sanyuan is rich in fertile land and its food industry has been renowned in Shaanxi throughout history. Private and township enterprises developed extensively after the 1980s, absorbing many surplus farmer labourers. However, per capita income for farmers in Sanyuan is still much lower than that for China as a whole. Sanyuan is regarded as one of the origins of the Han Chinese population and culture, and retains the core elements of the traditional Yellow River culture. Since it is located in a plain, large family clans exist in almost every village and are influential in village social life. Villages maintain a strict patriarchal family system with patrilocal marriage.

Lueyang county is a mountainous county located in the Qing mountains in the far south of Shaanxi province, far from Xi'an, on the borders of Sichuan and Gansu provinces. It has a relatively small population—about 200,000. Before the Tang dynasty, Lueyang was occupied mostly by minority Chinese, who were defeated by the Han Chinese, and most were forced to migrate to other places. In the late 17th century, the Qing government forced people in central China, where the population was dense, to migrate to southern Shaanxi, including Lueyang. Lueyang has been rich in natural resources, especially minerals and forests. Even though its per capita arable land is much higher than in Shaanxi and China, Lueyang is relatively underdeveloped compared with other counties in Shaanxi, and the standard of living is low. Most of its arable land is mountainous and infertile, so the population relies heavily on the forests, especially wild products. The difficult conditions require hard labour, mostly by men. In recent years, raising mushrooms has become the major cash generator. Since Lueyang is mountainous, villages are relatively small and the ancestors of most residents were immigrants from other areas, as a result of which, large family clans in Lueyang are few and unimportant to village life. The culture of Lueyang is that of southern Shaanxi, for which the patriarchal family system is somewhat weaker than in Sanyuan.

Songzi, located on the Jianghan Plain in southwest Hubei province on the southern side of the Yangtze River, had a population of 896,800 at the end of 1999. Songzi is a relatively well-developed agricultural county with fertile land that has produced the famous "Babao Cotton" for hundreds of years. The economy is based on this high quality cotton and related enterprises. The favourable local climate and many lakes and reservoirs support production of fruit and timber as well as aquiculture. In addition, enterprises run by villages or towns, and private ones run by farmers that involve agricultural by-products, textiles, and mining, are expanding. In the latter part of the Qing Dynasty, the swampland near a riverbank in the southeast of the county gradually became farmland, which attracted many new settlers. Thus, about half of the residents are outsiders whose arrival is recent enough to ensure that there are few dominant family clans. As a result, the patrilineal family system in Songzi is flexible.

Generally, minorities are rare in the three counties, whose ethnic composition is similar: the proportion of Han Chinese is overwhelming. As a result, the family planning policies in the three counties are also very similar. However, Lueyang and Songzi have fewer children per couple than does Sanyuan, and for years the former counties have been more advanced in terms of population control.

In the surveys, cluster sampling was used to select survey sites and couples. That is, in each county, we first selected one township, and then a number of neighbouring administrative villages within this township, until the number of couples was more than 1,500. In Sanyuan, about 1,567 households including 1,866 couples living in 23 natural villages were included in the survey. In Lueyang, about 1,364 households, including 1,581 couples living in 46 natural villages, were included in the survey. Many measures were used to ensure the quality of interviews. There were about 14 invalid questionnaires, and some responses had missing items, but not many. In addition, reinterviews were conducted by supervisors, and the overall proportion of inconsistent items was about 7.4 percent. The data were entered into a Foxpro database, and we randomly selected 10 percent of the data to check the accuracy and quality of data entry. The proportion of items with input error was less than one percent. Tests were also made for logical inconsistencies. In short, every measure was carefully implemented to ensure the quality of field interviews, data entry, and the data themselves. Errors were present but at an acceptable level, and the data can be regarded as reliable and of satisfactory quality.

Detailed information about the three counties and the surveys is available in Li et al. (1998) and Jin et al. (2002).

Table 1 reports the number of births and SRB by birth cohort in the three counties. Clearly, since the 1990s there has been strong son preference in Sanyuan but not in Lueyang or Songzi. Our surveys revealed that this SRB difference reflected differences in attitudes toward girl children.

The models of son preference described above and the survey data permitted estimates of son-preference potency and transmission of bias disaggregated by age group in Lueyang and Sanyuan. The estimated son-preference potencies for the two counties are shown in Table 2, while the coefficients of transmission of no preference are in Table 3 (Li , Feldman, and Li 1998).

Table 1. Number of births and sex ratio at birth, by birth cohort in the three counties.

	Sanyuan		Lueyang		Songzi	
Birth cohort	**Number**	**SRB**	**Number**	**SRB**	**Number**	**SRB**
1930s–1970s	2,394	92.4	1,854	98.5	2,024	107.8
1980s	1,324	102.8	1,039	106.6	991	106.0
1990s	891	119.5	678	96.5	589	89.4

Source: Li et al. (1998), Jin et al. (2002).

Table 2. Estimated son-preference potency, Sanyuan and Lueyang.

County	All women	15–29	30–44	45+
Sanyuan	0.44	0.50	0.40	0.35
Lueyang	0.08	0.06	0.03	0.20

Source: Li, Feldman, and Li (1998).

Table 2 shows that son-preference potency in Lueyang is extremely weak, but strong in Sanyuan. These are the basic causes of high SRB in Sanyuan and normal SRB in Lueyang.

The results reported in Table 3 are consistent with those of Table 2 and show that cultural transmission rates of no-son preference are higher for Lueyang than for Sanyuan. Again, this highlights that high SRB is a result of son preference, and this preference is the result of a process of transmission. The transmission rates of no-son preference are declining in both counties, i.e., transmission rates of son preference are increasing. Thus, under the recent and current regime of low fertility, the transmission of son preference in rural China has been accelerating, leading to an increase in son-preference potency and hence in SRB (Li, Feldman, and Li 1998).

An important issue that has also drawn extensive attention by the government and the general public, concerns future trends in son preference and SRB. To explore this question, we incorporated the estimated parameters from Sanyuan and Lueyang into our demographic model with cultural transmission of son preference for the whole of China, so as to predict the future of son preference and SRB in China (Li, N., Feldman, and Li, S. 2000). First, we used data from the 1990 Population Census to estimate various initial values of cultural and population variables to use in the models for China. Then, as in conventional high, medium, and low population projections, we assumed that a "good" situation of cultural transmission of son preference for China is similar to that in Lueyang, and a "bad" situation is similar to that in Sanyuan. That is, we used our estimated coefficients of cultural transmission of son preference in Sanyuan and Lueyang as the two possible scenarios for the whole of China. The details are described by Li, N., Feldman, and Li, S. (2000). The estimated trend for SRB is shown in Figure 1.

Table 3. Transmission coefficients of no-son preference, Sanyuan and Lueyang.

County	Coefficient	15–44	15–29	30–44
Sanyuan	C_v, C_o	0.50	0.24	0.59
	M_0	0.36	0.46	0.38
Lueyang	C_v, C_o	0.89	0.88	0.92
	M_0	0.94	0.79	0.97

Source: Li, Feldman, and Li (1998).

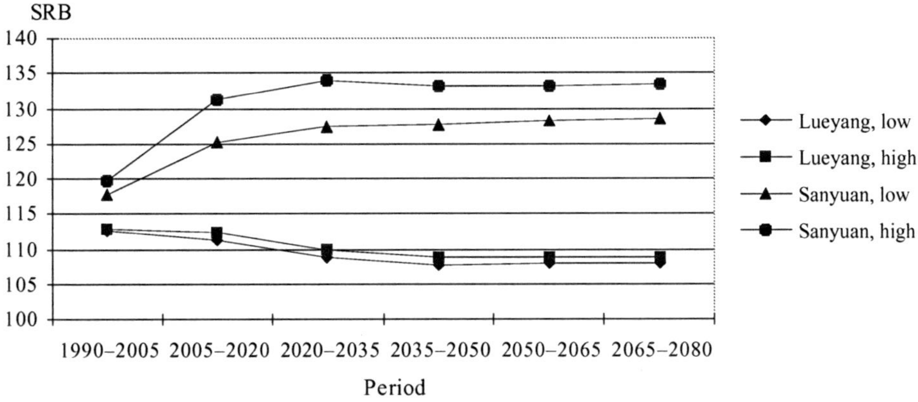

Figure 1. Simulated sex ratio at birth (SRB) for China's future (Li, Feldman, and Li 2000).

Figure 1 reveals that, if son preference in China were transmitted at rates characteristic of Lueyang, son preference would gradually decline and SRB would become normal at equilibrium. If son preference were transmitted at rates characteristic of Sanyuan, SRB would increase to 130–135 in the long run. Under the following two conditions, the predictions using Sanyuan's parameters seem more plausible for the future of the whole of China. The first is stable fertility and the second is that the current sex-selective abortion rate remains unchanged. Applying the transmission rates of Lueyang to China seems too optimistic, since it would require acceptance of uxorilocal marriage throughout rural China. However, if a reliable and broad pension system were established, or people were willing and able to save enough for their old age, it would be possible for SRB to approach the normal value in the long run. Under such circumstances, a son would no longer be necessary for the financial support of the parents; and as in Lueyang, incoming sons-in-law could provide old-age support for parents (Li, Feldman, and Li 1998).

3. Marriage

3.1. DIVERSIFIED MARRIAGE FORM

Why is son preference much stronger in Sanyuan than in Lueyang and Songzi under similar family planning implementation and similar levels of fertility? We claim the reason lies in the extent of acceptance and practice of uxorilocal marriage.

Marriages in rural China may be broadly classified into two categories: virilocal, where the wife leaves her natal family and moves to her husband's family; uxorilocal, where the husband leaves his natal family and moves to his wife's. Throughout the history of China, among Han Chinese virilocal marriage has been almost universal while uxorilocal has been rare. Location of postmarital coresidence reflects the core elements of the rural Chinese patrilineal family system, and virilocal marriage is perceived to ensure the purity, integrity, and continuity of the family (Li, S., Feldman, and Li, N. 2000). The family system and marriage customs have been changing to some extent, but virilocal marriage remains overwhelmingly dominant in contemporary rural China (Lavely and Ren 1992) and even in big cities such as Beijing (Guo and Chen 1999).

In China, social and political status determine the mode of mate selection in the process of union formation. For the former, the better educated are also likely to enjoy a higher degree of occupational prestige; for the latter, party members and those who come from "good" families have greater potential to succeed both professionally and economically. Although people voluntarily seek out those who are like themselves (homogamy), some heterogamy always exists (Xu, Ji, and Tung 2000). In a society where the patriarchal and patrilineal family system is practised, some hypergamy (in which women marry up in status) is expected. However, in uxorilocal marriage, the man, rather than the woman, usually marries up since the most common cause for him to enter his wife's family is poverty of his natal family.

With the change of assortment in the social and political status systems caused by the profound sociopolitical transformations under state socialism, patterns of assortative mating have changed (Xu, Ji, and Tung 2000). For those couples who married before 1977, assorting by political status might have been more usual than by social status; for those who married in the economic reform era after 1978, level of education has become increasingly important. In Songzi we found that, during the Cultural Revolution (between 1966 and 1976), many men whose family class origins were "bad" married up by adopting uxorilocal marriage in order to elevate their status. However, in recent years homogamy has become more common in uxorilocal marriage. More men with higher education or those who come from richer families adopt uxorilocal marriage, while families calling in sons-in-law are beginning to take the family status of the other party into account in the process of choosing husbands for their daughters.

Although cultural factors defining the patrilineal family system play a decisive role in determining marriage type, demographic and economic factors may also affect an individual family's decisions about marriage type, leading under some conditions to the occurrence of uxorilocal marriage. Uxorilocal marriage driven by demographic factors occurs in families without a son or in those that for various reasons are unable to adopt a son, and has been called "contingent" by Wolf (1989). To ensure continuity of the family lineage and for security in their old age, parents usually call in a son-in-law for one of their daughters, and either the son-in-law assumes his new family's surname as an adopted son, or one of the son-in-law's sons is assigned this family's surname to carry on the lineage. In this case, uxorilocal marriage is mainly preservative, to maintain the family lineage (Pasternak 1985), and the husband in such a uxorilocal marriage is in essence a kind of "male daughter-in-law." However, in a society governed by a rigid patrilineal family system, uxorilocal marriage usually entails high costs and stigma in terms of both individual and familial status and is therefore looked down upon by most rural communities (Pasternak 1985; Wolf 1989; Li, S., Feldman, and Li, N. 2000). Men who enter such a marriage are usually from very poor families that have more than two sons and cannot afford marriages for all their sons (Pasternak 1985). The general properties of the patrilineal family system, the high psychological costs of uxorilocal marriage, and low proportion of no-son families in populations with high fertility, have resulted in uxorilocal marriage being rare throughout rural China. In most of the country the frequency of demographically driven uxorilocal marriages is less than or equal to that of no-son families.

Uxorilocal marriage driven by other nondemographic, mostly economic, considerations occurs in families that have at least one son but face economic constraints, including a shortage of male labour, marriage costs such as bride price and dowry, or need for old age support, etc. (Li and Zhu 1999; Li, Feldman, and Li 2001). Uxorilocal marriage occurs in families deficient in male labour, where calling in a son-in-law is a means of increasing the number of male labourers in the family (Pasternak 1985; Li, Feldman, and Li 2000); or in relatively poor families, because the cost of uxorilocal marriage is inexpensive for all parties involved (Pasternak 1985); or in a family that badly needs a diligent worker who is predicted to acquire wealth, because a son-in-law coming from the outside is usually perceived as able to make more money than the natives of his adopted county (Li, Feldman, and Li 2001). In this case, uxorilocal marriage is mainly a practical resolution of economic and other difficulties in families that already have at least one son. Wolf (1989) calls uxorilocal marriage in these situations "institutional." In general, this kind of uxorilocal marriage is rare across rural China although it has been reported as prevalent in a few Chinese rural areas at different times. In these cases, the proportion of uxorilocal marriages is much higher than that of no-son families (Pasternak 1985; Wolf 1989; Han 1992; Chuang and Wolf 1995; Yan 1995; Yan and Shi 1996).

Our surveys show that the distribution of marriage forms differs in the three counties (Li et al. 1998; Jin et al. 2002). In Sanyuan, virilocal marriage is overwhelmingly dominant at about 90 percent while uxorilocal marriage is rare at about only 4 percent, close to the proportion of no-son families. In Lueyang, however, the proportion of virilocal marriage is 58 percent and that of uxorilocal marriage about 32 percent. In Songzi, the corresponding proportions are 75 and 21 percent, respectively. That the frequencies of uxorilocal marriage in Songzi and Lueyang exceed those of no-son families suggests that both contingent and institutional uxorilocal marriages occur in these counties.

3.2. UXORILOCAL MARRIAGE IN LUEYANG AND SONGZI

In Lueyang and Songzi, as has historically been the case in most rural areas in China, adoption of uxorilocal marriage usually requires a written or oral contract between the incoming son-in-law and his wife's parents before marriage. A typical contract contains four elements (Li and Zhu 1999). The first concerns the surname of the incoming son-in-law. Generally, his surname is changed to his wife's, and he is regarded as an adopted son. The second concerns his children's surname. Customarily, the first child will use its mother's surname and the second its father's surname, ensuring that the surnames of both families are passed on. The third concerns marriage costs including bride price and dowry, and here the bride price of an incoming son-in-law is less than usual or totally waived. The fourth addresses old age support of the wife's parents and inheritance of her family's property in the future. Conventionally, the incoming son-in-law is obliged to support his parents-in-law and has the right to inherit their family property. In practice, the contents of such contracts change with family structure and attitudes of family members, and differ among families

(Li and Zhu 1999). With recent socioeconomic changes, people do not care as much about the contract or change in the son-in-law's surname as in the past. Moreover, an incoming son-in-law and his children do not necessarily have to use his wife's surname. In other words, the practice of uxorilocal marriage has become more flexible than in the past.

We found that uxorilocal marriage in Lueyang and Songzi is not only a contingent variant of the Chinese patrilineal family system but also an institutional option for farmers—a phenomenon rarely observed in Han areas of rural China. In addition to the above-mentioned individual and family considerations, acceptance of uxorilocal marriage is also closely related to the community settings of Lueyang and Songzi (Li and Zhu 1999; Li, S., Feldman, and Li, N. 2000; Jin and Li 2003). First, uxorilocal marriage has historically been prevalent in Lueyang and Songzi. As an individual option for farmers, it has become part of the local community's cultural norms. Second, large family clans are not strong and their impact on members' marriage decisions is weak. Third, there are less community restrictions on adoption of uxorilocal marriage and less discrimination from the community against incoming sons-in-law. Further, that some infertile women in Lueyang's high mountainous areas are more likely to adopt daughters actually creates demand for uxorilocal marriage.

Individual and family factors underlying uxorilocal marriage are not the only reasons for the prevalence of uxorilocal marriage as an institutional option, since these factors also operate in other rural areas of China. The special community tolerance of uxorilocal marriage related to local sociocultural and geographic characteristics is also very important.

3.3. CULTURAL TRANSMISSION OF UXORILOCAL MARRIAGE

Data from the 1997 survey were used to estimate the extent of cultural transmission of uxorilocal marriage in Lueyang. Marriage cohorts in Lueyang were classified into two groups: before 1978 and after 1978, so as to investigate the impact of rural economic reform on the intergenerational transmission of uxorilocal marriage (Li, S., Feldman, and Li, N. 2000). Applying the vertical cultural transmission model developed by Cavalli-Sforza and Feldman (1981), we obtained estimates of transmission parameters shown in Table 4.

Table 4 shows that there exist positive transmission effects from parents to sons and daughters and positive nonparental transmission effects for the marriage cohort prior to 1978, and

Table 4. Vertical transmission of uxorilocal marriage-by-marriage cohort, Lueyang.

Husband's parents' marriage	Wife's parents' marriage	Before 1978		1978 and after	
		Frequency of pair	Transmission coefficient	Frequency of pair	Transmission coefficient
Uxorilocal	Uxorilocal	14	0.64	60	0.15
Uxorilocal	Nonuxorilocal	57	0.44	114	0.25
Nonuxorilocal	Uxorilocal	74	0.53	147	0.43
Nonuxorilocal	Nonuxorilocal	427	0.29	570	0.28

Source: Li, S., Feldman, and Li, N. (2000).

Table 5. Maximum-likelihood estimates of additive models of vertical transmission of uxorilocal marriage, Lueyang.

Marriage cohort	a_0	a_h	a_w	χ^2	P	Number of cases
Before 1977	0.29	0.15	0.24	0.06	>0.80	572
1978 and after	0.30	−0.11	0.08	9.89	<0.01	891

Source: Li, S., Feldman, and Li, N. (2000).

the latter are more important than the former. Since 1978, the coefficients of nonparental transmission have changed little while the coefficients of parental transmission have declined a lot. Further, for the marriage cohort after 1978, transmission effects from parents to daughters are positive but those from parents to sons are negative, and nonparental transmission effects are larger than parental transmission effects. Maximum likelihood estimates of coefficients of additive vertical transmission (Cavalli-Sforza and Feldman 1981) for marriage cohorts before and after 1978, including a_h for husband's parents, a_w for wife's parents, and a_0 for nonparents are shown in Table 5 and are consistent with those in Table 4.

Tables 4 and 5 suggest that rural economic reform has had little effect on nonparental transmission of uxorilocal marriage but has had strong effects on parental transmission in Lueyang. Rural economic reform since 1978 has resulted in the collective production team being replaced by the household as the basic production unit, as well as increased mobility of farmers, and an increased circle of potential marriage partners. These have produced dramatic changes in the individual and social life of farmers and have affected farmers' choice of marriage type and marriage arrangement. As a result, parental and nonparental transmission of uxorilocal marriage in Lueyang is affected (Li, S., Feldman, and Li, N. 2000).

3.4. COMPARISON OF DETERMINANTS OF UXORILOCAL MARRIAGE

Logistic regression analyses (see Table 6) showed that the determinants common to the two types of uxorilocal marriage are household demographic structure, membership in a large family clan, and attitude toward a man's uxorilocal marriage. For both contingent and institutional types, a husband with more brothers is more likely to engage in an uxorilocal marriage, while a wife with more siblings is less likely to engage in such a marriage. Membership in a large family clan reduces a couple's likelihood of accepting an uxorilocal marriage; a couple with a positive attitude toward uxorilocal marriage is also more likely to engage in such a marriage. The observed effects are stronger for contingent than for institutional uxorilocal marriage.

The most important contributor in the acceptance of institutional uxorilocal marriage in Lueyang is the parental marriage type. Table 4 showed that uxorilocal marriage in Lueyang is transmissible within a household and that the transmission itself has been affected by rural economic reform (Li, Feldman, and Li 2000). Actually only institutional uxorilocal marriage is transmissible within the household while a contingent one is not, and the transmission of institutional uxorilocal marriage is also affected by the changing

Table 6. Effects of household and individual determinants on the odds ratios of being in institutional and contingent uxorilocal marriage, Lueyang.

	Institutional			Contingent		
		Net effects			Net effects	
Explanatory variables	Gross effects	Model I	Model II	Gross effects	Model I	Model II
Household factors						
Parental marriage type						
Husband's parental marriage						
Uxorilocal	0.73	0.78	2.19*	1.05	1.13	1.16
Wife's parental marriage						
Uxorilocal	1.88***	1.80**	3.32***	1.28	1.30	1.32
Number of siblings						
Husband's brothers						
1	1.22	1.29	1.32	2.45*	2.65*	2.65*
2+	2.35***	2.49***	2.50***	3.59***	3.68***	3.69***
Husband's sisters						
1	0.85	0.81	0.81	1.02	0.98	0.98
2+	0.74	0.75	0.74	1.24	1.17	1.17
Wife's brothers						
2+	0.56***	0.56***	0.56***			
Wife's sisters						
1	0.54**	0.59**	0.59*	0.39**	0.45*	0.45*
2+	0.46***	0.51***	0.51***	0.27***	0.30***	0.30***
Large family clan						
Yes	0.49+	0.46+	0.44+	0.10*	0.09*	0.09*
Individual factors						
Marriage year						
1978+	0.74*	0.84	1.27	1.10	0.85	0.86
Age at marriage						
Husband						
20–24	0.85	0.75	0.78	1.25	1.15	1.15
25+	1.16	0.95	0.94	1.07	0.98	0.98
Wife						
18–22	0.64*	0.64*	0.66*	0.64	0.51*	0.51*
23+	0.50**	0.52*	0.55*	0.49+	0.57	0.57
Education						
Husband's education						
1–6	0.58**	0.71+	0.73	0.79	0.92	0.92
7–9	0.35***	0.53*	0.52*	0.54	0.73	0.73
10+	0.23***	0.33*	0.33*	0.43	0.42	0.42

Table 6. (*Continued*)

Explanatory variables	Institutional: Gross effects	Institutional: Net effects Model I	Institutional: Net effects Model II	Contingent: Gross effects	Contingent: Net effects Model I	Contingent: Net effects Model II
Wife's education						
1–6	1.16	1.18	1.15	1.88^{*}	1.92	1.92^{+}
7–9	1.53^{+}	1.56	1.55	2.46^{*}	4.22	4.23^{**}
10+	4.08^{***}	4.87^{***}	4.75^{***}	4.99	9.63^{*}	9.63^{+}
Adoption						
Husband's adoption						
Yes	0.42^{*}	0.47^{+}	0.44	0.65	1.00	1.00
Wife's adoption						
Yes	2.21^{**}	1.95^{*}	2.03^{*}	2.19^{**}	1.57	1.57
Marriage arrangement						
Self-arranged	0.81	0.87	0.88	0.55^{*}	0.33^{**}	0.33^{**}
Attitude toward a man's uxorilocal marriage						
Acceptable without a son	0.73^{*}	0.75^{+}	0.76^{+}	1.11	0.97	0.97
Unacceptable no matter what	0.34^{**}	0.46^{+}	0.48^{+}	0.30^{*}	0.21^{*}	0.21^{*}
Interactons						
HPMU[†] × M78+[‡]			0.21^{***}			0.95
WPMU[†] × M78+			0.39^{*}			0.98
−2LL		$1{,}086^{***}$	1066^{***}		394^{***}	394^{***}
Number of cases	1,108	1,108	1,108	343	343	343

Note: All variables are dummy categories. The mean equals percent relative to the omitted categories in the table, which are as follows: parental marriage type, nonuxorilocal marriage; number of siblings, 0; large family clan, no; marriage year, before 1978; age at marriage, 19 for husband and 17 for wife; education, 0; adoption, no; marriage arrangement, introduced; attitude toward uxorilocal marriage, acceptable no matter what.

$^{***}p < 0.001$; $^{**}p < 0.01$; $^{*}p < 0.05$; $^{+}p < 0.1$

[†]HPMU and WPMU, husband's or wife's parents are in uxorilocal marriage.

[‡]M78+, marriage cohort 1978 and after.

Source: Li, Feldman, and Li (2002).

social and household environment, as represented by the effect of rural economic reform on institutional uxorilocal marriage after 1978 in Lueyang. An institutional uxorilocal marriage is more likely to occur between a wife with a relatively high education level and a husband with a relatively low education level. Among couples in institutional uxorilocal marriages, the wife is more likely to have married early and to be an adopted daughter, and the husband is less likely to be an adopted son. These results also confirm that institutional uxorilocal marriage in Lueyang is mostly determined by individual and household status.

It should be noted that Lueyang is a special county, where both contingent and institutional uxorilocal marriages are common. However, the determinants of acceptance of contingent uxorilocal marriage reported in Lueyang are basically consistent with those found in Sanyuan, a typical Chinese county, where uxorilocal marriage has been rare, but where most uxorilocal marriages are contingent. With sustained low fertility and an increasing proportion of no-son families, contingent uxorilocal marriage is expected to increase greatly across rural China in the foreseeable future, in order to provide family support for the elderly who are not covered by any social security system.

4. Intergenerational Transfer

4.1. MEASUREMENT OF INTERGENERATIONAL TRANSFER

As mentioned before, the patrilineal family system in rural China shapes the gender-based pattern of children's support for their older parents. Under this system, intergenerational transfer between parents and sons is greater than between parents and daughters, and is characterized by reciprocity over a longer period (Greenhalgh 1985). Parents usually invest a lot in sons in terms of their education, marriage, etc., and expect later repayment by being supported by their sons in their old age. By contrast, parents usually invest less in daughters in terms of their education, marriage, etc., and expect daughters to repay them before marriage by doing housework, taking care of younger siblings, etc. While a daughter's moral obligations may continue after marriage, her formal obligations to her parents end at the time of marriage (Greenhalgh 1985; Yang 1996). Although dramatic social and economic transformations in rural society have brought about some changes in the traditional patrilineal family system and gender-based pattern of children's support for their older parents, the essential gender-dependent mode of old age support remains dominant. Currently, sons are expected to provide primary support for their older parents, while daughters tend to provide supplementary support for their parents through emotional connection, care in daily life, etc. This has been reported in a few studies of old age support in Mainland China and Taiwan (Freedman et al. 1978; Lee, Parish, and Willis 1994; Xu 1996; Yang 1996).

The intergenerational transfer addressed in our survey of Songzi includes only mutual financial transfers, housework, and agricultural labour assistance between children and their non-coresiding parents in 2000. The main reason for restricting attention to non-coresiding parents lies in the fact that in rural China, married children coresiding with and sharing the same household economy as their parents would automatically provide economic and noneconomic help for them. In the survey, children who have separated from parental families were no longer regarded as coresiding with parents, even though they may actually have resided with their parents within the same courtyard or under the same roof.

4.2. LIKELIHOOD OF CHILDREN'S HELP TO PARENTS

Logistic models were employed to analyse the likelihood that children of different gender and marriage forms would provide old age support for parents, where control variables include individual characteristics of children and parents (Li, Feldman, and Jin 2002c).

Table 7. Estimated child's odds ratios of giving various forms of help to parents in logistic model for Songzi.

Variable	Gross help			Net help		
	Financial	Housework	Agricultural fieldwork	Financial	Housework	Agricultural fieldwork
Child gender and marriage						
Son/virilocal	1.00	1.00	1.00	1.00	1.00	1.00
Son/uxorilocal	2.39**	1.30	0.72	1.67+	1.29	0.71
Daughter/virilocal	3.39***	1.68*	1.23	1.67**	2.01*	1.13
Daughter/uxorilocal	0.64	3.07**	1.08	0.64	3.25**	0.92
Child characteristics						
Age	1.02	0.98	0.99	1.01	1.01	1.00
Education	1.10*	0.99	0.98	1.09**	0.98	1.00
Number of brothers	1.16*	1.05	1.10	1.15**	1.08	1.08
Number of sisters	1.13*	0.96	1.01	1.19***	0.96	1.02
Household economy	1.11**	1.02	0.90*	1.11**	1.05	0.95
Number of offspring	0.93	0.66*	1.11	0.94	0.69+	1.06
Parent Characteristics						
Parent alive						
Both alive	1.00	1.00	1.00	1.00	1.00	1.00
Only father alive	1.10	1.25	0.73	1.09	1.52	0.92
Only mother alive	1.03	0.93	0.96	1.13	0.95	1.19
Age	0.99	0.99	1.00	1.01	0.98	1.01
Residence						
Local village	1.00	1.00	1.00	1.00	1.00	1.00
Local township	1.44	0.77	0.47***	1.40+	0.92	0.56**
Local county	1.68*	0.33***	0.21***	1.55*	0.40**	0.33***

(Cont.)

Table 7. (*Continued*)

Variable	Gross help			Net help		
	Financial	Housework	Agricultural fieldwork	Financial	Housework	Agricultural fieldwork
Other county	0.59+	0.21**	0.08***	0.67	0.25*	0.11***
Coresiding with other children	1.18	0.97	1.34	0.96	0.98	1.19
Children's main source of income	2.35***			2.09***		
Can do housework						
Yes		1.00			1.00	
Partly		0.90			1.08	
No		0.92			0.94	
Can do agricultural fieldwork						
Yes			1.00			1.00
Partly			0.60**			0.83
No			0.37***			0.50***
Giving financial help to child	3.98***	1.40+	1.15			
Giving housework help to child	1.31	3.30***	2.00***			
Giving agricultural fieldwork help to child	0.80	1.57*	5.52***			
Ever helped child with child care	1.28	1.47*	1.02	1.00	1.29	1.07
–2LL	1,128***	946***	1,127***	1,436***	768**	1,226***
Number of cases	1,152	1,152	1,152	1,152	1,152	1,152

*** $p < 0.001$; ** $p < 0.01$; * $p < 0.05$; + $p < 0.1$.
Source: Li S., Feldman, and Jin (2002c).

The results are shown in Table 7 and suggest that children's proximity to their parents greatly affects their likelihood and amount of old age support for their parents. Parents who are further away from their children tend to be less likely to receive housework and agricultural fieldwork help from their children. Parents residing within Songzi, but further from their children, are more likely to receive financial help from these children, though the amount of financial help is the same. Parents residing outside Songzi are the least likely to receive financial help from children, and the amount of financial help is lowest. It seems that the effect of proximity on provision of support for the elderly is very important in determining substitution among various modes of intergenerational transfer. This finding is consistent with those from studies on effects of living arrangements on children's support for the elderly (Yan, Chen, and Yang 2001).

An important finding is that while children's gender and marriage form have significant influence on their likelihood of providing financial and housework help for parents, they do not affect the amount of financial support from children for parents or children's likelihood of giving agricultural fieldwork help to parents. Specifically, sons in uxorilocal and daughters in virilocal marriages are more likely to provide financial support for older parents; and daughters, especially those in uxorilocal marriages, are more likely to provide housework help for older parents. It is difficult to determine from among all types of children who actually provides more total help for their parents. However, it is obvious that sons in virilocal marriages provide relatively less help for their parents than do other types of children.

Couples in virilocal marriages are more likely to provide financial and household support for the wife's parents. Couples in uxorilocal marriages are more likely to provide financial help for the husband's parents and housework support for the wife's parents. In other words, support from a couple in virilocal marriage is biased toward the wife's natal parents, while support from a couple in uxorilocal marriage is more balanced between both sets of parents. As to whether virilocal or uxorilocal couples are stronger in support of their non-coresiding parents, our results do not give a clear answer. However, our previous studies on family division showed that couples in uxorilocal marriages tend to coreside longer and leave the parents later than do couples in virilocal marriages (Li, Feldman, and Jin 2002a, 2002b). Considering that in rural China, couples who coreside with their parents usually share the same household economy, they would naturally provide necessary old-age support for their parents. In this context, provision of old-age support should not be worse for couples in uxorilocal marriages than for those in virilocal marriages. In fact, it may be better.

The final conclusion of our study on effects of children's gender and marriage form on their provision of old-age support to parents in Songzi is that daughter's support of their elderly parents is not weaker, but may be stronger, than that of sons. Similarly, old-age support to parents for couples in uxorilocal marriages is not weaker, but may be stronger, than that of couples in virilocal marriages. In other words, in Songzi, daughters and sons are equivalent; both can carry out the family obligation and social responsibility of providing old-age support for parents and carrying on the family name for generations.

5. Conclusion

Intergenerational transfer has been of three kinds in the studies reviewed here. First, we have cultural transmission of attitudes toward the sex of a child manifested most widely across China as son preference. Second, we have seen transmission of marital form—virilocal or uxorilocal—across generations. This transmission is influenced by socioeconomic reforms and by local social structures in the form of large family clans. The third class of intergenerational transfers reverses that traditionally studied by Western economists in that it concerns what children provide for their parents. There is a delicate balance between the kinds of marriage, and residential proximity in determining whether this upward transfer is in money or services. It is also interesting that reciprocity in transfers plays an important role beyond that of parents providing nurturance for growing children. Central to these considerations at least in the counties of China we have studied is the kind of marriage pattern approved of by the society.

There appear to be three family functions of uxorilocal marriage in Lueyang and Songzi: continuation of family lineage, old-age support, and provision of labour. In fact, virilocal marriage also has these functions, but the transfer of female labourers in virilocal marriage is not as significant as that of male labourers in uxorilocal marriage. Furthermore, virilocal marriage serves mainly the husbands' natal families, while uxorilocal marriage in Lueyang and Songzi is institutional, providing functions that are effective for both sets of natal families (Li and Zhu 1999; Jin and Li 2003). Uxorilocal marriage functions socially to promote social mobility and the flow of human resources, and also to reduce son preference within families and communities. This is because although son preference reflects women's low status in their families and communities under the patrilineal family system, it also reflects farmers' practical demand for continuing their family lineage, providing male labourers and old age security. The bilateral family functions of uxorilocal marriage to some extent resolve these practical problems. Thus, couples without a son may call in sons-in-law for their daughters to substitute for sons. Moreover, uxorilocal marriages are relatively egalitarian, which helps weaken son preference (Li and Zhu 1999; Jin and Li 2003).

Our findings have important implications for the current and future of rural Chinese society. While the sustained low fertility in rural China accelerates population ageing, rapid increase in the proportion of no-son families seriously challenges the traditional model of family support for the elderly that relies fundamentally on sons. Further, low fertility and concomitant demand for old age security intensifies the age-old son preference and decreases the survival rates of female children, as evidenced by a high sex ratio at birth and excess girl child mortality (Das Gupta and Li 1999). However, in Songzi and Lueyang, the above-mentioned problems are not so serious, even though fertility is low. Songzi, for example, has almost reached zero population growth and may be regarded as providing a glimpse of the future population in rural China. However, in these places where both virilocal and uxorilocal marriages have been historically accepted and practised, son preference is weak, sex ratio at birth is normal, and parents can receive adequate support for their old age from both sons and daughters in various marriage forms. In other words, no-son couples in Songzi can also count on their daughters and sons-in-law for their old-age security. Thus, promoting acceptance and practice of uxorilocal marriage on a large scale in areas with low fertility and a relatively highproportion of no-son families may be one of the important ways to resolve problems related to old-age security and son preference. Our viewpoints here are basically in agreement with those of other scholars on this issue (Yang 1998; Yan, Mao, and Lu 1999; Lu 2001).

Of course, Lueyang and Songzi are unique in rural China, and broader acceptance of uxorilocal marriage may take a long time. However, the increasing proportion of no-son couples resulting from low fertility indicates a great potential demand for uxorilocal marriage through which to provide old-age support. Sufficient increase in demand for uxorilocal marriage may overcome the resistance of the traditional patrilineal family system and community norms, and bring about important changes in marriage customs and the mode of family support for the elderly in rural areas. Community initiatives and governmental assistance could facilitate and accelerate this change (Yan, Mao, and Lu 1999).

References

Arnold, F. and Liu, Z. (1986), "Sex preference, fertility and family planning in China." *Population and Development Review*, 12:221–246.

Bongaarts, J. and Watkins, Z. (1996), "Social interactions and contemporary fertility transitions." *Population and Development Review*, 22(4):639–682.

Cavalli-Sforza, L.L. and Feldman, M.W. (1981), *Cultural Transmission and Evolution: A Quantitative Approach*, Princeton University Press: Princeton, NJ.

Cavalli-Sforza, L.L., Feldman, M.W., Chen, K.H. and Dornbusch, S.M. (1982), "Theory and observation in cultural transmission." *Science*, 218:19–27.

Chen, K.H., Cavalli-Sforza, L.L. and Feldman, M.W. (1982), "A study of cultural transmission in Taiwan." *Human Ecology*, 10(3):365–382.

Chuang, Y.C. and Wolf, A.P. (1995), "Marriage in Taiwan, 1881–1905." *The Journal of Asian Studies*, 54(3): 781–795.

Coale, A.J. and Banister, J. (1994), "Mortality, race, and the family: Five decades of missing females in China." *Demography*, 31(3):459–479.

Das Gupta, M. and Li, S. (1999), "Gender bias and marriage squeeze in China, South Korea and India 1920–1990." *Development and Change*, 30(3):619–652.

Davis-Friedmann, D. (1991), *Long Lives: Chinese Elderly and the Communist Revolution*. Stanford University Press: Stanford, CA.

Du, P. (1998), "Cohort analysis of changes of living arrangement of Chinese elderly." [In Chinese.] *Chinese Journal of Population Sciences*, 3:53–58.

Du, P. and Wu, C. (1998), "Analysis of main sources of income of the Chinese elderly." [In Chinese.] *Population Research*, 4:51–57.

Dyson, T. and Moore, M. (1983). "On kinship structure, female autonomy, and demographic behavior in India." *Population and Development Review*, 9(1):35–60.

Freedman, R., Moots, B., Sun, T. and Weinberger, M. (1978), "Household composition and extended kinship in Taiwan." *Population Studies*, 32(1):65–80.

Ganschow, T.W. (1978), "Aged in a revolutionary milieu: China." In: *Aging and the Elderly: Humanistic Perspectives in Gerontology*, S.F. Spicker, K.M. Woodward, and D.D. Van-Tassel, eds. pp. 303–320. Humanities Press: Atlantic Highlands, NJ.

Graham, M., Larsen, U. and Xu, X. (1998), "Son preference in Anhui province, China." *International Family Planning Perspectives*, 24(2):72–77.

Greenhalgh, S. (1985), "Sexual stratification: The other side of 'growth with equity' in East Asia." *Population and Development Review*, 11:265–314.

Gu, B. and Roy, K. (1995), "Sex ratio at birth in China, with reference to other areas in East Asia: What we know." *Asia-Pacific Population Journal*, 10(3):17–42.

Gu, S., Chen, X. and Liang, J. (1995), "Old-age support system and policy reform in China." *Paper presented at the International Conference on Aging in East–West*, Seoul National University: Korea, Seoul.

Guo, Z. and Chen, G. (1999), "On uxorilocal marriage in Beijing with data of 1995 one percent population survey." [In Chinese.] *Sociology Research*, 5:94–104.

Hammel, E.A. (1990), "A theory of culture for demography." *Population and Development Review*, 16(3): 455–485.

Han, C. (1992), "Uxorilocal marriage is good for family planning." [In Chinese.] *Population and Economics*, 6:17–19.

Hull, T.H. (1990), "Recent trends in sex ratios at birth in China." *Population and Development Review*, 16(1): 63–83.

Jin, X., Feldman, M.W., Li, S. and Zhu, C. (2002), "A survey of marriage and old-age support in Songzi, China." Revised edition. *Morrison Institute for Population and Resource Studies Working Paper 86*. Stanford University: Stanford, CA.

Jin, X. and Li, S. (2003), "Study on uxorilocal marriage in rural Chinese communities: Results from focus group discussions and in-depth interviews in Songzi." [In Chinese.] *Population and Economics*, 1:45–49.

Johansson, S. and Nygren, O. (1991), "The missing girls of China: A new demographic account." *Population and Development Review*, 17(1):35–51.

Kumm, J., Laland, K. and Feldman, M.W. (1994), "Gene-culture coevolution and sex ratios: The effects of infanticide, sex-selective abortion, sex selection, and sex-biased parental investment on the evolution of sex ratios." *Theoretical Population Biology* 46(3):249–278.

Laland, K. (1994), "The mathematical modeling of human culture and its implications for psychology and the human sciences." *British Journal of Psychology*, 84:145–169.

Laland, K., Kumm, J. and Feldman, M.W. (1995), "Gene-culture coevolutionary theory: A test case." *Current Anthropology*, 36(1):131–156.

Lavely, W., Li, J. and Li, J. (2001), "Sex preference for children in a Meifu Li community in Hainan, China." *Population Studies*, 55:319–329.

Lavely, W. and Ren, X. (1992), "Patrilocality and early marital co-residence in rural China, 1955–85." *The China Quarterly*, 130:378–391.

Lee, Y., Parish, W. and Willis, R. (1994), "Sons, daughters, and intergenerational support in Taiwan." *American Journal of Sociology*, 99(4):1010–1041.

Li, J. and Cooney, R.S. (1993), "Son preference and the one child policy in China: 1979–1988." *Population Research and Policy Review*, 12:108–115.

Li, N., Feldman, M.W. and Li, S. (1998), "Transmission of son preference: Estimates from a survey in two counties of China." *Morrison Institute for Population and Resource Studies Working Paper 76*. Stanford University: Stanford, CA.

Li, N., Feldman, M.W. and Li, S. (2000), "Cultural transmission in a demographic study of sex ratio at birth in China's future." *Theoretical Population Biology*, 58:161–172.

Li, N., Feldman, M.W. and Tuljapurkar, S. (1998), "A demographic model with cultural transmission of son preference."*Morrison Institute for Population and Resource Studies Working Paper 73*. Stanford University: Stanford, CA.

Li, N., Feldman, M.W. and Tuljapurkar, S. (1999), "Sex ratio at birth and son preference." *Mathematical Population Studies,* 8:91–107.

Li, S., Feldman, M.W. and Jin, X. (2002a), "Marriage form and duration of post-marital coresidence with parents in rural China: Evidence from Songzi." Revised edition. *Morrison Institute for Population and Resource Studies Working Paper 87*. Stanford University: Stanford, CA.

Li, S., Feldman, M.W. and Jin, X. (2002b), "Sons and daughters: Marriage form and family division in rural China." Revised edition. *Morrison Institute for Population and Resource Studies Working Paper 88*. Stanford University: Stanford, CA.

Li, S., Feldman, M.W. and Jin, X. (2002c), "Children, marriage and family support for the elderly in rural China." Revised edition. *Morrison Institute for Population and Resource Studies Working Paper 89*. Stanford University, Stanford, CA.

Li, S., Feldman, M.W. and Li, N. (2000), "Cultural transmission of uxorilocal marriage in Lueyang, China." *Journal of Family History*, 25(2):158–177.

Li, S., Feldman, M.W. and Li, N. (2001), "A comparative study of determinants of uxorilocal marriage in two counties in China." *Social Biology*, 48:125–150.

Li, S., Feldman, M.W. and Li, N. (2002). "Acceptance of two types of uxorilocal marriage in contemporary China: The case of Lueyang." Revised edition. *Morrison Institute for Population and Resource Studies Working Paper 85*. Stanford University: Stanford, CA.

Li, S., Feldman, M.W., Li, N. and Zhu, C. (1998), "A survey of transmission of son preference in two counties of China." *Morrison Institute for Population and Resource Studies Working Paper 75*. Stanford University: Stanford, CA.

Li, S. and Zhu, C. (1999), "Analysis of typical couples in uxorilocal marriage in Lueyang, China." [In Chinese.] *Population and Economics*, Supplement:85–93.

Li, S. and Zhu, C. (2001), *Research and Community Practice on Gender Differences in Child Survival in China*. Population Publishing Press: Beijing.

Lu, Y. (2001), "Childbearing interest: Rethinking of farmer's childbearing attitudes." [In Chinese.] *Population Research*, 2:17–27.

Park, C.B. and Cho, N.H. (1995), "Consequences of son preference in a low fertility society: Imbalance of the sex *ratio* at birth in Korea." *Population and Development Review*, 21:59–84.

Pasternak, B. (1985), "On the causes and demographic consequences of uxorilocal marriage in China." In: *Family and Population in East Asian History*, S.B. Hanley and A.P. Wolf, eds. pp. 309–334. Stanford University Press: Stanford, CA.

Pollak, R.A. and Watkins, S.C. (1993), "Culture and economic approaches to fertility." *Population and Development Review*, 19(3):467–496.

Poston, D.L., Liu, P., Gu, B. and McDaniel, T. (1997), "Son preference and the sex ratio at birth in China: A provincial level analysis." *Social Biology*, 44(1–2):55–76.

Shi, L.Y. (1993), "Family financial and household support exchange between generations: A survey of Chinese rural elderly." *The Gerontologist*, 33(4):468–480.

Skinner, G.W. (1997), "Family systems and demographic processes." In: *Anthropological Demography: Toward a New Synthesis*, D.I. Kertzer and T. Fricke, eds. pp. 53–95. University of Chicago Press: Chicago, IL.

Tuljapurkar, S., Li, N. and Feldman, M.W. (1995), "High sex ratios in China's future." *Science*, 267:874–876.

Wen, X. (1993), "Effects of son preference and population policy on sex ratios at birth in two provinces of China." *Journal of Biosocial Science*, 25:110–120.

Wolf, A.P. (1989), "The origins and explanations of variations in the Chinese kinship system." In: *Anthropological Studies of the Taiwan Area*, Chang et al., eds. pp. 241–260. National Taiwan University: Taipei, Taiwan.

Xie, Y. (1989), "Measuring regional variation in sex preference in China: A cautionary note." *Social Science Research*, 18:291–305.

Xu, Q. (1996), "A comparative study on son's and daughter's old-age support for parents." [In Chinese.] *Population Research*, 5:23–31.

Xu, Q. and Yuan, Y. (1997), "The role of family support in the old-age security in China." In *23rd IUSSP General Population Conference: Symposium on Demography of China*, China Population Association, ed. pp. 265–273.

Xu, X., Ji, J. and Tung, Y., (2000), "Social and political assortative mating in urban China." *Journal of Family Issues*, 21(1):47–77.

Yan, M. (1995), "An experimental study of the influences of marriage type on women's sex preferences in childbearing in rural China." [In Chinese.] *Population Science of China*, 5:11–16.

Yan, M., Mao, J. and Lu, J. (1999), "Exploring a fundamental way to reduce sex ratio at birth—reforms in marriage customs in Daye, Hubei." [In Chinese.] *Population and Economics*, 5:18–24.

Yan, M. and Shi, R. (1996), "The role of marriage type in fertility decline in rural China." [In Chinese.] *Population Science of China*, 5:32–38.

Yan, S., Chen, J. and Yang, S. (2001), "Impact of living arrangement on child's old-age support for parents." [In Chinese.] *Chinese Journal of Social Sciences*, 1:130–140.

Yang, H. (1996),"The distributive norm of monetary support to older parents: A look at a township in China." *Journal of Marriage and the Family*, 58:404–415.

Yang, Y. (1998), "On the positive impact of uxorilocal marriage on aging." [In Chinese.] *Population and Family Planning*, 6:31–34.

Yu, X. (1996), "Saving and population aging in China." [In Chinese.] *Population and Economics*, 3:10–17.

Zeng, Y., Tu, P., Gu, B., Xu, Y., Li, B. and Li, Y. (1993), "Causes and implications of the recent increase in the reported sex ratio at birth in China." *Population and Development Review*, 19(2):283–302.

CHAPTER 7. HEALTH AND LABOUR-FORCE PARTICIPATION OF THE ELDERLY IN TAIWAN

CEM METE AND T. PAUL SCHULTZ
Yale Univesity

1. Introduction

Economic performance of low-income countries may be affected by their system of health care. But there is no consensus regarding the optimum level of public health spending, or the efficiency and equity of various schemes for financing public and private health care. On the one hand, excessive government spending on health care in poor countries could divert resources from promising investment opportunities and thus translate into slower economic growth. Some countries have used public health subsidies with restraint at early stages of economic development, but at later stages public expenditure on health increase as a share of GDP—often with the goal of universal coverage of health care.[1] Reviewing the health care experiences of "successful" East Asian countries—Japan, Korea, Singapore, and Taiwan—Gertler (1998) notes that these countries achieved universal coverage of health care only after they had reached relatively high levels of income, were largely urbanized, and most of their workers were in the formal sector. Gertler cautions against early implementation of universal coverage because the resulting health subsidies encourage overuse of health care (i.e., moral hazard) and inefficient allocation of health goods and services. If cost inflation of medical curative care is partly borne by the private consumer, this may also deter the adoption of new medical technology until it is cost effective.

On the other hand, labour productivity and labour supply may positively respond to health improvements, creating economic gains to compensate for health subsidies (Costa 1996). Schultz and Tansel (1997) emphasize the positive effect of health status on workers' earnings due to increased productivity and decreased sickness-related absences from work. Strauss and Thomas (1998) maintain that the labour market consequences of poor health are likely to be more serious for the poor, who are more likely to suffer from severe health problems and to be working in jobs for which physical strength has a high payoff.[2]

---

[1] Newhouse (1993) discusses the evolution of the debate on universal coverage in the United States. Campbell and Ikegami (1998) focus on universal coverage by comparing the health systems of Japan and the United States. The Asian experience is summarized in Gertler (1998).

[2] It is also plausible that improved health status would improve the school performance of children (Rosso and Marek 1996). The improved school performance would be partly because of a less severe impact of sickness

Anne H. Gauthier et al. (eds.), Allocating Public and Private Resources across Generations, 163–200.

The provision of universal health care coverage may contribute to a healthier population through the use of more health care, and allocate more care to poorer segments of the population whose labour productivity might be more responsive to the provision of more health inputs. Conversely, such a national health policy might reduce labour-force participation and thereby erode the government's tax base and even reduce national income. Although the extension of health care coverage by a National Health Insurance (NHI) scheme might raise productive capacity through improvements in overall health, entitlement to the program would redistribute wealth toward those who were not currently working, reducing the motivation to work and to engage in precautionary savings to pay for unpredictable medical care for themselves and their families. This latter tendency would be stronger if the elderly worked in sectors that provided health insurance only to current employees, as was the case in much of Taiwan's economy in the 1980s. A growing literature comparing high income countries concludes that social security arrangements contribute to earlier retirement by taxing heavily the value of wages among the elderly, after adjustment for their loss of social-security wealth if the individual works beyond the age when pensions can be initiated (Krueger and Pischke 1992; Gruber and Wise 1999; Coile and Gruber 2000; National Research Council 2001; Chou and Staiger 2001). Thus, national health care programs may similarly allow the elderly to retire at an earlier age than they would otherwise, even when these programs contribute to improving the health and productive capacity of the elderly. But the direction and magnitude of the net effect of a NHI scheme on national income and welfare remains to be assessed. In the United States, medicare and social security are phased in approximately together, from age 59 to 65, providing only a short interval when pensions are available but medical insurance is not covered. In Taiwan, by contrast, retirement pensions are relatively smaller and rarer, and medical insurance for elderly nonworkers and dependents of workers was very limited until the NHI program was introduced in 1995.

In this paper we assess among elderly men and women how their health status affects their labour-force participation, and whether the national expansion in health insurance in 1995 encouraged earlier retirement and hence lower labour-force participation in 1996. We also seek to assess the effect on labour-force participation of self-reported health limitations and health status, recognizing that these health variables may be measured with error and may be endogenously affected by coordinated household behaviour. Many problems remain to be resolved with this form of analysis, including the development of more satisfactory methods for dealing with the endogeneity of household composition, the choice of living arrangements among the elderly, and their marital status.

2. Literature Review

Analyses of labour-force participation typically assume the demand of individuals for leisure (not working in the labour force) and market consumption goods depends on the wage they are offered in the labour force, their income without working in the labour force,

---

on the family budget. Higher educational attainment and/or better quality of education would, in turn, have a positive influence on economic growth in the long run.

and other-factors including exogenous health conditions (Killingsworth 1983; Ruhm 2001). Although this labour supply framework has been extended to study the household's coordination of the labour supply of all family members, it is not commonly employed to analyse labour-force participation among the elderly in low income countries. When it has been used to study the retirement decision in contemporary high income countries, administrative and tax provisions of the pension system exercise important empirical effects on the life-cycle timing of retirement (Gruber and Wise 1999). In less developed countries such as Taiwan, which have smaller and fewer pensions, the retirement decision may be more readily understood in terms of the standard labour-supply framework, including nonearned income, wealth, market wage offers, family support systems, and the evolving health status of the elderly. We first review two papers that use data from Taiwan: one focuses on the predictors of health status and the other one investigates the determinants of labour-force participation. A brief look at other empirical evidence follows.

Using data from the 1989 and 1993 Surveys of Health and Living Status (SHLS) of the Middle Aged and Elderly in Taiwan, Zimmer et al. (1998) find that educational attainment is associated with reduced likelihood of developing health-functional limitations in 1993, conditional on having no health limitation in 1989. For those who were limited in their health functioning in 1989, however, higher education had little influence on their health-functional transitions. It is difficult to interpret these findings, however, since social networks, health behaviour, and self-assessed health status are all treated as exogenous variables.

The effect of NHI on female labour-force participation in Taiwan is investigated by Chou and Staiger (2001) based on the Family Income and Expenditure Survey (FIES), and they find the availability of insurance for nonworkers (enabled by universal coverage) was associated with a four percentage point decline in married female labour-force participation. The authors conclude that countries considering universal health insurance should anticipate similar declines in labour-force participation. Even though there is theoretical justification for this outcome, the findings cannot be readily generalized, because the analysis focuses on a selected sample—married women aged 20–65, whose husbands are paid employees in the public or private sectors (women from agricultural families, as well as women whose husbands are self-employed or an employer are excluded); and the women must be a household head or married to a household head. Because the FIES do not have direct questions on health insurance status for each individual, Chou and Staiger distinguish between government employees' wives (who already had access to health insurance) and others—which may be a rough approximation to who had access to health insurance prior to the implementation of NHI. The exclusion of males and the elderly from the analysis also deserves reconsideration. Nonetheless, it is likely that the impact of NHI on labour-force participation would a priori be most substantial among married women and the elderly, and our analysis of the elderly based on the SHLS allows for a further examination of the labour-force participation effects of NHI in combination with detailed measures of health status.

We conclude this section by citing related evidence from the United States. Even though there is no universal health insurance in the United States, the studies investigating the relationship between social security benefits and retirement behaviour are relevant to this study (Gustman and Steinmeier 1994). This line of research, in general, has reached the conclusion that the level of social security benefits has a significant effect on the timing of retirement (Krueger and Pischke 1992; Gruber and Wise 1999; Coile and Gruber 2000). One possible limitation of this literature on the effect of social security benefits on labour supply is relevant to efforts to deduce the effect of NHI on labour supply; the cross-sectional estimation may be biased if unobserved individual heterogeneity which affects labour supply is also related to which persons benefit most from the NHI insurance coverage. Without controlling for individual heterogeneity, the changes in labour supply associated with the introduction of NHI may be due to other compositional changes occurring in the population or heterogeneity in the response to the treatment of insurance coverage.

3. Health System in Taiwan and its Reform

As a result of the sharp reduction in fertility and increase in life span, the share of elderly in the population of Taiwan is increasing: 8.7 percent of the population were aged 60 and over in 1987, and the estimate for year 2020 is 21 percent (Chang and Hermalin 1989). The implementation of the NHI from March 1995 is believed to have an especially large impact on the elderly, both because (i) eligibility for most health insurance programs prior to 1995 was dependent on employment status; and (ii) this group faces high medical expenditure (Republic of China–Taiwan 1997 Yearbook).

Prior to March 1995, 59 percent of Taiwan's population had health insurance under 13 public health plans. The three main insurance categories were Labour Insurance, Government Employee Insurance, and Farmers Insurance. Private health insurance serves a negligible fraction of the Taiwan population. NHI subsumed and extended the existing insurance schemes, but the old schemes were not abolished, for they continue to provide special benefits for extraordinary financial cases, e.g., the Labour Insurance program offers some benefits to workers under age 60 and the Farmers Insurance provides some special benefits to registered or working farmers (Department of Health 1992; Republic of China–Taiwan Yearbooks 1997 and 2000).

The beneficiaries of NHI, after paying their premium and obtaining NHI cards, are entitled to receive medical services including outpatient service, inpatient care, Chinese medicine, dental care, childbirth, physical therapy, preventive health care, home care, rehabilitation for chronic mental illness, etc. Although enrollment in NHI is compulsory, program coverage increased but was not immediately universal. At the end of 1998, 96 percent of the population participated in the program (up from around 90 percent during the latter half of 1995). By 1996 about 93 percent of medical institutions nationwide were participating in NHI. People aged 70 or older, as well as members of low-income households (as defined by the Social Support Law) pay no premium. Between 70 and 95 percent of hospitalization fees are also paid by the NHI program. Thus, NHI by 1996 covered the medical expenditures of a large proportion of the population who had no health insurance before 1995 (Republic of China–Taiwan 2000 Yearbook).

4. Labour-Force Participation, Health Status and Health Expenditures over Time

Figures 1, 2a and 2b depict health status and labour-force participation by age and sex in Taiwan. The data come from the 1989 and 1996 SHLS of the Middle Aged and Elderly in

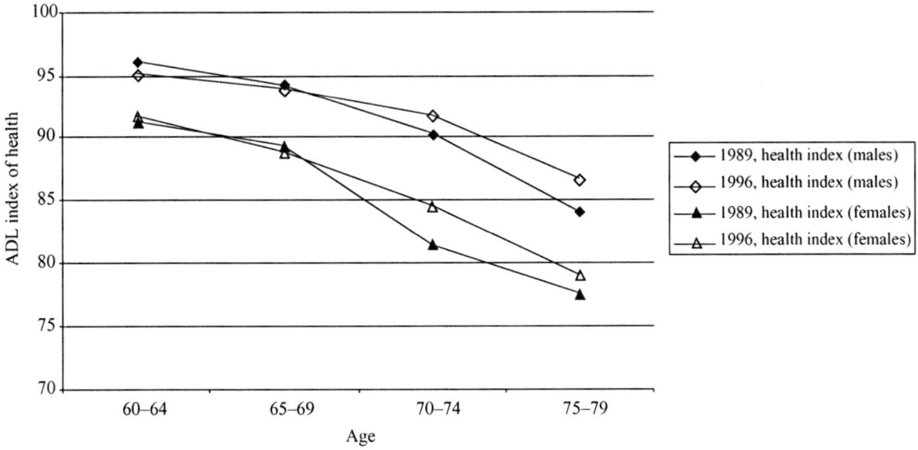

Figure 1. ADL index of health: by age, gender and year.

Taiwan.[3] Figure 1 is based on the activities of daily living index (using seven activities) ranging from 0 (cannot perform any of the seven activities listed) to 100 (no functional limitations).[4] Comparison of Activities of Daily Living (ADL) indexes for 1989 and 1996 suggest that improvements in health among both men and women aged 70 and older may be emerging even in this short span of years. There are significant differences between males and females, with females reporting more functional limitations. This finding is in line with the U.S. literature (Smith and Kington 1997; Verbrugge 1989).[5]

Elderly males in Taiwan are less likely to work in 1996, compared to 1989, as shown in Figure 2a.[6] The reduction seems to occur mostly through a reduction in part-time work. Among females the percentage working also declined from 1989 to 1996, but those working full-time increased at all ages, implying that the propensity to engage in part-time work has also declined for women (Figure 2b). The patterns of labour-force participation by sex are depicted in Table 1, using data from the FIES (various years). The FIES are not necessarily representative of the same population as the SHLS, but the FIES are useful both because of their larger sample size and because these surveys were conducted following

[3] In 1996, in addition to following-up the elderly interviewed in 1989, a new panel of individuals aged 50–66 was also surveyed, and as a result a representative sample of elderly aged 60 and more exist both for 1989 and 1996.

[4] Section 6 provides more information on the construction of this index.

[5] Sex differences in self-reported indicators of morbidity are generally attributed to (i) biological differences by sex, (ii) differences between males and females in perceiving and reporting health problems, (iii) differences in contacts with the health-care system, which could increase information and diagnosis of health conditions, and (iv) differences in mortality by sex, leading to a selection bias in the health status of survivors.

[6] In the United States, the spike in age pattern of retirement has been documented by a number of studies (Hurd 1990; Rust and Phelan 1997). The Taiwan data, however, do not show a sudden increase in retirement at a specific age, probably because pensions for the elderly replace only a small fraction of the wage received by most workers before retirement. Workers are also eligible for pensions in Taiwan, even if they continue to receive earnings from work.

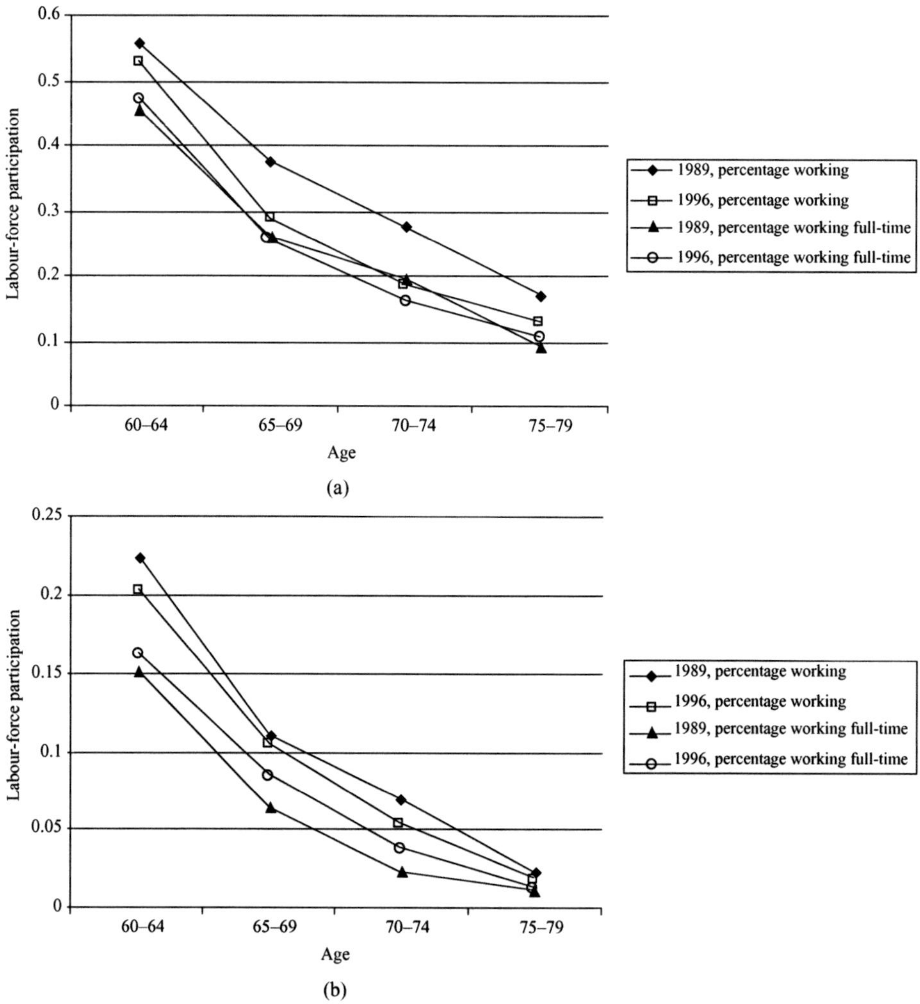

Figure 2. Labour-force participation of males by age, full/part-time and year.

a relatively consistent methodology since 1976 to develop price indexes and construct the national income accounts. The FIES also provide information on private discretionary expenditures on health, health insurance premiums paid by private households, and public subsidies for health insurance used by households. However, the questionnaires eliciting whether a worker is employed part-time or full-time appear to have changed in the FIES after 1995, introducing a possible discontinuity in the measurement of part-time workers as reported in Table 1.

The labour-force participation rate for males between the ages of 25–59 has declined gradually in Taiwan, at least from 1980, not unlike other countries that have been experiencing

Table 1. Proportion of age and sex group working in the labour force for selected years, 1976 to 1996, according to the FIES.*

Sex and age (proportion part time)	1976	1980	1984	1988	1992	1993	1994	1995	1996	1997	1998	1999
Men												
25–49	0.98	0.98	0.97	0.97	0.96	0.96	0.96	0.96	0.95	0.95	0.95	0.94
	(0.000)	(0.001)	(0.005)	(0.004)	(0.005)	(0.005)	(0.005)	(0.008)	(0.009)	(0.008)	(0.012)	(0.011)
50–59	0.91	0.93	0.90	0.90	0.89	0.88	0.88	0.86	0.85	0.86	0.85	0.85
	(0.002)	(0.023)	(0.052)	(0.041)	(0.016)	(0.020)	(0.019)	(0.023)	(0.015)	(0.020)	(0.022)	(0.027)
60–64	0.70	0.73	0.75	0.76	0.72	0.72	0.71	0.68	0.65	0.66	0.60	0.64
	(0.007)	(0.043)	(0.158)	(0.208)	(0.152)	(0.147)	(0.121)	(0.111)	(0.050)	(0.051)	(0.040)	(0.043)
65–69	0.53	0.52	0.54	0.60	0.57	0.58	0.59	0.55	0.53	0.51	0.50	0.48
	†	(0.125)	(0.258)	(0.274)	(0.331)	(0.402)	(0.413)	(0.371)	(0.227)	(0.198)	(0.175)	(0.155)
70–74	0.31	0.33	0.31	0.37	0.44	0.45	0.41	0.44	0.42	0.41	0.41	0.45
	(0.019)	(0.119)	(0.224)	(0.288)	(0.425)	(0.458)	(0.476)	(0.536)	(0.217)	(0.243)	(0.245)	(0.264)
75 or more	0.13	0.21	0.11	0.21	0.24	0.25	0.23	0.25	0.25	0.26	0.26	0.28
	(0.056)	(0.143)	(0.157)	(0.262)	(0.430)	(0.503)	(0.509)	(0.552)	(0.150)	(0.162)	(0.176)	(0.198)
Female												
25–49	0.44	0.49	0.53	0.58	0.61	0.63	0.64	0.64	0.66	0.65	0.65	0.67
	(0.003)	(0.009)	(0.042)	(0.040)	(0.028)	(0.031)	(0.028)	(0.031)	(0.022)	(0.019)	(0.020)	(0.018)
50–59	0.35	0.41	0.40	0.42	0.44	0.43	0.45	0.43	0.42	0.40	0.41	0.44
	†	(0.009)	(0.036)	(0.053)	(0.039)	(0.048)	(0.047)	(0.047)	(0.024)	(0.021)	(0.022)	(0.022)
60–64	0.18	0.31	0.25	0.28	0.32	0.30	0.32	0.31	0.27	0.27	0.24	0.26
	†	(0.005)	(0.025)	(0.108)	(0.043)	(0.084)	(0.082)	(0.096)	(0.027)	(0.022)	(0.031)	(0.030)
65–69	0.11	0.13	0.13	0.16	0.23	0.20	0.21	0.19	0.16	0.16	0.16	0.14
	†	(0.013)	(0.061)	(0.125)	(0.148)	(0.102)	(0.083)	(0.097)	(0.023)	(0.017)	(0.026)	(0.027)
70–74	0.029	0.065	0.029	0.073	0.12	0.12	0.11	0.11	0.10	0.10	0.084	0.095
	†	(0.040)	(0.067)	(0.237)	(0.069)	(0.134)	(0.169)	(0.221)	(0.023)	(0.020)	(0.035)	(0.021)
75 or more	0.0031	0.036	0.006	0.021	0.036	0.043	0.043	0.035	0.023	0.031	0.035	0.044
	†	†	(0.333)	(0.231)	(0.115)	(0.152)	(0.176)	(0.192)	(0.004)	(0.008)	(0.011)	(0.011)

*Labour force equal to persons working in a full-time or part-time job from 1976 to 1995, and reporting occupation and industry. Part-time reported from 1996 to 1999 in new questionnaire sequence (25), possibly related to sharp decline in part-time share after 1995.

†No part-time workers reported in marked cells.

substantial economic development (Durand 1975; Gruber and Wise 1999). Among men aged 60–69 participation rates first rise until 1988 and then begin to decline. Male participation rates for those aged 70–74 rise until 1993, and then stabilize, while there is no clear trend in the participation rates among males over 74. But it is notable that participation remains about a quarter at these advanced ages, much higher than in the OECD high-income countries. The proportion of each age group working part-time is reported in parentheses beneath the overall participation rates, and these part-time rates tend to increase through 1995, encompassing most of the period covered by our panel survey analysis. These data suggest that the increase in part-time jobs by the elderly may help to explain the rise until the early 1990s in the overall labour-force participation rates among males in Taiwan.

Labor force participation rates of women have increased gradually in the prime working ages in many parts of the world (Durand 1975; Schultz 1990), and this trend is evident in Taiwan for women aged 25–49 from 1976 to 1996. But in Taiwan there is in addition a large shift of female participation from work in agricultural self-employment and as an unpaid family worker to wage employment at the beginning of this period (Levenson 1996; Levenson 2000; Schultz 1999). For females aged 50–59 the participation rate peaks in 1994 at 45 percent and has nearly recovered this level again by 1999. Among older women the secular trend of increasing participation is evident until the early 1990s, after which the participation rates stabilize and in some cases fall slightly.

Table 2 reports the share of household total expenditures spent on discretionary health goods and services from the FIES, which decreased from 5.6 percent in 1992 to 3.2 percent in 1995 and 1996, possibly because the NHI was extended to all persons in March 1995.[7] When health insurance premiums are included in the household's health expenditure share after 1992, this share of all private health expenditure declined from 5.04 percent in 1994 to 3.69 in 1996.[8]

5. A Heuristic Framework for Studying Labour Supply and Health of the Elderly

An individual's single period utility function depends on leisure, l, consumption, C, of market goods after paying for health care, and health status, H:

$$U = U(l, C, H)$$

subject to a constraint in which time can be allocated either to market labour, L, or leisure, l:

$$T = L + l$$

---

[7] These estimates are prepared by the authors from the Family Income and Expenditure Survey (FIES) files, representing the Taiwan area of the Republic of China, collected by the Directorate General of Budget, Accounting, and Statistics, Executive Yuan.

[8] A public insurance subsidy for health services consumed appears to be attributed to household's in the FIES, based on the number of outpatient visits and days of inpatient (hospital) care reported by household members in the previous year. Including this public subsidy one obtains the last series of public and private expenditures on health care as a fraction of household total expenditure (and subsidies). The total share of household resources used for health, including this public subsidy for health insurance coverage, increased from 9.02 percent in 1993 to 12.04 percent in 1999 as shown by Table 2 (Cf. Gerdtham and Jönsson 2000).

Table 2. Percent of household total expenditure on health care.*

	1976	1980	1984	1988	1992	1993	1994	1995	1996	1997	1998	1999
Discretionary health expenditure	2.09	4.31	5.08	5.48	5.60	4.08	4.27	3.23	3.24	3.27	3.13	3.19
Expenditure plus health insurance premium	n.a.	n.a.	n.a.	n.a.	n.a.	4.79	5.04	3.57	3.69	3.78	3.74	3.98
Expenditure, insurance premium and National Health Insurance	n.a.	n.a.	n.a.	n.a.	n.a.	9.02	9.86	11.41	10.99	10.98	11.52	12.04

*Imputed share of NHI outlays. Data come from FIES, various years.
n.a.: health insurance premium not reported before 1993.

and consumption depends on market labour supply, market wage, W, and nonearned income inclusive of pensions that are not conditional on work or retirement, V, minus the relative price of health care P_H, which may be reduced or redistributed by a government subsidy, S_H:

$$C = WL + V - (P_H - S_H)$$

where all elements are expressed in real terms by dividing them by the price of consumption goods other than health.

The issue is how health status of the elderly and NHI affect the labour supply of the elderly. The market wage offer an individual receives depends on the individual's education, age, sex, health status, and other things:

$$W = W(E, A, H, e_1) \tag{1}$$

and labour supply could also be affected by the wage or implicitly by these same variables, plus nonearned income and the relative price of health care after the government subsidy:

$$L = L(W, E, A, H, V, (P_H - S_H), e_2) = L_{rf}(E, A, H, V, (P_H - S_H), e_1, e_2) \tag{2}$$

where the errors of measurement, functional form, and omitted variables are denoted by the e's, and the second expression represents a reduced form in which the wage (unobserved in our data) has been solved out. Some studies of the impact of health status on labour supply assume that health status is exogenous in equation (2), which implies that the errors in the wage and labour-supply equations are uncorrelated with the error in equation (3) determining health status (Schoenbaum 1995):

$$H = H(E, A, V, P_H - S_H, Z, e_3) \tag{3}$$

Because we anticipate that the errors affecting health status across elderly individuals could be related to those errors affecting wages and labour supply, we need an exclusion restriction if we are to identify the estimated effect of endogenous health status on labour supply in equation (2), or some variable(s) Z which are assumed to influence health, but not to directly affect wages or labour supply. We will assume that information regarding the individual's parents' longevity and ethnic origin, residential location of the individual at age 12, and proxies for regional nutritional variation at birthplace are valid instrumental variables Z, which help to predict health status, but are not correlated otherwise with the error in the labour-force participation equation. In other words, the identification restriction implied by the choice of Z allows one to estimate without bias the labour-supply equation (2). The use of such an instrumental variable method should also correct for the attenuation bias due to health being measured with error, which is likely to otherwise underestimate the effect of health status on labour supply (Schoenbaum 1995). With self-reported survey evaluations of an individual's health status, this problem of measurement error may be particularly serious.

6. The Available Data

Our analysis is based on the first three waves of the SHLS of the Middle Aged and Elderly in Taiwan (collected for the years 1989, 1993, and 1996), conducted by the Taiwan Provincial Institute of Family Planning and the Population Studies Center of the University of Michigan. The Round 1 survey sample included 4049 individuals aged 60 or older. These individuals were then contacted again in 1993 and in 1996. In 1993, 3449 individuals were alive, and 92 percent of these persons were successfully reinterviewed. In 1996, about 90 percent of the 2968 survivors were reinterviewed. In addition to reinterviewing the panel sample, the 1996 survey also included a new sample of individuals, aged 50–66. The sample of the elderly is nationally representative: all elderly, including the institutionalized, are sampled.[9] The health section of the questionnaire is particularly comprehensive, providing us with a variety of variables to measure health status. The main limitation of the surveys is that not much data is collected on wages, incomes, or assets; and response rates are low on the single income question. As a consequence, we initially estimate a reduced specification of labour-force participation in which we do not attempt to distinguish how poor health differentially affects worker productivity, wages, wealth, and labour supply or effort we instead estimate labour-force participation as a function of health status; we assume that adult health status is influenced by the longevity and other characteristics of the parents and family, as well as the location when the individual was a child.

With the exception of health status proxies, the variables in our labour-force participation model are relatively straightforward. The measurement of health status, however, deserves some explanation. Survey variables commonly used to measure health status include (1) subjective self-evaluation of how well the individual feels at the time of the survey, (2) a health index based on the daily functions which the individual is limited in performing, (3) height and weight, (4) self-reported specific health problems or illnesses, (5) clinically confirmed incidence of specific diseases or health conditions, and (6) number of days of work lost due to illness in a prior reference period. The alternative ways of measuring health have their distinctive strengths and shortcomings, and the preferred measure may depend on the research objective. The relationship between these alternative health measures is also of interest. To the extent that data exist on more than one measure, it may be useful to replicate the analysis using the alternative measures to examine the robustness of the results. We examine health status measures (1), (2), and (4).[10]

A serious problem with relying on self-reported health measures in order to investigate labour-market behaviour is that those who have a preference to retire early or enjoy leisure may overstate health problems that might limit their capacity to work, and those who have a preference to stay in the labour force may understate these health problems (Dwyer and Mitchell 1999; Bound 1991). In this case, one would expect a stronger inverse association between self-reported health and labour-force participation than would be estimated if an

---

[9] As described by the survey documentation, in Taiwan, the elderly (whether institutionalized or not) are included in the "household register". The surveys adopted sampling designs that were a probability sample of all individuals in the relevant age groups in the household register.

[10] In our SHLS data nonreporting of anthropometric information is substantial for (3), and (5) is reported also for a limited number of respondents, perhaps only those whose diagnosis was clinically confirmed.

objective measure of health status were available to relate to labour-force participation. Studies of U.S. data tend to support this hypothesis of a self-reporting bias, notably among those applying for disability insurance, but there is less agreement on the magnitude of the bias (Dwyer and Mitchell 1999; Johansson 1991). The number of days of work lost due to illness may also produce misleading results because individuals have some discretion as to whether to work (Bartel and Taubman 1979; Schultz and Tansel 1997). The problem with using self-reported specific illnesses involves the inability of most persons to self-diagnose their illnesses; and those who use medical care facilities tend to be richer, better educated, and possibly better insured, other things being equal. Those reporting a specific illness are therefore a self-selected sample of those who were diagnosed (or better informed) and had the illness, not a random sample of the ill. What is more, recent evidence using data from Canada suggests there could be significant discrepancies between medical records and self-reports (Baker, Stabile, and Deri 2001). Infrequent specific diseases and illnesses also require large samples before there are a sufficient number of each type of illness to support this form of statistical analysis.

Our preferred measure of health status is an index based on Activities of Daily Living (ADL);[11] but we also report models that use self-evaluated health (SEH) status and specific problems and illnesses. The self-assessment of health question provides five categories: poor, not so good, average, good, excellent. The specific illness indicators are constructed using questions on high blood pressure, diabetes, heart disease, stroke, respiratory problems, arthritis or rheumatism, ulcer, liver problem, cataract, glaucoma, kidney disease, anaemia, bone disease/fracture, prostate trouble (for males), sight problems, and hearing problems.

The ADL index is constructed as the sum of the capacities of the individual to do the following seven activities: squat, raise both hands over the head, grasp or turn objects with fingers, lift or carry something weighing 11–12 kg, walk up two or three flights of stairs, bath oneself, and walk for 200–300 metres. In order to capture the severity of each separate activity limitations, value 0 denotes no problem in doing it, 1 some difficulty, 2 much difficulty, and 3 if entirely unable to do the activity. After adding these codes for each activity, we convert this sum to a 0–100 scale, 100 indicating the worst health status observed in the data. The more objective multiple dimensions of ADL questions are thought to provide less subjective indicators of health status and ones that can be partly validated by the interviewer. Such ADL indexes are well replicated at the individual level over time, and have been systematically validated by clinical examinations (Stewart and Ware 1992). To describe improvements in health status or in physical functioning in ADL as an increase in the health index, we subtract this number from 100. Thus, an ADL index of 0 (100) indicates the worst (best) health status observed in the data. The health status indicator is also scaled from 0 (poor) to 4 (excellent).

The descriptive statistics for our health proxies are reported in Tables A. 1, A. –2, and A. 3. Table A. 1 reports for elderly men and women the mean values for all health variables that

[11] A daily activity health index may be of limited use in signalling the health status of younger individuals (most observations may be close to the best health). But for the elderly, as the earlier figures suggest, ADLs may be good proxies for health status (Stewart and Ware 1992; Strauss et al. 1995).

are used in our analyses. In 12 out of 15 of the specific morbidities for which men and women can be compared (unlike e.g. prostate troubles which are male specific), women over age 60 report illness or health problems more often than do men.

Table A. 2 reports the mean value of the ADL index for each self-assessed health category, separately for each gender. Females not only report more functional limitations than men (as shown in Table A. 1, index $85.5 < 93.2$), but they also report more functional disabilities than do men within each of the self-assessed health status categories.

To summarize the relationship between specific illnesses and functional limitations, the ADL index and self-evaluated health are regressed against all 16 specific illnesses outcomes. The OLS coefficients estimated on the specific illness indicators are reported in Table A. 3, where all statistically significant ($p < 0.05$) coefficients are negative. The first two columns report results for the ADL index, separately for men and women. The number of statistically significant estimates is 12 for males and 9 for females, but only in four cases are the estimates more significant for men than for women: diabetes, heart disease, stroke, and bronchitis and respiratory problems. The last two columns report results for self-evaluated health and this time we have 15 statistically significant estimates for men and 14 for women. In 11 cases the estimates for males are more significant than those for females (these 11 include 4 illnesses highlighted by the ADL index regressions mentioned above).

Table A 4 reports the means of the survey variables for the pooled sample of all three rounds of the SHLS, and for the three variables associated with the region of current residence and of birth. The working hypothesis is that the indicators of parent longevity, father's schooling, city/town/village residence at age 12, and vegetable and protein (pork) consumption per capita in the region of birth, are all conditions which are relevant to childhood nutrition and health care that should affect adult health status among the elderly, but do not directly affect current labour-force participation, except through the respondent's own education, age, and so on.

7. Empirical Analysis

The objective of this analysis is to investigate the determinants of labour-force participation of the elderly, paying special attention to the influence of health status on labour-force participation and the possible impact of the implementation of NHI starting from early 1995. First, the estimation of health status indicators is reported, followed by labour-force participation. Then, instrumental variable estimates are investigated where the family origin and status variables are expected to affect health status and thereby influence labour-force participation. Finally, estimates of the effect of the NHI program are obtained.

7.1. HEALTH STATUS

Health status determinants are estimated for males and females in Tables 3 and 4 using first the ADL index of health as the dependent variable, and then the self-evaluated health (SEH) indicator with its five categories of improvement. The explanatory variables include age, age squared, marital status, own schooling attainment, ethnicity, household possessions

Table 3. Males: determinants of health status. (|t|-ratios calculated using Huber–White variance estimator).

	OLS estimates			
	ADL index		Self-evaluated health	
Independent variables	Coefficient	|t|-ratio	Coefficient	|t|-ratio
Age	2.20	2.69	–0.051	1.18
Age squared/100	–1.93	3.36	0.027	0.89
Marital status (1 married)	–0.417	0.75	–0.049	1.31
Illiterate*	—	—	—	—
Literate	–0.190	0.20	–0.055	0.92
1–6 years of schooling	0.418	0.66	0.011	0.25
7 years of more schooling	1.48	1.92	0.214	4.05
Fuchien*	—	—	—	—
Hakka	–1.22	1.70	–0.016	0.33
Mainlander	–3.31	4.11	–0.135	1.29
Other	–5.87	2.80	–0.549	3.53
Household possessions index	0.611	4.16	0.105	9.99
Regional unemployment rate (by gender)	–0.832	1.93	–0.055	1.71
Year 1993 dummy	0.588	1.07	0.035	0.93
Year 1996 dummy	1.66	1.97	–0.159	2.75
Father's schooling Illiterate*	—	—	—	—
Missing	–0.558	0.66	0.016	0.27
Literate	1.25	2.19	0.009	0.23
Attended school	0.946	1.49	0.070	1.43
Birth place (1 if born in Taiwan)	–3.18	4.00	–0.271	2.61
Residence at age 12 Farm*	—	—	—	—
Town	–1.20	1.88	–0.100	2.44
City	–0.257	0.47	0.122	3.00
Mother died before age 60	0.248	0.48	–0.073	1.99
Father died before age 60	0.146	0.33	–0.003	0.09
Average vegetable consumption per capita (kg) in 1967, at the region of birth	0.097	2.67	0.007	2.50
Average hog consumption per capita (kg) in 1967, at the region of birth	0.051	0.84	–0.005	1.16
Constant	24.5	0.88	4.09	2.58
F-test (*p*-value): Joint significance of the identifier variables (bold above)	0.0001		0.0000	
Overall significance (*p*-value)	0.0000		0.08	
R^2	0.08		0.0000	
Number of observations	5080		5080	

*Marginal effects of the variables evaluated at the sample mean on full-time labour-force participation, and on part-time participation in brackets.

Table 4. Females: determinants of health status. (|t|-ratios calculated using Huber–White variance estimator).

	OLS estimates			
	ADL index		Self-evaluated health	
Independent variables	**Coefficient**	**|t|-ratio**	**Coefficient**	**|t|-ratio**
Age	2.22	2.22	−0.161	3.37
Age squared/100	−2.15	3.09	0.107	3.30
Marital status (1 married)	−0.276	0.43	0.011	0.30
Illiterate*	—	—	—	—
Literate	3.15	2.71	0.273	3.97
1–6 years of schooling	4.08	5.48	0.140	3.04
7 years of more schooling	6.71	6.58	0.472	5.81
Fuchien*	—	—	—	—
Hakka	−0.146	0.16	0.054	1.11
Mainlander	1.92	0.64	−0.018	0.11
Other	0.367	0.18	−0.256	2.11
Household possessions index	0.615	2.78	0.065	5.38
Regional unemployment rate (by gender)	1.73	3.70	0.041	1.47
Year 1993 dummy	0.782	1.05	0.086	2.06
Year 1996 dummy	−0.668	0.70	−0.199	4.04
Father's schooling Illiterate*	—	—	—	—
Missing	0.592	0.59	−0.061	1.15
Literate	2.73	3.45	0.148	3.24
Attended school	1.77	1.93	0.083	1.40
Birth place (1 if born in Taiwan)	3.35	1.07	−0.047	0.29
Residence at age 12 Farm*	—	—	—	—
Town	−2.11	2.45	−0.099	2.12
City	0.286	0.35	0.003	0.06
Mother died before age 60	−1.83	2.34	−0.103	2.41
Father died before age 60	1.20	1.78	0.060	1.60
Average vegetable consumption per capita (kg) in 1967, at the region of birth	0.071	1.42	−0.002	0.70
Average hog consumption per capita (kg) in 1967, at the region of birth	0.110	1.69	0.005	1.26
Constant	15.7	0.44	7.54	4.26
F-test (p-value): Joint significance of the identifier variables (bold above)	0.0001		0.0016	
Overall significance (p-value)	0.0000		0.0000	
R^2	0.13		0.07	
Number of observations	3736		3736	

*Marginal effects of the variables evaluated at the sample mean on full-time labour-force participation, and on part-time participation in brackets.

index (proxy for wealth), a regional unemployment rate[12] (by gender), year dummies, father's schooling (the Taiwanese surveys did not ask about mother's schooling), a dummy variable to indicate birth in Taiwan (versus in Mainland China), farm/town/city residence at age 12, dummy variables that indicate if the mother and father died before age 60, average vegetable consumption per capita (kg) and average pork consumption per capita (kg) in 1967 in the region of birth.[13]

According to the ADL index, health declines for older males and females (within the sample 60 or over), whereas the SEH indicator declines until about age 85 for males and until 75 for females, controlling for the other variables in the regression. Being married is not associated with significant differences in either measure of health for either gender, in contrast to some studies in the United States (Feinstein 1993). Own schooling is related to monotonic improvements in ADLs for women (in the SEH model 1–6 years of schooling ranks below literacy), but with a significant improvement for men only for those with 7 or more years of schooling, consistent with some analyses of mortality and health (Strauss et al. 1995; Kitigawa and Hauser 1973). Similarly, the index derived from household possessions, which is the survey's only proxy for wealth, is related to improvements in ADL measures of health for men and women, just as a growing number of studies indicate that wealth is related in high income countries to improved survival among the elderly. The SEH indicator of health is also reported as increasing with own schooling and household wealth (Hurd and McFadden 1999; Attanasio and Hoynes 2000; Attanasio and Emmerson 2001).[14] The relationship between unemployment rate and health status is not well understood (Chen, Wittgenstein, and McKeon 1996; Novo, Hammarstrom, and Janlert 2001); and our findings do not offer robust evidence either: higher unemployment rate is associated with poor health for males and good health for females. Father's schooling leads to a rise in the ADL health

---

[12] The unemployment rate data are drawn from the *Yearbook of Manpower Statistics Taiwan Area, Republic of China* (1989, 1993 and 1996). Table A. 5 presents these data (by gender and 21 regions).

[13] The data come from *Taiwan Agricultural Products Wholesale Market Yearbook*, 1967 Edition, Department of Agriculture and Forestry, Provincial Government of Taiwan. We would have liked to use the relative food prices in the region of birth, at the time of birth or shortly thereafter. But the earliest food data (by region) that we could find related to consumption patterns in 1967. Thus, we must assume that food consumption variations across regions were related to relative price patterns, which tended to be persistent over time and hence relevant for the elderly surveyed in 1989 when they were young children. For those born in mainland China, the lack of consumption series is compounded into the ethnic origin variable, so the consumption data only distinguish among health outcomes for those born in Taiwan.

[14] In an alternative specification, wealth effect is allowed to vary by age (not reported). Both for males and females, these models present evidence of diminishing wealth effects by age, although the overall wealth effect does not disappear until after age 82 for males and 86 for females (based on models using the ADL index—for self-assessed health models the threshold ages are well over 90). We also experimented with interacting own schooling variables with age, but this exercise did not produce clear evidence. For males when the interaction variables are included in the model with the ADL index, an F test of joint significance of the age/schooling interaction variables resulted in a p-value of 0.44. When the dependent variable is self-evaluated health, however, being literate or having 1–6 years of schooling (as opposed to being illiterate or having attended school for 7 or more years) has beneficial effects on health as a male gets older (p-value for the joint significance of interaction variables being 0.03). For females, the model with the ADL index suggests that schooling (monotonically) translates into better health as an individual gets older (p-value for the joint significance of interaction variables is 0.05). But this time, the model with self-evaluated health results in imprecise parameter estimates (p-value 0.85).

index, or an improvement in health free of functional limitations, with the patterns being stronger for females than males. Residing at age 12 in a small town is less healthy than on a farm or in a city for both genders. If a female respondent's mother dies before age 60, she tends to have worse health, according to the ADL or SEH indicators; however, her father's early death is associated with fewer health problems. The parent's longevity is surprisingly not associated with the ADLs of the males (in contrast with the reported pattern in the United States, e.g., Smith and Kington 1997), though it is for male SEH with respect to their mother's longevity. The nutritional indicators at birthplace are only significant in the case of vegetable consumption being associated with improved male health according to either indicator, and in the case of hog consumption being associated with improved female health according to ADLs. The identifying variables in Tables 3 and 4 are jointly significant in all four cases at better than 0.02 percent level. Although there is general consistency between the more objective ADL index and the more subjective SEH indicator of health, it is our belief that the ADL-based measure is less "culturally conditioned" (Johannsson 1991), and is therefore the more reliable indicator of objective health status across a society and possibly over time.

7.2. LABOUR-FORCE PARTICIPATION

The dependent variable is an ordered categorical variable, which distinguishes between full-time and part-time workers and those not working in the market labour force. The determinants of the labour-force categories are estimated by maximum likelihood methods by fitting an ordered probit model (Madalla 1983). The explanatory variables are alternative indicators of health status, age, married, own schooling, a household wealth index[15], ethnicity, plus regional unemployment rate (by gender), and year dummies, and the threshold parameters for part-time and full-time work. Separate models for men and women are estimated. None of the significant parameters reported here would change if a probit were fit for participation in both full-time and part-time work (i.e., all work), though we prefer to retain the plausible distinction that is available in the survey data. This is a parsimonious specification, which could be generalized to a multinomial probit, but at the cost of nearly doubling the parameters estimated with the same modest sized sample.

Three alternative specifications are reported for men and women in Tables 5 and 6: (i) a model excluding any indicator of health status as an explanatory variable; (ii) one that includes health status (either ADL index of SEH indicator) as if it were exogenous and measured without error; and (iii) one that treats either of the health status variables as an endogenous explanatory variable which may also be measured with random error. The last column (in both tables) presents marginal effects evaluated at sample means for full-time and in brackets part-time labour-force participation. An instrumental variable two-stage conditional maximum likelihood (2SCML) model of Rivers and Vuong (1988) is estimated in which health status is endogenous, and the models of health status reported in the previous section serve as the first-stage equations. The inclusion of residuals from

[15] The household possessions index is formed by the summation of the indicators for the availability of telephone, colour television set, refrigerator, washing machine, VCR, stereo, and air conditioner.

Table 5. Males: labour-force participation models (not-working, part-time work and full-time work). Ordered probit estimates. (|t|-ratios, calculated using Huber–White variance estimator, in parentheses.)

	Labour-force participation models					
		Health indicators		Health indicators as endogenous variables		
Independent variables	Basic equation	ADL index	SEH	ADL index	SEH	Marginal effects: endogenous ADL index specification*
ADL Index		0.036 (10.45)				
Self-evaluation of health			0.217 (12.3)			
Estimated ADL index				0.049 (2.62)		0.013 [0.003]
Estimated self-evaluation of health					0.661 (3.23)	
ADL index residual				0.036 (10.44)		
Self-evaluation of health residual					0.214 (12.09)	
Age	−0.206 (3.23)	−0.241 (3.64)	−0.213 (3.31)	−0.270 (3.45)	−0.190 (2.94)	−0.074 [−0.016]
Age squared/100	0.095 (2.13)	0.125 (2.70)	0.101 (2.23)	0.151 (2.55)	0.089 (1.98)	0.041 [0.009]
Marital status (1 married)	0.232 (4.67)	0.261 (5.14)	0.251 (5.00)	0.266 (5.18)	0.275 (5.33)	0.073 [0.015]
Illiterate[†]	—	—	—	—	—	
Literate	0.009 (0.13)	0.005 (0.07)	0.028 (0.37)	0.004 (0.06)	0.048 (0.63)	0.001 [0.000]

1–6 years of schooling	−0.026	−0.039	−0.025	−0.048	−0.036	−0.013
	(0.50)	(0.74)	(0.48)	(.88)	(0.69)	[−0.003]
7 years or more schooling	−0.056	−0.118	−0.112	−0.146	−0.232	−0.040
	(0.92)	(1.90)	(1.81)	(1.99)	(2.82)	[−0.008]
Fuchien†	—	—	—	—	—	
Hakka	0.097	0.127	0.108	0.137	0.117	0.038
	(1.79)	(2.27)	(1.98)	(2.36)	(2.14)	[0.008]
Mainlander	−0.011	0.001	−0.040	0.002	−0.096	0.001
	(0.23)	(0.03)	(0.82)	(0.04)	(1.75)	[0.000]
Other	−0.156	−0.009	−0.041	0.065	0.201	0.018
	(1.03)	(0.06)	(0.27)	(0.35)	(1.07)	[0.004]
Household possessions index	−0.037	−0.054	−0.060	−0.062	−0.107	−0.017
	(2.81)	(4.10)	(4.48)	(3.65)	(4.23)	[−0.004]
Regional unemployment rate (by gender)	−0.087	−0.073	−0.080	−0.061	−0.067	−0.017
	(2.20)	(1.78)	(1.99)	(1.68)	(1.66)	[−0.004]
Year 1993 dummy	−0.094	−0.108	−0.099	−0.116	−0.117	−0.032
	(2.04)	(2.30)	(2.12)	(2.40)	(2.48)	[−0.007]
Year 1996 dummy	−0.058	−0.084	−0.020	−0.107	−0.063	−0.030
	(0.77)	(1.08)	(0.27)	(1.26)	(0.74)	[−0.006]
Threshold 1 (part-time)	−9.43	−6.99	−9.16	−6.59	−7.30	
	(4.15)	(2.97)	(4.02)	(2.73)	(3.02)	
Threshold 2 (full-time)	−9.19	−6.74	−8.91	−6.33	−7.04	
	(4.04)	(2.86)	(3.89)	(2.63)	(2.91)	
Number of observations	5080	5080	5080	5080	5080	
chi squared (13, 14, 14,15, 15) in order	554.09	613.85	677.06	615.64	679.06	
p-value	0.0000	0.0000	0.0000	0.0000	0.0000	
Log likelihood	−3785.95	−3642.00	−3711.97	−3641.76	−3709.68	

*Marginal effects of the variables evaluated at the sample mean on full-time labour-force participation, and on part-time participation in brackets.

† Omitted category.

Table 6. Females: labour-force participation models (not-working, part-time work and full-time work). Ordered probit estimates. (|t|-ratios, calculated using Huber–White variance estimator, in parentheses.)

	Labour-force participation models					
		Health indicators		Health indicators as endogenous variables		
Independent variables	**Basic equation**	**ADL index**	**SEH**	**ADL index**	**SEH**	**Marginal effects: endogenous ADL index specification***
ADL index		0.022				
		(7.48)				
Self-evaluation of health			0.197			
			(6.78)			
Estimated ADL index				0.041		0.003
				(2.49)		[0.002]
Estimated self-evaluation of health					0.078	
					(0.24)	
ADL index residual				0.022		
				(7.43)		
Self-evaluation of health residual					0.198	
					(6.85)	
Age	−0.178	−0.170	−0.150	−0.211	−0.171	−0.016
	(1.45)	(1.29)	(1.22)	(1.56)	(1.27)	[−0.010]
Age squared/100	0.078	0.080	0.059	0.120	0.073	0.009
	(0.90)	(0.86)	(0.68)	(1.22)	(0.77)	[0.006]
Marital status (1 married)	0.135	0.139	0.133	0.142	0.135	0.011
	(2.17)	(2.18)	(2.11)	(2.24)	(2.15)	[0.007]
Illiterate†	—	—	—	—	—	
Literate	0.294	0.253	0.244	0.182	0.279	0.014
	(2.78)	(2.35)	(2.30)	(1.55)	(2.01)	[0.009]

1–6 years of schooling	0.086	0.031	0.052	−0.059	0.073	−0.004
	(1.09)	(0.39)	(0.65)	(0.55)	(0.78)	[−0.003]
7 years or more schooling	0.374	0.278	0.279	0.133	0.343	0.010
	(2.96)	(2.16)	(2.15)	(0.78)	(1.69)	[0.006]
Fuchien†	—	—	—	—	—	
Hakka	−0.062	−0.095	−0.077	−0.110	−0.066	−0.008
	(0.76)	(1.15)	(0.94)	(1.31)	(0.76)	[−0.005]
Mainlander	−0.379	−0.361	−0.388	−0.357	−0.381	−0.027
	(3.05)	(2.82)	(3.06)	(2.79)	(2.95)	[−0.017]
Other	0.211	0.213	0.293	0.210	0.258	0.016
	(1.07)	(1.06)	(1.51)	(1.05)	(1.17)	[0.010]
Household possessions index	−0.111	−0.126	−0.125	−0.137	−0.117	−0.010
	(5.35)	(5.87)	(5.90)	(5.78)	(3.79)	[−0.006]
Regional unemployment rate (by gender)	0.083	0.071	0.078	0.043	0.082	0.003
	(1.79)	(1.47)	(1.65)	(0.77)	(1.67)	[0.002]
Year 1993 dummy	0.114	0.103	0.092	0.088	0.102	0.007
	(1.58)	(1.40)	(1.26)	(1.17)	(1.30)	[0.004]
Year 1996 dummy	−0.120	−0.129	−0.089	−0.119	−0.111	−0.009
	(1.23)	(1.26)	(0.90)	(1.16)	(0.93)	[−0.006]
Threshold 1 (part-time)	−7.48	−4.90	−6.17	−4.33	−7.11	
	(1.75)	(1.07)	(1.43)	(0.94)	(1.42)	
Threshold 2 (full-time)	−7.20	−4.61	−5.90	−4.05	−6.83	
	(1.68)	(1.01)	(1.36)	(.87)	(1.37)	
Number of observations	3736	3736	3736	3736	3736	
chi squared (13, 14, 14,15, 15) in order	210.21	222.65	228.77	223.32	229.32	
p-value	0.0000	0.0000	0.0000	0.0000	0.0000	
Log likelihood	−1316.93	−1274.68	−1294.44	−1274.09	−1294.38	

*Marginal effects of the variables evaluated at the sample mean on full-time labour-force participation, and on part-time participation in brackets.

†Omitted category.

the first-stage equation provides a specification test for the null hypothesis that the health status variables are exogenous; and the coefficients on the predicted health status variable are the estimates of causal effects of health on labour-force participation.

First, it should be noted that both health indicators are positively associated with labour-force participation, whether treated as exogenous or endogenous, for both genders, though the estimate of the effect of SEH on labour-force participation is not statistically significant for women when treated as endogenous. However, for both men and women, the specification test indicates that the exogeneity of either health status variable can be rejected, leading to our preference for the 2SCML instrumental variable (IV) estimates in the last two columns.

The IV impact of the ADL index on labour-force participation is simulated by age and sex, evaluating model predictions at sample means allowing health status to vary. We consider the mean values of ADLs for health status (93.2 for males and 85.5 for females) as well as an ADL index of 100, which corresponds to no functional limitations, or perfect health. At age 65, the full-time labour-force participation for males would increase from 69 to 78 percent with an increase in their ADL index from 93.2 to 100. The overall male labour-force participation (including part-time work) increases from 77 to 85 percent as a result of this magnitude of improvement in ADLs. For women aged 65 an ADL increase from 85.5 to 100 is associated with a near doubling of their full-time labour-force participation from 25 to 45 percent, whereas their overall labour-force participation (including part-time work) would increase from 35 to 57 percent. The differences in labour-force participation rates due to this large improvement in ADLs become smaller among the eldest of the elderly, but the total effect on labour-force size remain substantial, at least for men, beyond age 70 in Taiwan.

In all specifications labour-force participation declines with age within the sample range for both men and women, and is higher for married than single persons. Own schooling tends to reduce participation among men, significantly for those with 7 or more years of schooling, if health status is controlled; whereas for women, being literate or with 7 or more years of schooling is associated with a greater likelihood of participating in the labour force than being illiterate or with only 1–6 years of schooling. The regional level of unemployment deters male labour-force participation, but is unexpectedly not significantly related to greater female labour-force participation.[16]

With reference to the overall impact of the introduction in the NHI in early 1995, it may be noted that female labour-force participation (including part-time work) among this elderly sample may have increased by about 1.1 percentage point in 1993 compared with 1989 (the omitted category), but decreased 1.5 percent in 1996 compared with 1989, based on

[16] The "displaced worker effect" could explain this trend, in that when husbands are unemployed the wives are more likely to work. This seems to be a plausible explanation: for when the female unemployment rates are replaced by male unemployment rates and the female labour-force participation models are rerun, in all specifications the estimated unemployment rate coefficients are positive and statistically significant at the 5 percent level.

the preferred endogenous ADL index specification (the reported numbers are obtained by adding the marginal effects for full-time and part-time work, which are presented in the last column of Table 6). A 2.6 percentage points decline for elderly women's participation from 1993 to 1996 would therefore have occurred in this time interval when NHI was introduced. For elderly males the participation rate is 3.9 percent lower in 1993 than 1989; whereas in 1996 the rate is 3.6 percent lower than in 1989, implying a 0.3 percent rise from 1993 to 1996 (the marginal effects are listed in Table 5, separately for full-time and part-time work). But this three-year change in labour-force participation rates could be explained by other developments than the introduction of NHI, and we will return to a more satisfactory estimate of the labour-supply effects due to the expansion of the NHI coverage in the next section.

The labour-force participation equations are reestimated in Table 7 by replacing the two overall indicators of health status by the 16 specific forms of morbidity or disease type. There are insufficient instrumental variables in our survey describing life-cycle conditions and parental longevity to identify all of these forms of morbidity simultaneously as endogenous determinants of labour-force participation. We therefore report only ordered probit models in which the specific morbidities are assumed to be exogenous and measured without error. Despite their collinearity, of the 16 morbidity indicators, 7 remain individually significantly associated at the 5 percent level with a reduction in male labour-force participation, and for women 5 out of 15 morbidities are individually statistically significant in their partial association with labour-force participation. It may be noted that all of the statistically significant partial associations are inverse as expected, with the largest coefficients associated with glaucoma and hearing difficulties for women, stroke and heart disease for men, diabetes for both sexes, etc.

7.3. NATIONAL HEALTH INSURANCE

To estimate with more precision the impact of National Health Insurance (NHI) on the labour-force participation of the elderly, we make the distinction between individuals who already had benefits similar to those provided by the NHI in 1993, and those who were covered by insurance only after the implementation of NHI in 1995. The approach is similar to that used by Chou and Staiger (2001), who focus on a sample of younger married women aged 20–65 whose husbands are wage earners. No information on health insurance was collected in the 1989 round of our SHLS survey; but in 1993 and 1996 questions were answered about health insurance. We assume that those who did not have health insurance in 1993, as well as the workers and their dependents covered by labourer, farmer, fisherman, and private insurances benefited from the expansion of NHI in March 1995. Government, military, and veteran workers and retirees in these three categories of employment, and their spouses, probably received expansions of their insurance in 1982 and thereafter; and may therefore have not materially gained benefits with the start of NHI in 1995. For example, labourer and farmer insurance schemes covered only workers, not retirees or dependents before the start of NHI.

Before moving to multivariate analysis, it is instructive to construct difference-in-difference estimates (Table 8). Males who benefited from the introduction of NHI (as defined in the previous paragraph) are more likely to be working relative to the others, both in 1993 and

Table 7. Labour-force participation (not-working, part-time work and full-time work) and specific diseases. Ordered probit estimates. (|t|-ratios calculated using Huber–White variance estimator.)

Independent variables	Labour-force participation models							
					Marginal effects*		Marginal effects*	
	Males		Females		Males		Females	
	Coef.	|t|-ratio	Coef.	|t|-ratio	Full-time	Part-time	Full-time	Part-time
High blood pressure	−0.119	2.55	−0.091	1.29	−0.035	−0.006	−0.008	−0.005
Diabetes	−0.185	2.37	−0.207	1.99	−0.054	−0.010	−0.017	−0.010
Heart disease	−0.273	4.98	−0.105	1.35	−0.079	−0.015	−0.009	−0.005
Stroke	−0.728	5.99	−0.288	1.39	−0.212	−0.039	−0.024	−0.014
Bronchitis, pneumonia, or other respiratory ailment	−0.019	0.38	−0.019	0.20	−0.006	−0.001	−0.002	−0.001
Arthritis or rheumatism	−0.022	0.47	−0.138	2.06	−0.007	−0.001	−0.011	−0.007
Gastric ulcer or stomach ailment	−0.129	2.52	−0.081	0.98	−0.038	−0.007	−0.007	−0.004
Liver or gall bladder disease	−0.099	1.20	−0.218	1.40	−0.029	−0.005	−0.018	−0.011
Cataract	−0.104	1.84	−0.271	3.31	−0.030	−0.006	−0.022	−0.013
Glaucoma	−0.032	0.21	−0.970	2.36	−0.009	−0.002	−0.080	−0.048
Kidney disease	−0.160	1.78	−0.044	0.38	−0.047	−0.009	−0.004	−0.002
Anemia	−0.027	0.34	−0.124	1.37	−0.008	−0.002	−0.010	−0.006
Bone diseases/fracture	−0.002	0.03	−0.007	0.09	−0.001	0.000	−0.001	−0.000
Prostate trouble	−0.148	2.05	—	—	−0.043	−0.008	—	—
Sight difficulties	−0.123	1.99	−0.054	0.73	−0.036	−0.007	−0.004	−0.003
Hearing difficulties	−0.085	1.27	−0.336	2.54	−0.025	−0.005	−0.027	−0.016

Age	−0.191	3.00	−0.159	1.27	−0.056	−0.010	−0.013	−0.008
Age squared/100	0.086	1.92	0.067	0.76	0.025	0.005	0.006	0.003
Marital status (1 married)	0.243	4.84	0.150	2.38	0.071	0.013	0.012	0.007
Father's schooling	—	—	—	—	—	—	—	—
Illiterate†								
Literate	0.016	0.21	0.241	2.29	0.004	0.001	0.020	0.012
1–6 years of schooling	−0.029	0.54	0.037	0.45	−0.008	−0.002	0.003	0.002
7 years or more schooling	−0.072	1.14	0.287	2.23	−0.021	−0.004	0.024	0.014
Fuchien†	—	—	—	—	—	—	—	—
Hakka	0.103	1.87	−0.090	1.09	0.030	0.006	−0.007	−0.004
Mainlander	0.020	0.40	−0.329	2.60	0.006	0.001	−0.027	−0.016
Other	−0.182	1.19	0.259	1.33	−0.053	−0.010	0.021	0.013
Household possessions index	−0.045	3.40	−0.113	5.33	−0.013	−0.002	−0.009	−0.006
Regional unemployment rate	−0.085	2.09	0.121	2.51	−0.025	−0.005	0.010	0.006
Year 1993 dummy	−0.075	1.58	0.127	1.64	−0.022	−0.004	0.010	0.006
Year 1996 dummy	−0.047	0.61	−0.152	1.50	−0.014	−0.003	−0.012	−0.007
Threshold 1 (part-time)	−9.08	4.00	−6.95	1.58				
Threshold 2 (full-time)	−8.83	3.89	−6.67	1.52				
Number of observations	5080		3736					
chi squared (29, 28) in order	695.29		237.10					
p-value	0.0000		0.0000					
Log likelihood	−3703.29		−1280.61					

*Marginal effects of the variables evaluated at the sample mean on full-time and part-time labour-force participation.
†Omitted category.

Table 8. Differences between labour-force participation rates between the individuals who were eligible for supplemental NHI benefits and others and over time.*

	1993 differences in labour-force participation (eligible for supplemental NHI benefits after 1995–others) Single differences (D)		1996 differences in labour-force participation (eligible for supplemental NHI benefits after 1995–others) Single differences (D)		1996–1993 Difference in differences Double differences (DD)	
	Male	**Female**	**Male**	**Female**	**Male**	**Female**
Work	0.176	0.029	0.077	0.027	−0.099	−0.002
	(7.56)	(1.41)	(3.33)	(1.54)	(1.37)	(1.01)
Full-time work	0.078	0.015	0.059	0.019	−0.019	0.004
	(3.77)	(0.915)	(2.71)	(1.28)	(1.36)	(0.543)

* |t|-ratios, based on probit models to fit the contingency table for labour-force participation, in parentheses.

in 1996 (the single differences, D, are statistically significant at the 5 percent level). For males the difference-in-differences (DD) estimate is large and in the expected direction (suggesting a 1.9 percentage point decline in full-time work, 9.9 percentage points overall decline when part-time work is also included), but not statistically significant at the 10 percent level. For females, neither in 1993 nor in 1996 do we detect significant differences in labour-force participation patterns between those who "benefited from the introduction of NHI" and others. The DD estimates are negligible (signalling a 0.4 percentage point *increase* in full-time work and a 0.2 percentage point decline in labour-force participation when part-time work is also considered), and not statistically significant at the 10 percent level for females.

Table 9 provides the ordered probit estimates of the effect of NHI on labour-force participation for the group eligible to benefit; the participation model without the health index is reported first, followed by the specification with the ADL index treated as endogenous (Table A. 6 reports the descriptive statistics on this reduced sample of individuals observed in 1993 and 1996.). The critical estimate in this table is that of the first coefficient reported on the interaction between the year (1996) and the NHI beneficiary group. For males, when both full-time and part-time work is considered, the impact of the NHI on the NHI beneficiary group's labour-force participation appears to be sizable (a 3.9 percent decline in labour-force participation, considering the marginal effects for both full-time and part-time work for the endogenous ADL index specification, as reported in Table 9); but the estimated coefficient is not significantly different from zero. For females, the coefficient on the interaction term is unexpectedly positive (suggesting a 1.6 percent increase in labour-force participation—1.0 due to full-time participation and 0.6 to part-time participation—marginal effects being listed in Table 9); but this time too, the estimate is not statistically significant. We conclude from our analysis that those most likely to gain health insurance coverage for themselves and their dependents from the expansion of NHI did not change their likelihood of labour-force participation in 1996. But a cautionary note is also in order. Only 8 percent (186 observations) of this elderly female sample is working and 26 percent (744 observations) of the male sample is working, which underscores the need for a larger survey and a better method of inferring who is eligible to benefit from the NHI program. We are doubtful whether the expansion of the coverage by NHI to the elderly was a large factor in reducing the size of the labour force; whereas the program may have had an important effect in equalizing the economic burden of health care among this elderly population in Taiwan.

8. Conclusions

Poor health status among the elderly, as summarized by the ADL index and SEH indicator reported in a Taiwan Survey of Health from 1989 to 1996, are associated with reduced participation in the labour-force for both elderly men and women. These health effects on labour supply and on the postponement of retirement of individuals aged 60 and over are substantial in this rapidly growing middle-income country. The econometric specification tests reject the exogeneity of these health status indicators, and a set of characteristics of the respondent's residence at age 12 and at birth, as well as parents, provides a basis for identifying the endogenous effects of health status on current labour-force participation

Table 9. The impact of the NHI program on labour-force participation (not working, part-time work and full-time work). Ordered probit estimates. |t|-ratios calculated using the Huber–White variance estimator. Marginal effects for full-time work (in brackets) and part-time work (parentheses) are reported for the interaction term.

	Labour-force participation							
					Endogenous ADL index		Endogenous ADL index	
	Males		Females		Males		Females	
Independent variables	**Coefficient**	**|t|-ratio**	**Coefficient**	**|t|-ratio**	**Coefficient**	**|t|-ratio**	**Coefficient**	**|t|-ratio**
Interaction term between year 1996 dummy and Eligible for supplemental NHI benefits after 1995	−0.179 [−0.045] (−0.010)	1.57	0.067 [0.006] (0.003)	0.29	−0.136 [−0.031] (−0.008)	1.16	0.164 [0.010] (0.006)	0.69
Eligible for supplemental NHI benefits after 1995	0.630	7.33	0.229	1.60	0.604	6.94	0.153	1.03
Estimated ADL index	—	—	—	—	0.018	0.85	0.040	1.80
ADL Index residual	—	—	—	—	0.034	7.14	0.025	5.78
Age	−0.127	1.01	−0.272	1.29	−0.112	0.79	−0.286	1.26
Age squared/100	0.039	0.46	0.141	0.97	0.028	0.28	0.167	1.06
Marital status (1 married)	0.266	3.79	0.153	1.78	0.303	4.20	0.193	2.11
Illiterate*	—	—	—	—	—	—	—	—
Literate	−0.017	0.16	0.142	0.92	−0.026	0.24	0.009	0.05

1–6 years of schooling	−0.003	0.05	0.078	0.71	−0.008	0.11	−0.092	0.59
7 years or more schooling	−0.050	0.56	0.384	2.16	−0.087	0.87	0.182	0.80
Fuchien*	—	—	—	—	—	—	—	—
Hakka	0.055	0.75	−0.098	0.88	0.100	1.30	−0.131	1.15
Mainlander	0.292	3.75	−0.154	0.93	0.295	3.64	−0.204	1.16
Other	−0.439	1.93	−0.186	0.59	−0.438	1.68	−0.147	0.45
Household possessions index	−0.037	2.03	−0.114	3.92	−0.044	2.19	−0.137	4.08
Regional unemployment rate (by gender)	−0.111	1.84	0.116	1.55	−0.101	1.64	0.085	1.09
Year 1996 dummy	0.183	1.38	−0.292	1.34	0.159	1.17	−0.348	1.55
Threshold 1 (part-time)	−6.03	1.32	−10.83	1.42	−3.80	0.82	−7.25	0.86
Threshold 2 (full-time)	−5.80	1.27	−10.60	1.39	−3.56	0.77	−6.99	0.83
Number of observations	2906		2196		2906		2196	
chi squared (14,14,16,16) in order	264.47		98.25		305.75		108.94	
p-value	0.0000		0.0000		0.0000		0.0000	
Log likelihood	−1916.05		−672.62		−1844.48		−646.23	

*Marginal effects of the variables evaluated at the mean sample on full-time labour-force participation, and on part-time participation in brackets.

in part-time or full-time work. The IV estimates of the endogenous effect of health are larger than the estimates predicting that health is exogenous, with the exception of women with the self-evaluation indicator for which the IV coefficient is not significant. Both men and women reduce their labour-force participation when the household's wealth proxy is higher, and both are more likely to work if they are married and younger. The respondent's own schooling, at least 7 years or more, is associated with lower participation for the male and higher participation for the female, suggesting that women's labour supply responds positively to their wage opportunities as proxied by their education, compared with men whose labour supply decreases with their wage opportunities in these elderly years.

The establishment of a NHI program in Taiwan in March 1995 raises the prospect that offering health insurance to the population, whether they are currently working in the labour force or not, could reduce the incentive to work, particularly among secondary workers who are not likely to be covered in their health-care expenditure unless they themselves are working. One might expect this health insurance subsidy to be particularly valuable to the poor and to most elderly Taiwanese who could not obtain private health insurance before 1995. Economic theory would lead us to expect that the NHI could have contributed to a reduction in labour supply, particularly among secondary workers, such as women. The analysis reported here did not find a statistically significant response; and indeed the direction of the response among those most likely to benefit from NIH was to increase labour-force participation for women and decrease participation for men by 4 percentage points, although neither estimate is statistically significant. The 4 percent decline in labour-force participation by a selected sample of married women aged 20–65 reported by Chou and Staiger (2001) using another survey data source is a challenge to replicate here. We suspect that a change in the questionnaire in their survey between 1994 and 1995 may have been responsible for the reported decline of part-time workers after 1995. But it is also possible that the NIH contributed to the sudden decline in part-time work, which is noted in Table 1. To distinguish between these explanations may require additional data or a different way of interpreting the various survey and administrative information.

Appendix

Table A. 1. Means and standard deviations of the health indicators.*

	Males		Females	
Diseases	**Mean**	**Std. dev.**	**Mean**	**Std. dev.**
High blood pressure	0.260		0.323	
Diabetes	0.076		0.132	
Heart disease	0.180		0.262	
Stroke	0.050		0.034	
Bronchitis, pneumonia, or other respiratory ailment	0.182		0.133	
Arthritis or rheumatism	0.219		0.345	
Gastric ulcer or stomach ailment	0.177		0.192	
Liver or gall bladder disease	0.061		0.047	
Cataract	0.167		0.275	
Glaucoma	0.020		0.029	
Kidney disease	0.056		0.085	
Anaemia	0.064		0.150	
Bone diseases/fracture	0.129		0.207	
Prostate trouble	0.089		—	
Sight difficulties	0.133		0.287	
Hearing difficulties	0.099		0.100	
ADL index	93.2	15.3	85.5	19.7
Self-evaluated health	2.44	1.10	2.00	1.04
Number of observations	5080		3736	

*For a binary dummy variable with a mean of m, the standard deviation is (m(1 – m))exp 1/2.

Table A. 2. The relationship between the ADL index and self-evaluated health status.

	Males ADL index		Females ADL index	
Self-evaluation of health status	**Mean**	**Frequency (%)**	**Mean**	**Frequency (%)**
Poor	56.8	186 (3.6)	53.8	208 (5.6)
Not so good	85.6	826 (16.3)	77.5	1037 (27.7)
Average	94.2	1756 (34.5)	89.3	1415 (37.9)
Good	97.2	1223 (24.1)	93.3	704 (18.9)
Excellent	99.1	1089 (21.5)	96.8	372 (9.9)
Total	93.2	5080	85.5	3736

Table A. 3. The relationship between the ADL index, self-evaluated health, and specific diseases. (|t|-ratios, calculated using Huber-White variance estimator, in parentheses)

OLS estimates	ADL index		Self-evaluated health	
Specific conditions	Males	Females	Males	Females
High blood pressure	−0.504 (1.10)	−0.841 (1.30)	−0.170 (5.24)	−0.148 (4.38)
Diabetes	−4.06 (5.56)	−2.58 (2.95)	−0.383 (7.39)	−0.299 (6.52)
Heart disease	−2.44 (4.67)	−1.26 (1.81)	−0.382 (10.3)	−0.242 (6.62)
Stroke	−23.4 (25.9)	−23.1 (14.27)	−0.732 (11.5)	−0.693 (8.18)
Bronchitis, pneumonia, or other respiratory ailment	−4.00 (7.92)	−2.26 (2.58)	−0.462 (12.9)	−0.245 (5.32)
Arthritis or rheumatism	−2.26 (4.76)	−3.30 (5.28)	−0.221 (6.58)	−0.193 (5.88)
Gastric ulcer or stomach ailment	0.308 (.60)	0.043 (0.06)	−0.283 (7.82)	−0.226 (5.68)
Liver or gall bladder disease	1.36 (1.66)	1.53 (1.09)	−0.278 (4.80)	−0.176 (2.40)
Cataract	−1.47 (2.79)	−2.14 (3.16)	−0.073 (1.96)	−0.116 (3.26)
Glaucoma	0.577 (0.42)	−1.07 (0.61)	−0.004 (0.04)	−0.086 (0.93)
Kidney disease	−1.67 (1.96)	−4.12 (3.80)	−0.275 (4.55)	−0.343 (6.02)
Anaemia	−2.16 (2.71)	−1.50 (1.80)	−0.381 (6.73)	−0.297 (6.78)
Bone diseases/fracture	−3.06 (5.31)	−5.07 (6.92)	−0.185 (4.52)	−0.171 (4.44)
Prostate trouble	−1.85 (2.72)	—	−0.167 (3.46)	—
Sight difficulties	−6.54 (11.13)	−8.88 (13.2)	−0.533 (12.8)	−0.395 (11.2)
Hearing difficulties	−5.37 (8.25)	−10.1 (10.2)	−0.345 (7.47)	−0.218 (4.21)
Constant	98.7 (338)	94.4 (190)	3.01 (145)	2.60 (99.6)
R^2	0.22	0.20	0.25	0.21
Number of observations	5080	3736	5080	3736

Table A. 4. Means and standard deviations of the variables examined in the models investigating the determinants of health status and labour-force participation of the elderly.*

	Males		Females	
Variables	**Mean**	**Std. Dev.**	**Mean**	**Std. Dev.**
Labour-force participation				
Full-time	0.247		0.061	
Part-time	0.076		0.037	
Not-working	0.677		0.902	
Age	69.8	6.09	70.9	6.28
Married	0.758		0.522	
Own schooling Illiterate	0.199		0.651	
Literate	0.102		0.069	
1–6 years of schooling	0.409		0.213	
7 years or more schooling	0.290		0.067	
Ethnicity Fuchien	0.516		0.726	
Hakka	0.141		0.168	
Mainlander	0.330		0.089	
Other	0.013		0.018	
Household possessions index	4.81	1.63	4.82	1.46
Father's schooling Illiterate	0.463		0.535	
Missing	0.077		0.114	
Literate	0.292		0.223	
Attended school	0.169		0.128	
Mother died before age 60	0.222		0.192	
Father died before age 60	0.314		0.279	
Year 1989	0.413		0.399	
Year 1993	0.325		0.329	
Year 1996	0.262		0.272	
Born in Taiwan	0.653		0.906	
Residence at age 12 Farm	0.610		0.593	
Town	0.155		0.174	
City	0.235		0.233	
Average vegetable consumption per capita (kg) in 1967, at the region of birth	87.0	5.77	86.8	6.80
Average hog consumption per capita (kg) in 1967, at the region of birth	33.8	4.02	33.7	4.71
Regional unemployment rate	1.84	0.774	1.87	0.693
Number of observations	5080		3736	

*For a binary dummy variable with a mean of m, the standard deviation is (m(1 − m))exp 1/2.

Table A. 5. Unemployment rates and food consumption by region.

	1989 Male unemp. rate	1993 Male unemp. rate	1996 Male unemp. rate	1989 Female unemp. rate	1993 Female unemp. rate	1996 Female unemp. rate	1967 Vegetable cons. (kg) per capita	1967 Hog cons. (kg) per capita
Taipei municipality	1.71	1.65	3.09	1.77	1.96	2.66	84.1	38.7
Keelung city	3.28	3.06	3.92	4.37	3.92	3.28	84.1*	29.4
Hsinchu city	2.23	1.19	1.91	2.03	1.76	1.75	88.5	38.7*
Taichung city	2.05	1.70	2.99	1.92	2.24	2.68	78.0	33.0**
Chiayi city	2.32	2.19	2.22	2.11	2.06	1.57	84.6	34.2*
Tainan city	2.48	1.77	3.86	2.59	1.53	3.12	79.5	34.2
Kaohsiung municipality	1.66	1.24	3.03	2.02	1.83	3.15	84.6	26.8
Ilan Hsien	1.78	1.52	3.39	1.58	2.57	3.01	88.5*	21.1*
Taipei Hsien	1.50	1.37	2.83	1.47	1.43	2.30	84.1*	38.7*
Taoyuan Hsien	1.17	1.23	1.94	1.17	1.45	2.35	88.5*	38.7*
Miaoli Hsien	1.34	1.22	2.26	1.49	1.69	2.46	88.5*	38.7*
Taichung Hsien	0.83	0.99	2.65	0.99	1.51	2.28	78.0	33.0**
Changhwa Hsien	0.98	1.01	2.12	0.86	1.12	2.08	95.4	33.0**
Nantou Hsien	1.72	1.52	2.54	1.20	1.29	2.67	95.8	33.0**
Yunlin Hsien	1.05	0.85	1.89	0.72	1.15	1.82	90.3	33.0**
Chiayi Hsien	1.38	0.86	2.26	0.73	0.72	1.99	84.6	34.2*
Tainan Hsien	1.93	1.50	2.21	1.35	1.27	2.18	79.5	34.2
Kaohsiung Hsien	1.78	1.34	3.41	1.95	1.77	1.99	84.6	26.8
Pingtung Hsien	1.24	1.10	2.80	1.40	1.21	2.36	101	33.7
Taitung Hsien	2.06	1.67	3.30	2.34	1.34	3.60	81.0	20.8
Hualien Hsien	2.08	1.32	3.88	1.83	1.80	2.65	100	57.9

The unemployment rate data come from the *Yearbooks of Manpower Statistics Taiwan Area, Republic of China* (1989, 1993 and 1996). The regional grouping used in these surveys is identical to that of the *Survey of Health and Living Status of the Middle Aged and Elderly in Taiwan*. The vegetable and hog consumption data come from the *Taiwan Agricultural Products Wholesale Market Yearbook* (1967), which employs a different grouping with some commonalities. In some cases this later reference provides food consumption data for a broader region, and in some other cases data exist for a town that is located in one of our regional categories. The cases where there is not a perfect match among region names are marked with *. Also, the vegetable and hog consumption data were not necessarily reported for the same regions, and so in some cases there is a match for vegetable consumption but not for hog consumption and vice versa. For five of the 21 regions (marked with **), there is no data on hog consumption: average hog consumption per capita of all cities/towns listed in the *Taiwan Agricultural Products Wholesale Market Yearbook* (1967), 33 kg, is assigned to these cases.

Table A. 6. Means of the variables examined in the model investigating the impact of the National Health Insurance program on labour-force participation: data comes from the 1993 and 1996 waves of the survey. For continuous variables, standard deviations are reported in parentheses beneath the means.

Variables	Males	Females
Labour-force participation		
Full-time	0.194	0.052
Part-time	0.062	0.028
Not-working	0.744	0.920
Age	71.6	72.6
	(5.42)	(5.68)
Married	0.763	0.514
Illiterate	0.192	0.639
Literate	0.097	0.070
1–6 years of schooling	0.418	0.222
7 years or more schooling	0.293	0.069
Fuchien	0.514	0.725
Hakka	0.146	0.170
Mainlander	0.328	0.088
Other	0.012	0.017
Household possessions index	4.84	4.83
	(1.60)	(1.44)
Regional unemployment rate	2.01	2.01
	(0.860)	(0.638)
Year 1993	0.568	0.559
Year 1996	0.432	0.441
Eligible for supplemental NHI benefits after 1995	0.618	0.786
Interaction term: Year 1996 ∗ Eligible for supplemental NHI benefits after 1995	0.268	0.344
Number of observations	2906	2196

References

Attanasio, O.P. and Emmerson, C. (2001), "Differential mortality in the U.K." National Bureau of Economic Research *Working Paper 8241*, Cambridge, MA.

Attanasio, O.P. and Hoynes, H.W. (2000), "Differential mortality and wealth accumulation." *Journal of Human Resources*, 35(1):1–29.

Baker, M., Stabile, M. and Deri, C. (2001), "What do self-reported, objective, measures of health measure?" National Bureau of Economic Research *Working Paper No. 8419*, Cambridge, MA.

Bartel, A. and Taubman, P. (1979), "Health and labor market success: The role of various diseases." *The Review of Economics and Statistics*, 61(1):1–8.

Bound, J. (1991), "Self-reported versus objective measures of health in retirement models." *Journal of Human Resources*, 26(1):106–138.

Campbell, J.C. and Ikegami, N. (1998), *The Art of Balance in Health Policy. Maintaining Japan's Low-Cost, Egalitarian System.* Cambridge University Press: New York, NY.

Chang, M.-C. and Hermalin, A.I. (1989), *1989 Survey of Health and Living Status of the Elderly in Taiwan: Questionnaire and Survey Design.* University of Michigan, Population Studies Center, Ann Arbor, MI.

Chen, L.C., Wittgenstein, F. and McKeon, E. (1996), "The upsurge of mortality in Russia: Causes and policy implications." *Population and Development Review*, 22(3):517–530.

Chou, Y.J. and Staiger, D. (2001), "Health insurance and female labor supply in Taiwan." *Journal of Health Economics,* 20(2):187–211.

Coile, C. and Gruber, J. (2000), "Social security and retirement." National Bureau of Economic Research *Working Paper No. 7830.*

Costa, D.L. (1996), "Health and labor force participation of older men, 1900–1991." *Journal of Economic History*, 56(1):62–89.

Department of Health. (1992), *Public Health in Taiwan Area, Republic of China.* Executive Yuan: Taipei.

Directorate General of Budget, Accounting and Statistics (various years), Personal Income Distribution Survey (Microdata files called Family Income and Expenditure Survey). Executive Yuan, Taipei, Taiwan Area of the Republic of China.

Durand, J.D. (1975), *The Labor Force in Economic Development.* Princeton University Press: Princeton, NJ.

Dwyer, D.S. and Mitchell, O.S. (1999), "Health problems as determinants of retirement: Are self-rated measures endogenous?" *Journal of Health Economics*, 18(2):173–193.

Feinstein, J. (1993), "The relationship between socioeconomic status and health: A review of the Literature." *The Milbank Quarterly*, 71(2):279–322.

Gerdtham, U.-G. and Jönsson, B. (2000), "International comparisons of health expenditure." In: *Handbook of Health Economics*, Vol. 1, Chap. 1, A.J. Culyer and J.P. Newhouse, eds. Elsevier Science: Amsterdam.

Gertler, P.J. (1998), "On the road to social health insurance: The asian experience." *World Development*, 26(4):717–732.

Gruber, J. and Wise, D. (1999), *Social Security and Retirement Around the World.* University of Chicago Press: Chicago.

Gustman, A. and Steinmeier, T. (1994), "Employer provided health insurance and retirement behavior." *Industrial and Labor Relations Review*, 47(3):124–140.

Hurd, M. (1990), "Research on the elderly: Economic status, retirement, and consumption and savings." *Journal of Economic Literature*, 28(2):565–637.

Hurd, M. and McFadden, D. (1999), "Predictors of mortality among the elderly." National Bureau of Economic Research *Working Paper 7140*, Cambridge, MA.

Johansson, S.R. (1991), "The health transition: The cultural inflation of morbidity during the decline in mortality." *Health Transition Review*, 1(1):39–68.

Killingsworth, M.R. (1983), Labor Supply, Cambridge UK: Cambridge University Press.

Kitigawa, E. and Hauser, P. (1973), *Differential Mortality in the United States: A Study in Socioeconomic Epidemiology*. Harvard University Press: Cambridge.

Krueger, A.B. and Pischke, J.-S. (1992), "The effect of social security on labor supply: A cohort analysis of the notch generation." *Journal of Labor Economics*, 10(4):412–437.

Levenson, A.R. (1996), "Do consumers respond to future income shocks? Evidence from social security reform in Taiwan." *Journal of Public Economics*, 62:275–295.

Levenson, A.R. (2000), "The role of agricultural and female labor mobility in Taiwan's industrialization: 1976–1991." *Review of Development Economics*, 4(1):101–119.

Madalla, G.S. (1983), *Limited-dependent and Qualitative Variables in Econometrics*. Cambridge University Press: Cambridge.

National Research Council (2001), Preparing for an Aging World: The Case for Cross-National Research, Washington D.C., National Academy Press.

Newhouse, J.P. et. al. (1993), *Free for All? Lessons from the RAND Health Insurance Experiment*. Harvard University Press: Cambridge, MA.

Novo, M., Hammarstrom, A. and Janlert, U. (2001), "Do high levels of unemployment influence the health of those who are not unemployed? A gendered comparison of young men and women during boom and recession." *Social Science and Medicine*, 53:293–303.

Republic of China, Various Years, *Report on the Survey of Personal Income Distribution in Taiwan Area of the Republic of China*. Directorate General of Budget, Accounting and Statistics, Executive Yuan, Taipei.

Republic of China (1997), *Statistical Yearbook of the Republic of China*. Directorate General of Budget, Accounting and Statistics, Executive Yuan, Taipei.

Rivers, D. and Vuong, Q. (1988), "Limited information estimators and exogeneity tests for simultaneous probit models." *Journal of Econometrics*, 39(3):347–66.

Rosso, J.M. and Marek, T. (1996), "Class action: Improving school performance in the developing world through better health and nutrition." *Directions in Development*. The World Bank: Washington, DC.

Ruhm, C.J. (2001), "Economic Expansions are Unhealthy: Evidence from Microdata," National Bureau of Economic Research *Working Paper No. 8447*, Cambridge, MA.

Rust, J. and Phelan, C. (1997), "How social security and medicare affect retirement behavior in a world of incomplete markets." *Econometrics*, 65(4):781–831.

Schoenbaum, M. (1995), "*The Health Status and Labor Force Behavior of the Elderly in Taiwan*." Manuscript, University of California at Berkeley.

Schultz, T.P. (1990), "Women's changing participation in the labor force," *Economic Development and Cultural Change*, 38(3):599–634.

Schultz, T.P. (1999), "Income Inequality in Taiwan, 1976–1995," In: *Political Economic of Taiwan's Development in the 21st century*, G. Ranis, S.-C. Hu and Y.-P. Chu, eds. E. Elgar Publishers: England.

Schultz, T.P. and Tansel, A. (1997), "Wage and labor supply effects of illness in Côte d'Ivoire and Ghana: Instrumental variable estimates for days disabled." *Journal of Development Economics*, 53:251–286.

Smith, J.P. and Kington, R. (1997), "Demographic and economic correlates of health in old age." *Demography*, 34(1):159–170.

Stewart, A.L. and Ware, J.E. Jr. (eds.) (1992), *Measuring Functioning and Well-Being*. Duke University Press: Durham, NC.

Strauss, J., Gertler, P., Rahman, O. and Fox, K. (1995), "Gender and life-cycle differentials in the pattern and determinants of adult health." In: *Investments in Women's Human Capital*, T.P. Schultz, ed. University of Chicago Press: Chicago.

Strauss, J. and Thomas, D. (1998), "Health, nutrition, and economic development." *Journal of Economic Literature*, 36:766-817.

Verbrugge, L. (1989), "The twain meet: Empirical exporations of sex differences in health and mortality." *Journal of Health and Social Behavior*, 30(3):282–304.

Zimmer, Z., Liu, X., Hermalin, A. and Chuang, Y.-L. (1998), "Educational attainment and transitions in functional status among older Taiwanese." *Demography*, 35(3):361–375.

CHAPTER 8. SOCIAL EXPENDITURES ON CHILDREN AND THE ELDERLY IN OECD COUNTRIES, 1980–1995: SHIFTING ALLOCATIONS, CHANGING NEEDS

JANET C. GORNICK

Baruch College and The Graduate Center, City University of New York

1. Introduction

In the middle 1980s, two interrelated areas of inquiry emerged within the field of comparative welfare state research, and both are ongoing. First, after three decades of theory-building and empirical research on the initiation and expansion of the welfare state, scholars began to focus on social policy restructuring and retrenchment (Rein, Esping-Andersen, and Rainwater 1987; Munday 1989; Mishra 1990; Pierson 1994; Ploug and Kvist 1996). At the same time, new questions arose about intergenerational differentials, i.e., about the well-being of the elderly versus children, especially in relation to the relative allocation of resources between these two groups (O'Higgins 1988; Smeeding, Torrey, and Rein 1988; Pampel 1994). To a large degree, the second inquiry emerged from the first, in that concerns about welfare state cutbacks motivated questions about which groups' interests would prevail as public resources grew increasingly scarce.

This chapter assesses the allocation of public social welfare expenditures on the elderly versus children during a 15-year period—1980–1995—a period widely understood to be one of social policy restructuring and change across the industrialized countries. The analysis takes as its starting point a 1984 article by the American demographer Samuel Preston. In an influential article, Preston argued that, in the two prior decades, as the elderly share of the U.S. population grew and the child population fell, public resources shifted toward the elderly, because the elderly's swelling numbers granted them increased political leverage—or as popularly coined, "gray power." In this study, I revisit the Preston thesis, by considering trends in social expenditures on the elderly and children, against the backdrop of graying populations in a group of 14 industrialized countries, including the United States.

In the next section, I briefly review the literature on welfare state retrenchment and on shifting allocations between the elderly and children. In Section 3, I present my analytic approach and lay out a series of five research questions. Section 4 summarizes data and methods. Empirical results are presented in Section 5, and conclusions in Section 6.

Anne H. Gauthier et al. (eds.), Allocating Public and Private Resources across Generations, 201–226.

2. Background and Literature

A renewed focus on social policy retrenchment followed the tumultuous 1970s, when oil price shocks catalysed widespread increases in unemployment across the western industrialized democracies, in turn putting pressure on welfare state efforts. The burden of increasing unemployment was compounded by other factors—especially a rising old-age dependency ratio (i.e., elderly persons to nonelderly adults), increasing transnational competition, and a rash of conservative electoral victories. As a result, welfare state restructuring, along with labour market deregulation, was pushed onto political agendas across the industrialized countries, especially in Europe. In the last two decades, much comparative welfare state research has been preoccupied with the question of current and future welfare state reversals.

A primary focus of retrenchment scholarship has been the identification of causal factors that underlie social policy rollbacks and/or resilience (for a comprehensive review, see, Pierson's edited volume, *The New Politics of the Welfare State*, 2000). Overall, scholars working in this area operate within the frameworks of earlier literature on the determinants of welfare state expansion, including, initially, economic development, industrialization, and urbanization and, later, the political dominance of social democratic and labour parties. At the same time, a central insight in the retrenchment literature is that the determinants of growth and decline are not necessarily the same; in other words, the downward path is not a mirror image of the upward path. Pierson (1994), especially, has stressed the resilience of many welfare state programs, explaining their staying power in largely political terms; politicians want credit for program growth but seek to avoid blame for program cutting. Furthermore, policy provisions often create their own constituencies, rendering many social policies, in effect, self-perpetuating, as they are no longer dependent on the original catalytic factors.

A core question—one that is fundamentally descriptive but crucial—underpins this literature: to what extent *have* welfare states actually been altered, and in what way? In short, while public discourse, especially in the United States, stresses widespread retrenchment across the major welfare states of Europe (see, e.g., Gornick and Meyers 2001), the dominant conclusion in the academic literature has been that real policy change since the 1970s has actually been fairly limited. While total social expenditure growth has slowed and considerable reform has taken place—several countries have enacted a range of programmatic rule changes—overall, the research literature concludes that the welfare state is far from dismantled and the essential architecture remains intact (Pierson 1994, 2000; Daly 1997; Esping-Andersen 1999; Kamerman and Kahn 1999; Myles and Pierson 2000; Ploug 1999).

While the finding of overall welfare state resilience is widespread, the process of measuring and analysing recent welfare state changes up through the end of the 1990s remains unfinished. In particular, there are two principal weaknesses in the literature on shifting social policy outputs. One is that little research has taken into account changing levels of underlying need. As cross-national welfare state outputs have shifted over time, an array of demographic and economic factors have changed in tandem. In most of the Organization for Economic Cooperation and Development (OECD) countries,[1] after the 1970s, populations "grayed," unemployment rose, and market wages stagnated or fell; at the same time, married women's labour-force participation increased and, in some countries, family market income rose as a result. Yet, the extent to which welfare state outputs have changed relative to various indicators of underlying need—with some needs rising and some falling—remains largely unexplored.[2] While welfare state investments may be resilient, relative to total public spending or GDP or even per capita, it is possible that when we account more fully for changes in economic well-being (e.g., family market income) that social welfare efforts relative to need may have fallen, perhaps in some subgroups if not overall.

A second weakness in the retrenchment literature is the lack of research on shifting allocations, across groups, within welfare states. While much research focuses on trajectories in total outputs (i.e., the size of the pie), few studies assess changes in allocations among groups being served (i.e., how the pie is being re-sliced). Although many questions about welfare state rollbacks remain unanswered, there is little doubt that the postwar era of rapid welfare state growth is over, and that public social investments are under scrutiny and constrained. That raises the logical question: are some recipient groups holding their ground, or gaining, relative to others? In particular, are the elderly—whose population share is growing—gaining at the expense of families with children?

2.1. INTERGENERATIONAL RESOURCE ALLOCATIONS: THE POWER OF THE ELDERLY

Nearly two decades ago, Preston (1984) launched this question in the United States in his foundational article, "Children and the elderly: Divergent paths for America's dependents." Preston argued that in the years after the United States' War on Poverty, up through the early-middle 1980s, America's elderly population grew while its child population fell; simultaneously, resources shifted from children to the elderly, and their relative well-being changed as a result. Most notably, elderly poverty rates fell while child poverty grew more prevalent. Preston notes that the ballooning elderly population relative to the shrinking child population could have produced the reverse—fewer resources per elderly person and more per child—if each group's resources were, in fact, somehow fixed. "[T] his view would be characteristic of those who see the world through a Malthusian lens and find the main social drama to be the pressure of numbers on some kind of inelastic resource (p. 435)."

[1] The OECD is an intergovernmental organization with 30 members—primarily, the market-based economies of Europe, Australasia, and North America. Its primary purpose is to support information sharing and consultation across these countries.

[2] One exception is Clayton and Pontusson (1998), who studied the ratio of social expenditures to the pre-tax-and-transfer poverty rate; they also captured the ratio of expenditure to the sum of the number of elderly and unemployed persons. In the four countries they analyse—Germany, Sweden, the United Kingdom, and the United States—they find a greater degree of change in social welfare commitments than has been reported by other scholars: the growth of social spending per poor person, for example, failed to keep up with the growth of GDP per capita.

But, as Preston documents, a Malthusian drama did not unfold, surely not in the United States during the 1970s and early 1980s. As the elderly's population share grew, investments per elderly person did not fall nor did they simply remain constant—in fact, they grew. The elderly's slice of the social welfare pie expanded more quickly than did its population share. Preston concluded that, during the years in question, "many public programs benefiting children have been rolled back while programs targeted to the elderly have been maintained or expanded (p. 437)." His explanation is political: "... the changing numbers of young and old have altered the environment for public policy decisions. In a modern democracy, public decisions are obviously influenced by the power of special interest groups, and that power is in turn a function of the size of the groups, the wealth of the groups, and the degree to which that size and wealth can be mobilized for concerted action. In all of these areas, interests of the elderly have gained relative to children (p. 446)." In a powerful example, Preston reports that, during the 1970s—across the 50 U.S. states—the growth rate of the elderly population was negatively associated with the growth rate of teachers' salaries; in other words, where elderly populations grew the most, investments in children, via investments in teachers, grew the least.

2.2. CROSS-NATIONAL RESEARCH

While Preston's argument was forged with respect to the United States, it clearly had implications for other industrialized countries, because populations were graying in many countries. Not long after Preston's 1984 article, comparative researchers reported that similar shifts in relative well-being, and in resource allocation, were underway in other countries as well. Several contributions in a 1998-edited volume, *The Vulnerable*, addressed the question from a cross-national perspective. Smeeding, Torrey, and Rein (1988) reported that in Canada and the United Kingdom, between 1970 and 1984 the post-tax-and-transfer income of the elderly increased faster than that of the general population, while the incomes of single-parent families increased more slowly or fell. O'Higgins (1988) analysed per capita public spending on pensions and family allowances across the OECD countries between 1960 and 1984, and found a varied pattern: in some countries, family allowances rose in relation to pensions, while in others they fell. When he considered spending in other categories as well—especially public education—he concluded that, overall, resources were shifting toward the elderly in most countries. Like Preston, he noted that resources per elderly person grew alongside elderly population shares: i.e., rising investments in pensions were "not simply due to the fiscal consequence of demographic pressures; it also reflect[ed] decisions to increase pensions (p. 225)."

In her recent review of the literature, Lynch (2001) observes that research on the question of public investments in the elderly versus children has been fairly limited since Preston (1984) and *The Vulnerable* (1988). "While the concept of intergenerational justice has proved fertile ground for political and social theorists, very few empirical studies exist that would allow comparisons of how social provisions for different age groups vary across welfare state types, across countries, or across time." Nevertheless, a small cross-national literature focuses on this question and, overall, empirical support for Preston's prediction that elderly spending would crowd out spending on children is mixed at best.

Pampel (1994) studied spending trends in 18 countries between 1959 and the middle 1980s and concluded that, in many countries, ageing populations did not drive up spending relative to children. In a paper originally titled "Do the old eat the young?", Ghilarducci (1997) assesses the elderly/child tradeoff via a cross-sectional regression analysis of spending in 47 countries; she also finds that generous pension spending, as of the late 1980s, did not, in fact, lead to lower spending on children. Esping-Andersen and Sarasa (2000) considered social spending trends in 12 OECD countries, between the middle 1980s and middle 1990s. Adjusting for population shares, they found that in the majority of countries, transfers to families with children rose, while transfers to aged families fell—clearly not a story of elderly dominance in policy development.

A related literature analysed trajectories in public spending on children. While they did not address expenditure on the elderly, Gornick and Meyers (2001) assessed public expenditure on cash benefits for families in 14 European countries between 1980 and 1995, and found an overwhelming pattern of increase; per-child spending rose, on average, by over 50 percent during those years. They also found that per-child spending (via cash benefits) even rose sharply in Preston's central case, the United States, during that 15-year period.[3]

A growing *qualitative* literature on recent and ongoing policy reforms (i.e., rule changes) across the major welfare states lends further support to the possibility that children-related programs are, in fact, growing, while elderly targeted programs are being trimmed. During the 1980s and 1990s, old age pensions, along with disability and unemployment benefits, were reformed in nearly all of the OECD countries. Myles and Pierson (2000) report, for example, that since 1980 at least six countries—Australia, Canada, Denmark, Finland, New Zealand, and Sweden—have adopted some form of targeting to reduce formerly universal flat-rate benefits for the high-income elderly.

In contrast, there is accumulating evidence that programs targeted on families have been protected or expanded in the last two decades. Gauthier (1996, 1999) and Kamerman and Kahn (1999) conclude that cash benefits for families (typically, family allowances) were largely unchanged and, in many cases, increased during the 1980s and 1990s. Gauthier finds that, in general, the value of family allowance benefits, for those who received them, rose slightly between 1975 and 1990. Kamerman and Kahn (1999) concur with these conclusions about the 1980s and report additional evidence of the resilience of family allowances since that time. "In the 1990s," they note, "when many countries actually set out to curtail social expenditures, and targeted pension policies especially, . . . child and family benefits appear to have been protected in most countries" (Kamerman and Kahn 1999:24).

3. Analytic Approach and Research Questions

This study assesses public social expenditures between 1980 and 1995 across a group of 14 OECD countries. I aim, specifically, to address two of the weaknesses in the retrenchment literature outlined above: the lack of attention paid to allocations across groups, and the neglect of shifting levels of underlying need. Using Preston's work as a starting point, I consider his contention that as population composition has shifted, the elderly and children,

[3] In the United States, per-child spending increased from $304 to $575 (in 1990 dollars), as a result of expansions in the Earned Income Tax Credit.

as groups, have experienced "divergent paths." For both subgroups, I assess welfare state changes in the context of population ageing as well as shifts in mean family income—two measurable dimensions of need.

I extend Preston's 1984 findings by assessing a group of welfare states—including Preston's case, the United States—and by considering a subsequent time period. For most of the analyses, I emphasize change in these countries *as a group*, rather than cross-national variability among them, as preliminary analyses revealed more commonality than variation across countries (including the United States) on several indicators, especially on the direction of change (see Appendix 2). I turn to variation across countries to test two predictions that flow from Preston: one, that public resources would shift to the elderly (per elderly person) more in countries where the ratio of the elderly to child population grew the most; and two, that public resources would shift to the elderly (again, per elderly person) more in countries where total social spending resources became relatively more scarce (as competition would be heightened).

3.1. RESEARCH QUESTIONS

The empirical analysis of shifting allocations to the elderly and children during the 1980s and first half of the 1990s is organized around a series of five research questions. The first four questions concern these countries as a group, focusing on cross-national averages. The final question assesses variation across these countries in order to explore two predictions based on Preston's thesis that the elderly's growing population share will shift policy trajectories in directions that favour the elderly at the expense of children.

First, what happened to total social spending (i.e., the size of the pie) in relation to national economic capacity? And how did expenditures on the elderly and children change in relation to that total (i.e., their slices of that pie)? This opening question addresses the direction and magnitude of overall welfare state change during the study period, as well as the shifting shares of our two subgroups.

Second, how has spending targeted on the elderly and children fared relative to *other* spending categories that faced expansion pressures during these years, such as active or passive labour market programs and/or health expenditures? While we often frame this intergenerational equity question as one of the elderly versus children, it is also the case that investments targeted on both groups could grow (or shrink) as a share of the whole, if other expenditure categories change substantially.

Third, what happened to social spending *per capita*—in particular, spending on the elderly per elderly person, and spending on children, per child? Specifically, how are the findings about changing expenditures on the elderly and children, as shares of the whole, altered when we incorporate the ongoing ageing of the population? As Preston predicted, did elderly outlays rise *per elderly person* and child outlays fall *per child*—presumably, as the elderly's political power grew in relation to children's?

Fourth, what happened to spending on both the elderly and children, relative to shifting levels of economic need among elderly headed families and families with children? During the 1980s and 1990s, household earnings in both elderly families and families with children were subject to contradictory pressures. In elderly headed families, employment rates rose in some countries and fell in others. In Europe, several countries actively lowered the retirement age, mostly to free jobs for younger workers, while other countries aimed to increase elderly employment, under the rubric of social inclusion for all age groups.[4] At the same time, nonelderly families—for whom earnings constitute the lion's share of the income package—faced a range of deteriorating conditions. Household earnings were diminished by both high unemployment, especially in Europe, and stagnant wages, although those were counterbalanced to some extent by rising maternal employment nearly everywhere.

As Preston laid out, several change scenarios are possible with respect to public resources targeted on these two groups. The share of resources allocated to the elderly (or to children) could be fixed over time; it could keep pace with changes in the population share; or it could outpace changes in the population share. Preston reported that the third scenario had taken place in the United States—to the advantage of the elderly and the disadvantage of children—and he interpreted that as evidence that the elderly were influencing redistribution in their own favour, through their increased political power. By extension, a scenario in which resources per elderly person increased relative to the economic needs of the elderly—especially if need is falling—could be interpreted as further evidence of the elderly's political strength.

Fifth, did public resources shift to the elderly (per elderly person) more in countries where the elderly-to-child population ratio grew the most? And, did public resources shift to the elderly (again, per elderly person) more in countries where total social spending resources grew relatively more scarce?

Preston contended that the elderly will command more public resources relative to children where their relative population share increases the most (as he found vis-à-vis teachers' salaries in the United States). His framework also suggests that the elderly would exercise their growing political power vis-à-vis children more intensely where resources are growing more scarce; where social welfare outlays are the least constrained, the elderly face less need to compete with (and draw resources away from) children. Both of these predictions can be assessed, using variation across these 14 countries in rates of change in (1) elderly to child population ratios, (2) total social expenditures, and (3) social outlays to these subgroups.

4. Data and Methods

4.1. EXPENDITURES AS AN INDICATOR OF SOCIAL WELFARE EFFORTS

Policy change can be studied by tracking policy via social expenditures, or policy reforms (rule changes), or both. In this study, I rely on the expenditure approach, as it allows quantified measures of state effort, and is best suited for comparing policy shifts across

[4] See, for example, a recent European Commission (1999) report, *Towards a Europe for all Ages*, which lays out a strategy aimed at raising the employment rate of older workers. Kalisch et al. (1998) summarize recent policy reforms aimed at promoting longer employment.

subgroups as well as to changing levels of need. While the use of expenditure data to capture welfare state effort provides a parsimonious cross-national portrait, it does have several important limitations (see Dixon 1998 for a discussion). First, expenditure changes are a blunt measure of effort, in that they mask institutional changes; specifically, it is difficult to separate changes driven by demographic and economic factors from those put in place intentionally. And, since Preston's main contention concerns the effect of elderly political power, which would be exercised by influencing policy rules, it is necessary to capture intentional changes, if only indirectly. Furthermore, all expenditure datasets suffer from inevitable cross-national incomparabilities. In addition, expenditure data may be further limited by lags between rule changes and program spending levels. For example, changes in old age programs, such as decreases in pension benefits for future recipients, may not have measurable effects on social spending for a number of years. In this study that could cause a bias, in that changes in family cash programs would be expected to show up in expenditures almost immediately; as a result, the relative resilience of policies that serve the elderly could be overstated.

Although these limitations are important, social spending does provide a valuable vantage point on welfare state development. As Swank (2000) argues, aggregate spending—across locales and over time—is in fact highly correlated with theoretically and substantively important outcomes, such as income distribution. In addition, some of the core limitations are overcome to some extent in this study. As Dixon (1998) notes, the limitations of cross-country incomparability are much lessened when spending indicators are used to track change over time. Furthermore, the use of per-person measures of annual social spending (as a proxy for per-recipient) allows us to capture changes in annual expenditures that, in general, signal rule changes. Controlling for measures of need further allows us to infer rule changes from expenditure trends.

4.2. EXPENDITURE DATA: OECD SOCIAL EXPENDITURE DATABASE

The data on social welfare expenditures are from the first edition of the OECD *Social Expenditure Database: 1980–1996* (SOCX). The database contains annual data from 27 of the 29 OECD[5] countries, for the years 1980 up to and including 1995 (in some cases, 1996).

The present study, which covers the time period 1980–1995, includes 14 countries: Australia, Belgium, Canada, Denmark, Finland, France, Germany, Italy, Luxembourg, Netherlands, Norway, Sweden, the United Kingdom, and the United States. The selection of countries was driven by the availability of corresponding microdata for the appropriate years (for the market income and poverty analyses) through the Luxembourg Income Study (LIS).

While a second edition of SOCX (2001) brought spending data up to 1997 in most countries, the LIS data were available, for the most part, only up to and including 1995. Fortunately, the observed period contains the years during which social welfare retrenchment would be expected to be the most severe. After 1995 many of these economies shifted toward recovery and, in several cases, labour or social democratic governments replaced conservative governments.

---

[5] Hungary and Poland are not included in SOCX.

The SOCX database includes social expenditures paid and controlled by central, state, and local governments, including social insurance funds. A cross-country file provides *total social spending* data in 13 categories: (1) old-age cash benefits, (2) disability cash benefits, (3) occupational injury and disease benefits, (4) sickness benefits, (5) services for the elderly and disabled, (6) survivors benefits, (7) family cash benefits, (8) family services, (9) active labour market programs, (10) unemployment benefits, (11) health benefits, (12) housing benefits, and (13) assistance for contingencies (such as benefits for immigrants). Moreover, individual-country data files provide detailed spending breakdowns within these 13 program areas.

To assess overall welfare state trends, I begin by analysing *total social spending* (the sum of the 13 spending areas), and then turn to two particular components: expenditures on *old-age cash benefits* and *family cash benefits. Old-age cash benefits* include old age pensions, old age civil service pensions, veteran's old age pensions, early retirement pensions, and other old-age cash benefits. *Family cash benefits* include traditional cash transfer programs targeted on families (family allowances for children, family support benefits, and lone parent cash benefits), as well as paid family leave and refundable tax credits for families. Expenditures are captured in 1990 US dollars, adjusted for variation in purchasing power parities (PPPs).[6]

The SOCX data have some well-known limitations for comparing social expenditures on the elderly versus children (see Kamerman and Kahn 1997 for a discussion of OECD spending data). The key weakness is that programs not specifically associated with the elderly or with families—such as unemployment insurance, disability compensation, public health insurance programs, and programs for low-income persons—clearly serve the elderly and children as well as the working-age population; but it is not possible to separate those expenditures into elderly and child components. At the same time, fortunately, in countries where large means-tested programs are targeted on the elderly (e.g., old age pensions in Australia) or on families (e.g., the Aid to Families with Dependent Children (AFDC) program in the United States), OECD typically allocates those expenditures to the two focal categories: old-age cash or family cash spending. The most substantial problem lies in the inability to separate health spending between the elderly and children, as it constitutes, on average, a quarter of total social expenditures in these countries.

A second weakness concerns data on services for the elderly and for children. Spending on services for the disabled (of all ages) and the elderly are combined and cannot be disaggregated: so to include services for families—a separate item—would create a problematic incomparability. As a result, I omit services and focus on expenditures on cash benefits only. Fortunately, services represent a relatively small share of public investments for these two groups. That is especially true for the elderly, where—on average, across these 14 countries—only 4 percent of total social spending goes to services (for the elderly and disabled combined) compared to nearly 30 percent for old-age cash benefits. When we

[6] The use of PPP-adjusted dollars means that expenditure levels across countries should correspond to equivalent levels of aggregate purchasing power. The use of constant PPP-adjusted dollars thus controls both for cross-national variation in the cost of living and for inflation.

turn to families with children, less than 3 percent of total social spending goes to services, compared with 7 percent allocated to cash programs.

A third weakness of the SOCX data is the exclusion of most tax expenditures, which provide crucial benefits in some countries. At the same time, refundable tax credits, such as the United States' Earned Income Tax Credit (EITC), are generally included, either under family cash benefits or elsewhere in the database (e.g., in the "other contingencies" category). In this study, family cash benefits in the United States were adjusted to include expenditures on the EITC.

4.3. POPULATION INDICATORS

To capture changes in expenditures in relation to population shifts, the SOCX data are combined with population data, from OECD's *Labour Force Statistics* (various years). In the population-based indicators, the elderly are defined as aged 65 or older, and children as below age 15.

4.4. INCOME INDICATORS: THE LUXEMBOURG INCOME STUDY

Country-level indicators of market income and post-tax-and-transfer poverty, for the elderly and children, come from the LIS, an archive of comparable microdata from a large number of industrialized countries. The LIS datasets, primarily based on household surveys, contain demographic, labour market, and income data at the individual and household level. This study uses two datasets from each of these 14 countries, one from the "second wave" of the LIS data (1984–1987) and one from the "fourth wave" (1994–1997).[7]

For the poverty analysis, the poverty rate is defined as the percentage of persons living in families with post-tax-and-transfer income (what LIS calls "disposable personal income" or DPI) falling below 50 percent of their country's median, with family income adjusted for family size.[8]

In the analyses of families' mean market income, elderly headed families are defined as those with household heads aged 65 and older, and families with children include those with dependents below age 18. "Market income" is defined as the sum of gross wages and salary, self-employment income, and cash property income; LIS terms this "factor income" or FI.

5. Empirical Results

The first research question—what happened to magnitude of the pie, and the slices allocated explicitly to the elderly and children?—is addressed in Figure 1. On average, total social

---

[7] For information and documentation about the LIS microdata, visit the LIS web site: http://www.lisproject.org

[8] Disposable personal income includes all market income plus all social transfers, less employee payroll contributions and income taxes. As is standard in cross-national comparisons, families are defined as poor if their income falls below 50 percent of the median income in the country in which they reside. Income is adjusted for family size, where adjusted income equals income divided by the square root of family size.

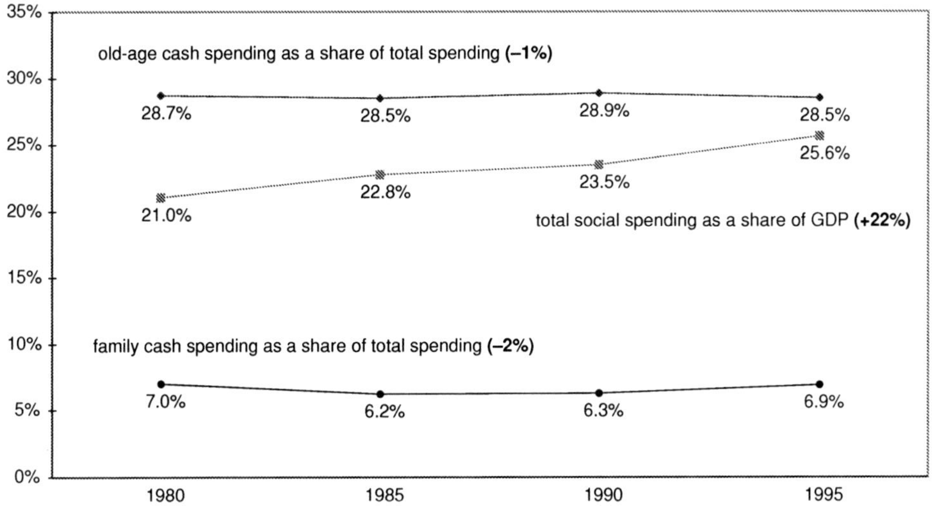

Note: Numbers in parentheses refer to percent change between 1980 and 1995.

Figure 1. Social spending as shares: total social spending, old-age cash spending, family cash spending (14-country averages, 1980–1995).

spending as a share of GDP grew substantially during the study years, from 21 to 26 percent of GDP—a remarkable 22 percent increase. As reported in Appendix 2, there was an overwhelming pattern of commonality across these countries. Total social spending grew everywhere with one exception, the Netherlands, where expenditures relative to GDP fell by 3 percent. The answer to the opening question—were total social welfare outlays rolled back during these years?—is clearly no.

Figure 1 also reports the shares of total spending allocated to old-age cash spending and to family cash spending during the study years. Spending on cash benefits for the elderly and families were remarkably stable, i.e., they were virtually unchanged between the beginning and end of the study period. Expenditures on cash benefits for the elderly remained about four times as high as cash benefits for families—29 percent versus 7 percent of total spending, respectively. The far higher share of social spending distributed to the elderly is not surprising. Although there are more children than elderly persons everywhere (with the exception of Italy as of 1995), public income transfers constitute a much larger share of the elderly's income package; families with children rely far more on market income. In any case, we see no evidence that the elderly gained at children's expense during the 1980s and early 1990s, at least not from this vantage point.

The second research question—what about the remainder of the pie?—is assessed in Figure 2, which reduces the 13 SOCX spending categories into seven. Clearly, three categories of spending constitute the other "big ticket items:" labour market programs (which include active labour market programs and unemployment compensation); health benefits; and a "catch-all" category that includes low-income, disability, occupational injury, sickness,

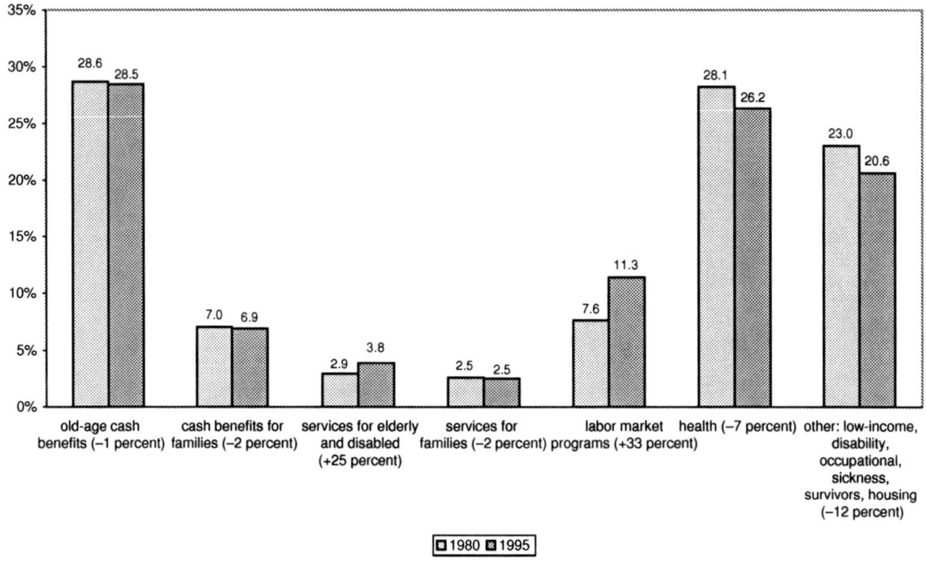

Figure 2. Major categories of spending as a share of total social spending (14-country averages, 1980 and 1995).

survivors, and housing programs. During the study years, spending on labour market programs grew substantially (by one-third, on average): that result is not surprising because unemployment rates, on average, doubled across these countries. Average spending in the other two large categories fell, such that average total spending on the three categories together remained unchanged (totalling 53–54 percent). The two focal categories—old-age cash benefits and family cash benefits—were, overall, the most stable program categories, along with services for families.

Further results indicate, again, an overall pattern of commonality, although the United States is somewhat of an outlier. Unlike the common pattern, the unemployment rate in the United States fell during the study years; accordingly, it was one of the few countries where the share of spending allocated to labour market programs did not increase. The United States is also an exception with respect to health spending: among these countries, during the study years, health spending substantially increased only in the United States.

Figure 2 also shows that spending on services for the elderly and disabled rose markedly (by 25 percent), although from a low base (3 percent of total social spending). Nevertheless, this finding is meaningful in a particular way. As noted earlier, when we use expenditures to compare changes in public supports for the elderly versus families, we face the problem that there may be a long lag time between the policy change and the point at which its consequences are evident in spending levels. That is more likely to be true with programs for the elderly (where changes often affect future recipients) than for families (where changes generally affect current recipients). This means that old age programs would

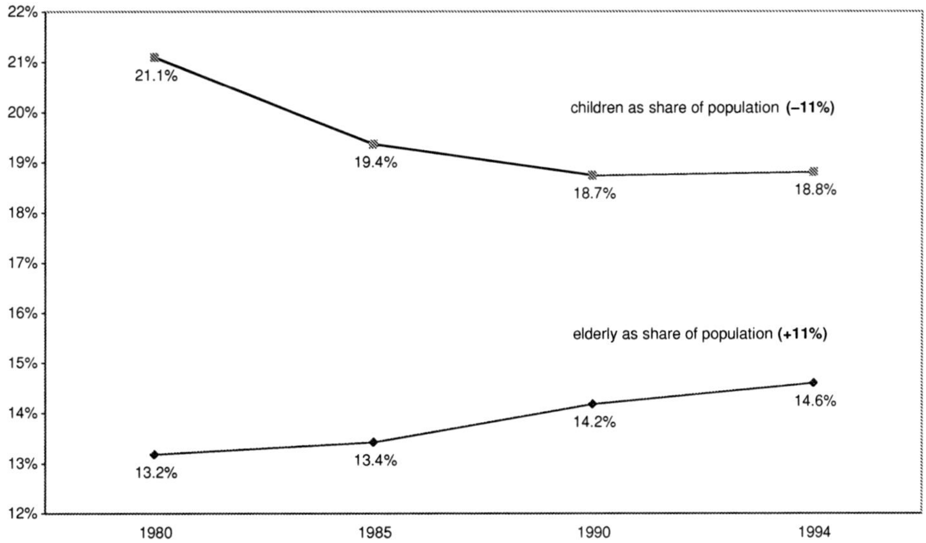

Note: Numbers in parentheses refer to percent change between 1980 and 1995.

Figure 3. Population shares: under age 15 and over age 64 (14-country averages, 1980–1995).

appear more stable than they are, relative to current policy activity; in other words, the stability that we see may mask underlying cuts that are not yet apparent. The results in Figure 2 suggest that our omission of services that are provided for the elderly works in the counterbalancing direction: i.e., their omission understates growth in programs that serve the elderly during the study period.

The third research question—what happened to social spending when we take population ageing into account?—is addressed in Figures 3 and 4. During the study years, total populations increased in all 14 countries, and by 11 percent on average (results not shown). Overall, there was relatively little variation in total population growth, with the exception of Germany, where unification increased the total population by a third. Furthermore, nearly everywhere, the working-age population (aged 15–64) grew at nearly the same rate as the total population.

Population ageing characterized all of the countries studied. The elderly share of the population (aged 65+) increased in all of them; the average increase was 11 percent (Figure 3). Simultaneously, the average child share of the population (less than age 15) fell by the same amount: 11 percent. (The child share fell in all countries, except in the United States, where it remained constant.) The ratio of elderly-to-children rose substantially everywhere, increasing, on average, from 0.64 to 0.80. Clearly, throughout this group of welfare states, the elderly population grew and the child population fell, resulting in a clear pattern of graying, albeit at somewhat varying rates. (We return to that variation later.)

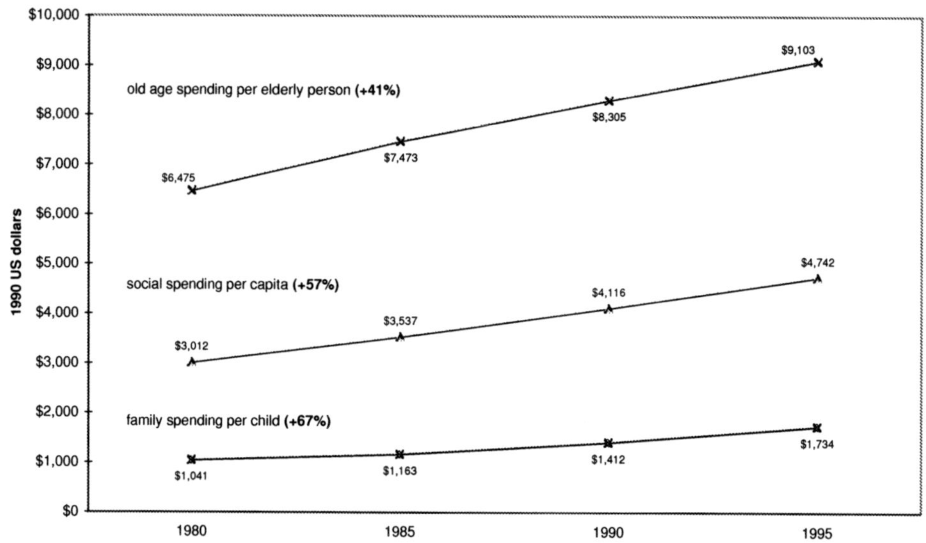

Note: Numbers in parentheses refer to percent change between 1980 and 1995.

Figure 4. Social spending per person: total social spending, old-age cash spending, family cash spending (14-country averages, 1980–1995).

Figure 4 reveals the interaction of Figures 1 and 3. Recall that, on average, total social spending increased as a share of GDP (by 22 percent). Because GDP growth outpaced population growth everywhere, social spending *per capita* grew even more sharply—by an average of 57 percent, as shown in Figure 4. Undoubtedly, despite all the demographic and economic pressures facing these welfare states, social investments per capita continued to rise substantially after 1980.

On average, what happened to our focal categories of spending—old-age and family cash benefits—per potential recipient, i.e., per elderly person and per child, respectively? Although old age spending held steady as a share of the total pie, the expanding pie meant that real old-age cash layouts increased. What is important for our purposes is that, as Preston predicted, real expenditures on the elderly increased more than did the elderly population. Thus, old age spending *per elderly person* increased—and by a substantial 41 percent. (Per elderly spending grew in all countries, with one exception: Australia, where it declined slightly; see Appendix 2.)

And what about children? Similarly, children's share of the pie remained constant, but the expanding pie, combined with the falling child population, meant that family cash expenditures per child rose even more sharply than did old age expenditures per elderly—by an extraordinary 67 percent. Considering these welfare states as a group, the graying of the population did not set in motion a process that resulted in a reduction of real public resources for children—as Preston predicted—surely not as captured by this indicator. Furthermore, as shown in Appendix 2, old age spending per elderly person and family cash spending per child *both grew* in 10 of the 14 countries. The exceptions were Australia (where elderly spending declined slightly), and Germany, Italy, and the Netherlands (where family spending declined).

Thus, the "Preston scenario"—where old age spending per elderly grew while family cash spending per child fell—was seen in 3 of the 14 countries: Germany, Italy, and the Netherlands. Further study of these three continental European countries would be fruitful. Particularly interesting is the contrast between Germany and Italy. Germany is the only country that experienced an absolute *increase* in the child population (+16 percent) as well as in the elderly population (+35 percent)—due to unification—whereas Italy saw the largest decrease in the number of children (−25 percent), the consequence of its dramatically declining fertility rate. Thus, the per-child decline in Germany may be specific to that historical transition; while the decline in investments in Italy reflects a substantial withdrawal of public resources allocated to family cash programs. Orsini (2001) describes the Italian case as unusual, relative to Europe as a whole, as well as relative to fairly similar welfare states. "Italy differs from the other ideal-typical examples of the conservative welfare regime (principally France and Germany): the extremely high percentage of total social transfers allocated to pensions crowds out social expenditure targeting younger households." The question of why the intergenerational outcome in Italy is so unusual, in cross-national terms, identifies it as a valuable case for further study.

The fourth research question—what happened to spending relative to shifting levels of underlying economic need?—is addressed in Figures 5 and 6. Need is captured here in

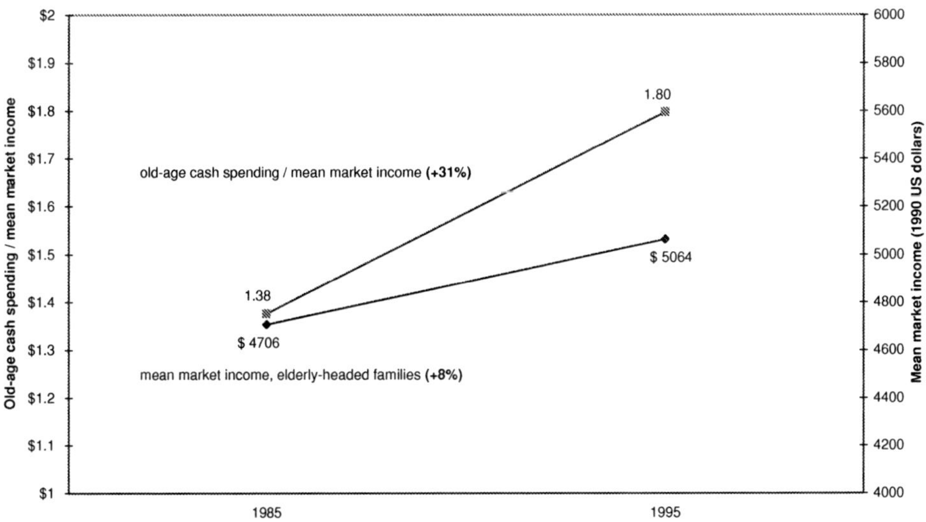

Note: Numbers in parentheses refer to percent change between 1985 and 1995.

Figure 5. Old-age cash spending relative to mean market income (14-country averages, 1985 and 1995).

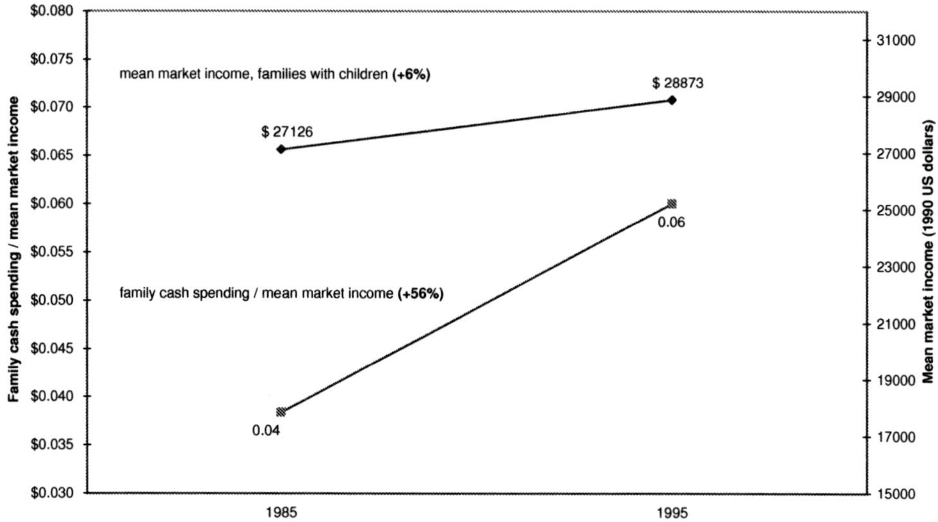

Note: Numbers in parentheses refer to percent change between 1985 and 1995.

Figure 6. Family cash spending relative to mean market income (14-country averages, 1985 and 1995).

terms of mean family market income, i.e., pre-tax-and-transfer income.[9] The figures indicate that across these countries, as a group, mean market income increased modestly, signalling small reductions in pretransfer need, in both elderly headed families and in families with children (with two-parent and single-parent families combined). In 1990 US dollars (PPP-adjusted), on average, elderly families' mean market income rose from $4,706 to $5,064 (an 8 percent increase), whereas the mean market income of families with children rose from $27,126 to $28,873 (a 6 percent increase). Results (not shown) indicate that, for the elderly, increases in market income were mostly due to rising average numbers of earners in elderly headed families, whereas, in families with children, the numbers of earners tended to rise as did average income per earner.

Figures 5 and 6 also show that, in both groups, growth in cash benefit expenditures per person grew much faster than did market income. Old-age cash expenditures per elderly person (which increased by 41 percent) also increased per dollar of market income received by elderly headed families (by 31 percent). Likewise, family cash expenditures per child (which grew by 67 percent) increased per dollar of family market income (by 56 percent). In other words, for each dollar earned in an elderly headed family in 1985, public expenditures on old-age cash benefits (per elderly person) amounted to $1.38; by 1995, that rose to

[9] Market income is post-taxes in a few countries: Belgium, France, Italy, and Luxembourg. That diminishes average family market income in these countries and, in turn, in the 14-country average. At the same time, these analyses are largely over time, so this incomparability has virtually no effect on the results.

\$1.80.[10] For each dollar earned in a family with children, family cash benefits (per child) totaled \$.04 in 1985, rising to \$.06 in 1995.

These findings have two implications. First, on average, these two categories of welfare state expenditure grew during the study years—per potential recipient in both age groups, and on top of modestly rising family market income in both age groups.

Second, although cash benefit expenditures *per dollar earned* are far less for families with children than for elderly families, expenditures per market dollar increased in both groups—and somewhat more so for children and their families. The resilience of family cash spending, relative to market income, is somewhat surprising, because—unlike most old-age cash benefits—a substantial share of family benefits in all countries are income-tested with respect to current income. On the other hand, some family cash benefits, such as paid maternity and parental leave, would increase as more women enter the labour market; rising female employment would both raise family income and push social expenditures up as well. In any case, across these countries as a group, there is scant evidence that the elderly's increasing advantage squeezed out spending on children, even when we take into account shifting levels of need.

A core component of Preston's argument rested on intergenerational poverty trends. He reported that during the 1970s and early 1980s, elderly poverty fell while child poverty rose; and, noting rising maternal employment, he attributed that largely to a withdrawal of public resources from children. Figure 7 reports average poverty rates for the elderly and children, across these countries, at 1985 and 1995. These are post-tax-and-transfer poverty rates, so they reflect a combination of market income and income from the public cash benefits programs at the core of this study. Here we see that—very much as Preston predicted—elderly poverty fell during the study years and child poverty rose, until they met in the middle, at 11.2 percent.

How is the finding of rising child poverty, alongside decreasing elderly poverty, reconcilable with the accumulating evidence that public expenditures on family cash benefits grew during the same years (by 67 percent), as did mean market income among families with children (by 6 percent)? Rising child poverty can be reconciled with these other trends for several reasons. First of all, post-tax-and-transfer income in families with children is primarily shaped by employment patterns and labour market conditions, rather than by public cash benefits; that is not the case for most elderly, who are far more shielded from current market conditions. And two trends during the 1980s and early 1990s raised children's risk of poverty in several of these countries: an increasing share of children lived with one parent, which reduced family market income, and labour market conditions declined, as joblessness rose and wages stagnated or fell. At the same time, in several of these countries, the bottom of the earnings distribution fell further away from the middle. That could result in rising child poverty rates—both absolute poverty, as Preston reported,

---

10 The elderly have two other substantial sources of income as well—income from assets, and from private pensions and annuities. While the LIS data do not include the former, they do include the latter, constituting a fruitful area for further study.

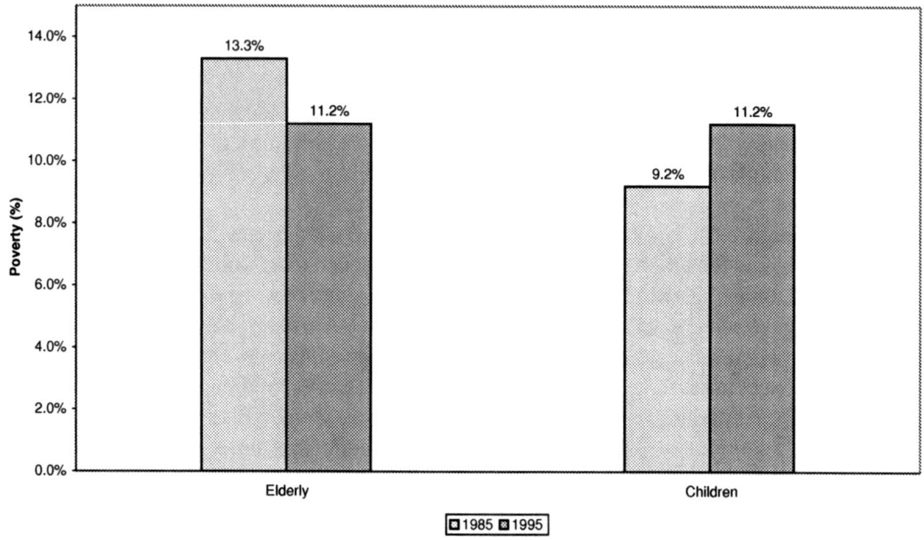

Figure 7. Post-tax-and-transfer poverty rates: elderly versus children (14-country averages, 1985 and 1995).

and relative poverty, as reported here—alongside rising market income at the mean and growing expenditures on family cash benefits (much of which goes to nonpoor families). The implication of this is that the poverty trajectory that Preston reported, and seen here as well, may be driven more by declining labour market conditions and changing family structures than by intergenerational competition and its effect on the allocation of public social welfare resources.

Figures 8 and 9 address the fifth and last research question, actually a pair of questions: Did public resources shift to the elderly (away from children) more in countries where the elderly-to-child population ratio grew the most? And, did public resources shift toward the elderly more where competition for resources was the most heightened? Preston's scenario of demographic "crowding out" would predict that children would lose the most (in resources per child) where the elderly gained the most in population, and he reported evidence of that result across the 50 U.S. states. Furthermore, his framework would suggest that as public social investments grew relatively scarcer, the elderly's gains would come more at the expense of children.

The first of these questions is addressed in Figure 8; the horizontal axis reports change in the ratio of the elderly population to the child population, and the vertical axis reports change in the ratio of old-age cash spending per elderly person to family cash spending per child.[11] (See Appendix 1 for a key to the country abbreviations that appear in Figures 8 and 9.) The

[11] For example, in Belgium, the ratio of the number of elderly to the number of children rose from 0.72 to 0.88, an increase of 23 percent [the x-axis]; the ratio of old-age to family cash spending, per person, increased from 2.95 to 3.96, an increase of 34 percent [the y-axis].

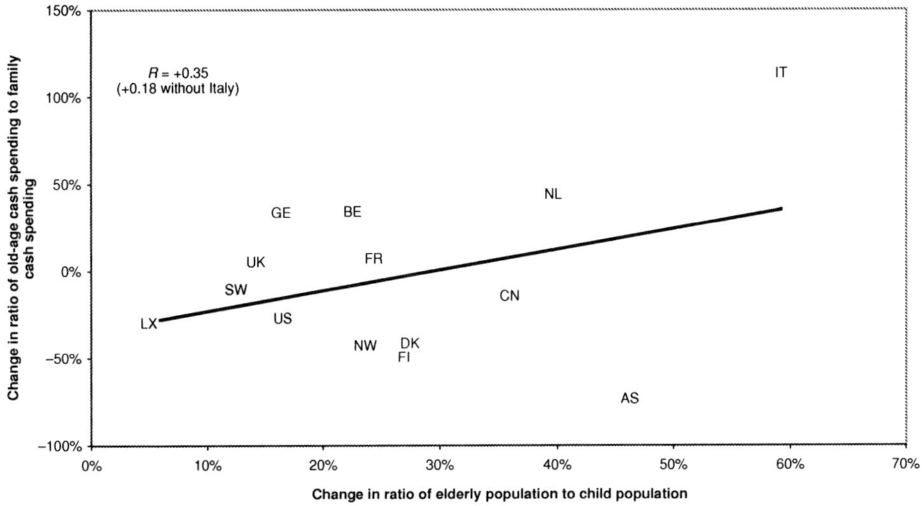

Figure 8. Change in ratio of elderly population to child population and change in ratio of old-age cash spending per elderly person to family cash spending per child (1980–1995).

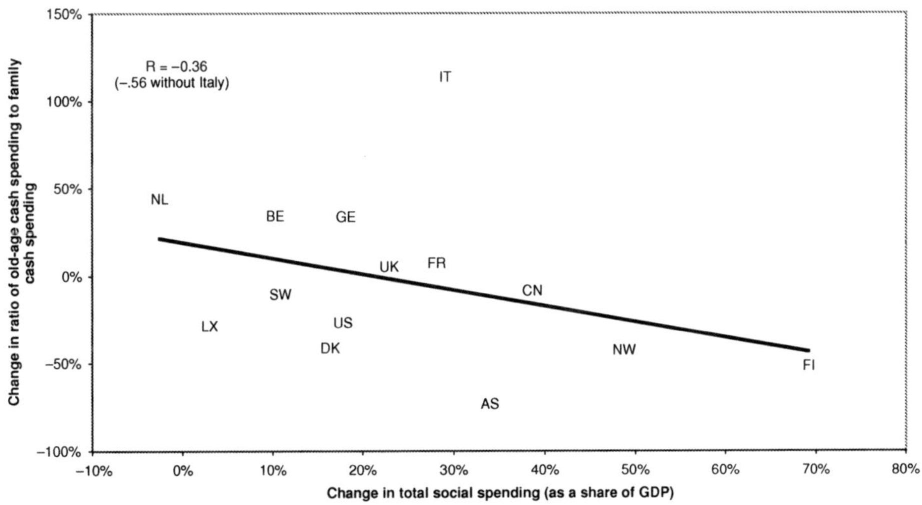

Figure 9. Change in total social spending (as a share of GDP) and change in ratio of old-age cash spending per elderly person to family cash spending per child (1980–1995).

Table 1. Effects of change in population ratio and in total spending on change in ratio of elderly-to-child expenditures.

Dependent variable:	Change in the ratio of old-age cash spending per elderly person to family cash spending per child	coefficient (SE)
Independent variables:	Change in the ratio of the elderly population to the child population	+1.574 (0.648)*
	Growth in total social spending as a share of GDP	−1.208 (0.845)*

$^*N = 14$, $p < 0.10$, $R^2 = 0.34$.

upward slope in Figure 8 lends some support to Preston's underlying thesis: spending per elderly person did, in fact, increase more, relative to spending per child, in those countries where the elderly gained the most in population. (Italy, in the upper right corner, might be the extreme "Preston case"; its elderly-to-child population ratio skyrocketed as did its elderly-to-child spending ratio.) The overall pattern seems, to some extent, to reflect the exercise of political power in the distribution process, as these spending changes reflect changing resources *per person*. While each group's share of total spending might increase more or less automatically with population share, per person spending is more likely to reflect changes in policy rules.

Did public resources shift to the elderly—again, per elderly person—more in countries where total welfare resources grew relatively scarcer, and thus competition was more intense? In Figure 9, the horizontal axis indicates growth in total social spending (as a share of GDP), and the vertical axis is the same as in Figure 8: change in the ratio of old-age cash spending per elderly person to family cash spending per child. The downward slope indicates that spending on the elderly did, in fact, increase more (relative to per child spending) where total social spending was more constrained (i.e., where it grew less).

An ordinary least squares (OLS) regression suggests that the two factors (in Figures 8 and 9) have independent effects (see Table 1).

Furthermore, the two together explain a third of the variation across these 14 countries in the change in elderly-to-child spending ratios ($R^2 = 0.34$).[12]

6. Conclusion

During the 1980s and early 1990s, the industrialized countries of Europe, North America, and Australasia saw their populations age rapidly. Elderly populations grew both relatively and absolutely while child population shares fell, in many cases so sharply that the absolute number of children actually declined. At the same time, these countries' social welfare

[12] Clearly, with an N of 14, regression results are suggestive at best. This regression was also estimated without Australia and Italy, two of the more unusual cases, and the results were largely the same. The coefficient on change in total social spending was again negative and significant ($p < .05$) and the coefficient on change in population ratios was again positive, but no longer significant.

systems operated under tremendous political and fiscal pressure, set in motion during the economic downturn of the early 1970s. According to Preston, the American demographer, when a similar scenario was played out within the United States in the 1970s and early 1980s, the elderly exercised their political power effectively—so effectively that public resources per elderly person increased while expenditures per child fell. Each group's resources turned out not to be fixed but instead elastic, and the elderly's share of resources rose with their population share, in fact *more than* their population share.

This chapter assessed whether a "Preston scenario" unfolded, across a group of 14 welfare states, during a subsequent time period, 1980–1995. Overall, these 14 countries were assessed *as a group*, parallel, to some extent, to Preston's treatment of the United States: he assessed trends in the nation as whole and then turned to variation across its 50 states. Clearly, working with cross-national averages is less standard, but I chose it as an analytic strategy because preliminary analyses revealed more commonality than variation across these countries, even including the United States.

A core finding of this study is that, during the 1980s and early 1990s, a "Prestonian scenario" unfolded—but only partially—across this group of welfare states as they negotiated dramatic population ageing alongside multiple economic and political challenges to their welfare states. On the one hand, there is evidence in favour of such a scenario. Spending on cash benefits for the elderly increased *per elderly person*, even as their numbers rose so sharply. In addition, spending targeted on the elderly gained on child spending more where elderly populations grew the most (and presumably their political power as well), and more where the need to compete for resources was higher. Furthermore, elderly post-tax-and-transfer poverty decreased while child poverty increased.

On the other hand, other findings argue against the conclusion that the elderly pulled resources away from children. On average, spending on family cash benefits *per child* grew as well—and quite dramatically—even as mean market family income rose modestly. In half of the countries, the ratio of old-age cash spending to family cash spending *fell*. Furthermore, while there is evidence that the magnitude of the population shift matters, as well as the overall degree of spending constraint (see Table 1), the majority of the variation in spending patterns remained unexplained.

Another central finding partially accounts for the absence of a clear distributional shift from the elderly to children. Despite widespread perceptions that welfare state rollbacks have been severe in recent decades, total social expenditures actually grew steadily throughout the period across the industrialized countries—both as a share of GDP (by 22 percent) and per capita (by 57 percent). During the study years, social expenditures continued to expand, allowing the elderly to gain more resources—perhaps as their numbers inflated their political power—while state spending on children grew as well.

6.1. FUTURE RESEARCH

Several interrelated lines of inquiry would be especially fruitful for demographers and welfare state researchers. First, while this paper stressed trajectories within this group of welfare states as a whole, much remains to be understood about variation across welfare state models, as well as across individual welfare states. Pampel (1994) concluded that a growing elderly population will be able to translate its numbers into advantageous policy shifts only in relatively pluralist political systems; in more strongly class-based systems, including several in place in Europe, universalist social policy principles blunt the power of the elderly cohort. At the same time, later research finds no consistent relationship between extant welfare state models and the age-orientation of social policies—perhaps due, in part, to the weakening of class-based social welfare politics over time in several welfare states. Lynch (2001) concluded that while countries tend to demonstrate consistent age biases across policy areas, "the age-orientation of social policy as a dimension of distributive politics ... is not captured by other welfare state typologies, suggesting a need to develop new accounts of the development of welfare states that include the dimension of age." Further research on the interplay of welfare state model and resource reallocation would also be enhanced by comparative, historical research focused on the outlying cases—beginning with, perhaps, Italy and Australia, as suggested by Figure 8 in this chapter.

Second, scholars have yet to adequately integrate conclusions about the demographic and political effects of ageing on public resource allocations with new lessons from the welfare state retrenchment literature. As Pierson (1994, 2000) and others have demonstrated, social policies are resistant to change and in complex ways. As a result, welfare state restructuring, and benefit and eligibility cuts in particular, operate under logics unlike those operating during expansionary periods. How the politics of ageing interacts with the dynamics of retrenchment—especially politicians' need to avoid taking blame—remains an open question. It is possible that the two key constituencies exercise a different balance of power in times of contraction than they did during more expansionary years.

Finally, since the 1980s, an extensive theoretical and empirical literature has developed that seeks to integrate gender into welfare state research. Most of it assesses the effects of welfare state features on women and families; a more limited strand analyses the role that gender has played in welfare state development. What is relevant here is that the gendered aspects of intergenerational competition for resources remains largely unexplored—which is particularly problematic, given that women are not only the key actors in family politics, they are also disproportionately represented among the elderly in all industrialized countries. Welfare state scholars working on gender and those concerned with intergenerational competition should move toward inquiries that aim to integrate these two areas of study.

Appendix 1. Notes on Figures.

All Figures:	Country abbreviations are as follows: AS = Australia; BE = Belgium; CN = Canada; DK = Denmark; FI = Finland; FR = France; GE = Germany; LX = Luxembourg; NL = Netherlands; NW = Norway; SW = Sweden; UK = United Kingdom; US = United States.
	Throughout this study, Germany refers to West Germany from 1980 through 1990, and unified Germany after 1990. Percent change always calculated as: (Time 2 – Time 1)/Time 1.
	All currency amounts are expressed in 1990 U.S. dollars, adjusted for purchasing power parities (PPPs). OECD PPPs are available in the SOCX database.

Appendix 1. (*Continued*)

Figure 1: Source: SOCX database. Total social expenditures include old-age cash benefits, disability cash benefits, occupational injury and disease benefits, sickness benefits, services for the elderly and disabled, survivors benefits, family cash benefits, family services, active labour market programs, unemployment benefits, health benefits, housing benefits, and assistance for contingencies (such as benefits for immigrants).

Old-age cash spending includes old age pensions, old age civil service pensions, veterans' old age pensions, early retirement pensions, and other old-age cash benefits.

Family cash spending includes traditional cash transfer programs targeted on families (family allowances for children, family support benefits, and lone parent cash benefits), as well as paid family leave and refundable tax credits for families.

Figure 2: Source: SOCX database.

Figure 3: Source: *Labour Force Statistics*, various years.

Figure 4: Source: Expenditures from SOCX database; population from OECD *Labour Force Statistics*, various years.

Figures 5, 6: Source: Expenditures from SOCX database; mean market income from Luxembourg Income Study (LIS), various datasets.

In the LIS database, time points within the waves vary slightly across countries. The data points presented in Figures 5 and 6 (both the SOCX expenditure data and the LIS microdata) are as follows:

	approximately	
	1985	*1995*
AS	1985	1994
BE	1985	1996
CN	1987	1994
DK	1987	1995
FI	1987	1995
FR	1984	1994
GE	1984	1994
IT	1986	1995
LX	1985	1994
NL	1987	1994
NW	1986	1995
SW	1987	1995
UK	1986	1995
US	1986	1994

Figure 7: LIS, various datasets. Poverty is defined as household income below 50 percent of the median, where a single median is applied for both the elderly and children. Income adjusted for family size is computed as family income divided by the square root of family size.

Denmark is excluded from Figure 7 due to data irregularities.

Figure 8, 9: Source: Expenditures from SOCX database; population from OECD *Labour Force Statistics*, various years.

Appendix 2. Percent Change, 1980–1995, Four Outcome Variables

	Social spending as a share of GDP (%)	Social spending per capita (%)	Old-age spending per elderly person (%)	Family spending per child (%)
Australia	34	71	−4	259
Belgium	10	37	36	1
Canada	39	66	44	68
Denmark	16	54	61	174
Finland	69	109	73	240
France	28	55	49	39
Germany	18	32	13	−16
Italy	29	70	65	−23
Luxembourg	3	83	76	145
Netherlands	−3	23	9	−24
Norway	49	112	69	193
Sweden	11	30	31	46
UK	23	65	62	54
US	18	51	16	59
change in cross-country average	**22**	**57**	**41**	**67**

References

Clayton, R. and Pontusson, J. (1998), "Welfare state retrenchment revisited: Entitlement cuts, public sector restructuring, and inegalitarian trends in advanced capitalist societies." *World Politics*, 51 (October 1998), 67–98.

Daly, M. (1997), "Welfare states under pressure: Cash benefits in European welfare states over the last ten years." *Journal of European Social Policy*, 7(2):129–146.

Dixon, J. (1998), "Comparative social security: The challenge of evaluation." *Journal of Comparative Policy Analysis: Research and Practice*, 1(1):61–95.

Esping-Andersen, G. and Sarasa, S. (2000), "Distributional dilemmas in 21st century welfare states." Unpublished manuscript.

Esping-Andersen, G. and Sarasa, S. (2002), "The generational conflict reconsidered." *Journal of European Social Policy*, 12(1):5–21.

Esping-Andersen, G. (1999), *Social Foundations of Post-Industrial Economies*. Oxford University Press: Oxford, UK.

European Commission (1999), *Towards a Europe for All Ages: Promoting Prosperity and Intergenerational Solidarity*. Commission of the European Communities: Brussels.

Gauthier, Anne H. (1996), *The State and the Family: A Comparative Analysis of Family Politics in Industrialized Countries*. Oxford University Press: Oxford.

Gauthier, Anne H. (1999), "Historical trends in state support for families in Europe (post-1945)." *Children and Youth Services Review*, 21(11/12), 937–965.

Ghilarducci, T. (1997), "Do the old eat the young? Intergenerational equity and public pensions." *Paper presented at the Second International Conference on Social Security, sponsored by the International Social Security Association*, January 23–28, 1998.

Gornick, J.C., and Meyers, M.K. (2001), "Lesson-drawing in family policy: Media reports and empirical evidence about European developments." *Journal of Comparative Policy Analysis: Research and Practice*, 3(1):31–57.

Kalisch, D.W., Aman, T. and Buchele, L.A. (1998), "Social and health policies in OECD countries: A survey of current programmes and recent developments." *Labour Market and Social Policy* Occasional Papers No. 33. Organization for Economic Cooperation and Development, Directorate for Education, Employment, Labour and Social Affairs, Paris.

Kamerman, S. and Kahn, A. (1999), "Child and family policy in an era of social policy retrenchment and restructuring." *Paper prepared for presentation at the Luxembourg Income Study Conference on Child Well-Being in Rich and Transition Countries*, September 30–October 2, Luxembourg.

Kamerman, S. and Kahn, A. (1997), "Investing in children: Government expenditures for children and their families in Western industrialized countries." In: *Child Poverty and Deprivation in the Industrialized Countries, 1945–1995,* Giovanni Andrea Cornia and Sheldon Danziger, eds. Oxford University Press: New York.

Lynch, J. (2001), "The age-orientation of social policy regimes in OECD countries." *Journal of Social Policy*, 30(3):411–436.

Mishra, R. (1990), *The Welfare State in Capitalist Society: Policies of Retrenchment and Maintenance in Europe, North America, and Australia*. University of Toronto Press: Toronto.

Munday, B. (ed.) (1989), *The Crisis in Welfare: An International Perspective on Social Services and Social Work*. St. Martin's Press: New York.

Myles, J. and Pierson, P. (2000), "The comparative political economy of pension reform." In: *The New Politics of the Welfare State*, P. Pierson, ed. pp. 305–333. Oxford University Press: Oxford, UK.

OECD. (various years), *Labour Force Statistics*. Organization for Economic Cooperation and Development: Paris.

OECD. (1999), *Social Expenditure Database, 1980–1996*. Organization for Economic Cooperation and Development: Paris.

O'Higgins, M. (1988), "The allocation of public resources to children and the elderly in OECD countries." In: *The Vulnerable,* J.L. Palmer, T. Smeeding, and B. Torrey, eds. pp. 201–228, Urban Institute Press: Washington, DC.

Orsini, K. (2001), "Yet the poorest, relatively speaking: Italian poverty rates in international perspective." *Luxembourg Income Study Working Paper 261*. Luxembourg Income Study: Luxembourg.

Pampel, F. (1994), "Population aging, class context, and age inequality in public spending." *American Journal of Sociology*, 100:153–195.

Pierson, P. (1994), *Dismantling the Welfare State? Reagan, Thatcher, and the Politics of Retrenchment*. Cambridge University Press: New York.

Pierson, P. (ed.) (2000), *The New Politics of the Welfare State*. Oxford University Press: Oxford.

Ploug, N. (1999), "Cuts in and reform of the Nordic cash benefit systems." In: *Nordic Social Policy: Changing Welfare States*, M. Kautto, M. Heikkilä, B. Hvinden, S. Marklund, and N. Ploug, eds. Routledge: London.

Ploug, N. and Kvist, J. (1996), *Social Security in Europe: Development or Dismantlement?* Kluwer Sovac Series on Social Security, Vol. 3. Kluwer Law International: The Hague.

Preston, S. (1984), "Children and the elderly: Divergent paths for America's dependents." *Demography*, 21(4):435–457.

Rein, M., Esping-Andersen, G. and Rainwater, L. (eds.) (1987), *Stagnation and Renewal in Social Policy*. M.E. Sharpe: Armonk, NY.

Smeeding, T., Torrey, B. and Rein, M. (1988), "Patterns of income and poverty: The economic status of children and the elderly in eight countries." In: *The Vulnerable*, J.L. Palmer, T. Smeeding, and B. Torrey, eds. pp. 89–120. Urban Institute Press: Washington, DC.

Swank, D. (2000), "Political institutions and welfare state restructuring: The impact of institutions on social policy change in developed democracies." In: *The New Politics of the Welfare State*. P. Pierson, ed. pp. 197–237. Oxford University Press: Oxford.

CHAPTER 9. CONSEQUENCES OF EDUCATIONAL CHANGE FOR THE BURDEN OF CHRONIC HEALTH PROBLEMS IN THE POPULATION

MARK D. HAYWARD
The University of Texas

EILEEN M. CRIMMINS
University of Southern California

ZHENMEI ZHANG
Bowling Green State University

Changes in the public and individual burden of chronic health problems have significant implications for the allocation of public and private resources across generations. Preston (1984) noted almost two decades ago that population ageing in the United States was accompanied by the rapid expansion of public programs benefiting the health of elders while public programs benefiting children's education contracted. Health care is the principal public service provided to the elderly while education is the counterpart for children.

Within a historical time period, political choices about the funding of age-targeted service programs have an urgency that oftentimes sweeps aside the fact that investments in children's well-being pay substantial dividends decades later when children become the elders of a population. In large part, this reflects a lack of attention both by policy makers and by demographers of these long-run associations. Here, we provide new insights into the long-run consequences of investments in children for the burden of chronic health problems by conducting a thought experiment in which we simulate how sweeping historical changes in a population's educational achievement potentially alters active life expectancy and the prevalence of functioning problems in the population.

Our thought experiment is based on a multistate life table model that first documents the educational disparities in active life expectancy for males and females with 0–6 years of education compared to persons with 12 years of education. We use morbidity and mortality data from a major longitudinal study of morbidity experience in a population, the Health and Retirement Study (HRS), as inputs for the life table model. An advantage of the model is that it explicitly takes into account how education shapes active life expectancy, as well as the prevalence of functioning problems, through its associations with functional

Anne H. Gauthier et al. (eds.), Allocating Public and Private Resources across Generations, 227–242.

changes and mortality. We then use simulations to illustrate how changes in the aggregate educational attainment and mortality of a population potentially alter active life expectancy and the prevalence of functional problems. The simulations combine historical information on American mortality since 1900 from the U.S. Social Security Administration with information on educational disparities in morbidity and mortality incidence derived from the HRS. We simulate in a backward fashion how changes in the burden of chronic health problems in a population over a 90-year period are associated with declining levels of mortality and higher levels of educational attainment.

Other nations are at various stages of economic development and the spread of mass education. At the same time that national populations are being transformed in terms of their characteristics, the "quiet" demographic revolution of population ageing is drawing policymakers' attention to the possible dilemma of coping with the burden of chronic health problems for burgeoning elderly populations while dealing with the exigencies of economic development. This study attempts to underscore the idea that while these may appear to be zero sum short-run decisions, investments in children potentially have substantial payoffs both in terms of reducing the burden of elders' health problems at the societal level and in improving the quality of life for individuals.

1. Implications of Educational Attainment for Population Health

A consistent finding in the population sciences is the strong association between educational attainment and adult health in cross-sectional data. This relationship holds across the life-cycle even into old age, although some research suggests that the association may diminish at advanced ages as genetic factors take on a more significant role in influencing health (Preston and Taubman 1994). Typically, the association has been gauged in terms of mortality and self-reported health, but research also shows a strong association between education and functional limitations (Crimmins, Hayward, and Saito 1996; Freedman and Martin 1999). The consistency of the findings across numerous studies is not surprising given the strong association between education and the risk of a range of chronic diseases—the primary source of mortality and functional limitations at older ages in developed nations (Hayward et al. 2000).

The association between education and adult health highlights the importance of understanding how shifts in the educational composition of a population are associated with trends in the health problems of a population. Assuming that the association between education and health is relatively stable over time, increases in educational attainment in a population ought to foster improvements in the health of a population (e.g., lower levels of mortality, the postponement of chronic diseases to older ages, and lower levels of functioning problems). Empirical evidence on historical trends in the association between greater levels of educational attainment in a population and health is relatively sparse, limited to relatively short historical periods, and largely restricted to developed nations. Some evidence points to a widening of educational disparities in mortality as average levels of education have increased in Western developed nations (Preston and Taubman 1994). More recent evidence for the United States points to a temporally stable association (Manton and Stallard 1997; Freedman and Martin 1999). There is also evidence of a strong association between education and functional limitations in a developing nation with low average levels of education among its elderly population (Zimmer et al. 1998) and between education and chronic diseases in a sample of U.S. Civil War veterans (Costa 1999). Available evidence, therefore, points to persistent educational disparities in health even in the face of economic development and the institutionalization of mature health care systems. For purposes of our thought experiment, we *assume* that the association between education and adult health remains constant over a relatively lengthy historical period—almost a century. The implications of imposing this assumption for distorting our analysis must await additional empirical studies.

The association between education and chronic health problems is, in large part, a reflection of the creation of health capital over the individual life-cycle. Greater levels of education indicate greater human capital. And, as a relatively exogenous factor in the life-cycle, education fosters good health by improving economic resources over the life-cycle, access to health care, and home and work environments relatively free of risk (Caldwell 1979; Behrman et al. 1991; Feinstein 1993; Freedman and Martin 1999; Hayward et al. 2000; Ross and Wu 1996; Zimmer, Hermalin, and Lin 2002). Education, because of its association with delayed gratification or because educated people are better consumers of biomedical information, is also associated with the avoidance of health risk behaviours such as smoking, obesity, and substance abuse (Winkleby et al. 1992; Brunner et al. 1996; Lynch, Kaplan, and Salonen 1997). Education also reduces the risk of some chronic diseases by fostering psychological resources including a sense of autonomy and control over one's surroundings and social support (Elo and Preston 1992; House et al. 1994; Ross and Mirowsky 1999).

Viewed historically, improvements in a population's level of education reflect improvements in individuals' capacity to reduce risks having a negative impact on long-term health. Not surprisingly, this type of long-term life course perspective has become an important organizational theme at the World Health Organization (WHO). A major program of WHO's Department of Noncommunicable Disease Prevention and Health Promotion (NPH), for example, is *Ageing and the Life Course*. This program emphasizes the importance of making significant investments in children, including improving education, as a primary prevention strategy in reducing chronic diseases in later life (http://www.who.int/hpr/ageing/lifefactsheet.pdf).

The association between education and the levels of adult health in a population is also a reflection of macro level factors embodied in aggregate measures of education (Palloni 1981). The level of educational attainment in a society reflects not only the stock of human capital but also the capacity of a social system to address societal health needs (Easterlin 1997). At the earlier historical stages of rising educational levels, the growth in mass education is made possible by the proliferation of schools. The spread of schools typically coincides with other health-related infrastructural changes in communities such as the introduction of the spread of modern sanitation practices, health care services, and medical technologies—the latter has been posited as having a significant influence on changes in adult health (Preston 1975; Palloni 1981).

Evaluating how *improvements* in a population's educational level lead to *improvements* in health is an indeterminate problem. In all likelihood, the association represents multifarious and complex mechanisms. A useful way to think about a population's educational level is in terms of the "social capacity" for population health—that is, the confluence of individual life course and institutional conditions favourable to improvements in health. Here, we attempt to gain insights into the consequences of improvements in the social capacity for population health by posing the following questions. What is the burden of disease in a population that has relatively low levels of education and high levels of mortality compared to an educated population with low mortality? Does an increasingly educated older population result in improvements in life expectancy and a relatively compressed period of morbidity, or are improvements in life expectancy accompanied by an extended period of morbidity? How are these changes in morbidity and mortality reflected in the prevalence of functional problems in the population?

In large part because of the absence of data, the body of empirical evidence addressing these questions is quite small. A recent study by Freedman and Martin (1999), however, provides some important clues about how population health changes as educational levels rise in a population. Their study examined changes in the prevalence of several major functional limitations in the 65 years of age and older population in the United States between 1984 and 1993. Using a decomposition approach, Freedman and Martin's analysis showed that improvements in educational attainment over the 9-year period were strongly associated with the declines in the prevalence of functional limitations—indeed, education had the greatest effect among a range of demographic and socioeconomic factors. Based on projected prevalence rates, Freedman and Martin also posit that the United States will experience continued declines in the prevalence of major functional limitations with rising educational levels of the future birth cohorts entering old age. If we extend the logic of Freedman and Martin's analysis, their results point to a pattern in which rising educational levels, an indicator of improved social capacity for population health, are accompanied by increased life expectancy, a compression of the years spent with functional problems, and a lower prevalence of functional limitations in the population.

2. Historical and International Differences in Educational Attainment

Americans born in the first half of the 20th century grew up in an era in which mass education became a reality. The median number of years of completed schooling for males 25–34 years of age in 1910 was 7.4; the median for females was 8.1 (Gustavus and Nam 1968). Only 15–16 percent of men and women had completed a high school education or more. Fifty years later, the median years of completed schooling for males and females 25–34 years of age had jumped to over 12.

The educational attainment of the older population in the United States during the latter part of the 20th century largely reflects the lagged effect of the educational attainment of young adults in the first half of the century. Among persons of 70–79 years of age, for example, over 80 percent of males had completed 8 years of schooling *or less* in 1940 (estimates calculated by authors using U.S. Census data). By 1992, this dropped to less than 25 percent (estimates calculated using data from the Assets and Health Dynamics Study). Females experienced a similar improvement in educational attainment. In a historical sense, this change represents the spread of mass education—especially at the secondary level—in the older population of the United States. The effects of mass education were relatively small until after 1960 when the older population showed dramatic improvements in educational attainment. Future improvements in educational attainment have been projected for the next 30 years (Day and Bauman 2000).

Global data show that the levels of educational attainment in developing nations fall far below current levels shown for the United States and other industrialized countries (Ahuja and Filmer 1995; Wils and Goujon 1998). Recent estimates of adult literacy rates in sub-Saharan Africa, Arab countries, and Southern Asia are approximately 60 percent for men and 30–40 percent for women (Wils and Goujon 1998), with lower literacy rates corresponding to larger gender gaps. Although adults in much of the developing world have very low levels of educational attainment—levels comparable to that of the U.S. older population at the start of the 20th century—most regions of the world have experienced significant growth in the educational attainment of young age groups. Southern Asia, the Arab countries, and Latin America are now approaching universal primary education (the exception is a worrisome trend in sub-Saharan Africa) (Wils and Goujon 1998). Ultimately, this trend toward mass education—at least at the primary level—points to gradual increases in the educational attainment of the adult population as members of recent birth cohorts age into adulthood. Globally, this signals an increase in the social capacity for improvements in population health at the older ages (Easterlin 1997).

3. Approach, Data and Measures

3.1. APPROACH

The analytic problem for this study is how to gauge changes in the burden of disease in the older population associated with improvements in education. Our simulation approach to the problem is built around a multistate life table model of active life expectancy in which persons can experience the onset of functional limitations, functioning can be regained, and mortality can occur from either of the two health states. Functional limitations are problems in performing basic physical and cognitive activities used in daily life. Our choice of functional limitations as the measure of the burden of disease is based on the idea that functional problems are the downstream outcomes of fatal and nonfatal chronic disease processes. Functional limitations are also the target of lengthy personal and medical care. In this sense, disease burden takes on a distinct social quality.

The multistate life table model imposes the need for longitudinal data to identify transitions on the state space—in this case, between health states and from the two functioning states to death. At the same time, our analytic goal focuses our attention on *changes* in the burden of disease associated with *improvements* in education. In the ideal world, we would use historical data referencing the morbid and mortal transitions of the type in the life table model along with data on changes in the level of educational attainment. In the absence of such data, we first make use of recent data on educational disparities in the active life expectancy process. Based on the idea of extrapolating historical trends from current observed levels of inequality, educational disparities provide us with a crude approximation of how population health changes as a country moves from a regime of low education to high education.

Our simulation approach also combines information about educational disparities in the active life expectancy process with historical information on mortality changes at the older ages. Our intent is to illustrate the consequences for the burden of disease of a regime of low education and relatively high mortality. As described later, for example, we calculate a life table model for persons aged 50 years and older who have 0–6 years of completed schooling and are subject to the mortality experience of the 1900 and 1940 U.S. population. The 1900 simulation yields a rough approximation of a population's burden of disease when the social capacity for population health is low—e.g., at the early stages of the transition to universal education. We also calculate a life table model for same-aged persons with 12 years of completed schooling subjected to the mortality experiences of the 1992 U.S. civilian noninstitutionalized population. This model provides a rough approximation of the burden of disease after the transition to universal education under a regime of low mortality—a model reflective of the social capacity of most developed nations. We calculate a third model holding the 1900 education and morbidity transition rates constant, but in which 1900 mortality rates are presumed to decline to the levels in 1940. This scenario is used to illustrate what might happen to the burden of disease when mortality in a population declines but improvements in education lag behind—a situation that is likely to occur relatively early in the process of modern disease fighting. We compare these models to evaluate how educational improvements are associated with the burden of disease in the population.

3.2. DATA AND MEASURES

Information on educational differences in the morbidity and mortality experiences of persons aged 50 years and older is based on the first two waves of the Health and Retirement Survey (HRS) and the Assets and Health Dynamics Survey (AHEAD). The HRS and AHEAD are nationally representative, longitudinal surveys of middle-aged and older Americans. The HRS is representative of persons 51–61 years of age and their spouses in 1992. The AHEAD is representative of persons 70 years of age and older and their spouses in 1994. Wave 2 surveys for both the HRS and AHEAD occurred approximately 2 years after the baseline survey. Age-ineligible spouses have no sample weights.

We have pooled the two surveys in order to estimate morbidity and mortality incidence from middle age into advanced old age ($N = 19{,}797$).[1] We ignore the sampling weights for the present analysis, because we use information on age ineligible spouses to estimate the incidence rates for the ages not covered by the two companion surveys (ages 50 and 62–69 at baseline).

Functional limitations are measured by a series of overlapping items in both the HRS and AHEAD. We use five measures to identify whether a person has functional limitations. The measures are:

- Walking several blocks
- Picking up a dime from a table

[1] The total sample size is 19,799. Two respondents did not provide sufficient information to classify their functioning.

- Climbing stairs without resting
- Lifting or carrying weights over 10 pounds
- Pulling or pushing large objects like a living room chair.

Mobility functioning is referenced by walking several blocks and climbing stairs. Strength measures include lifting weights, and pulling/pushing large objects. The measure, picking up a dime from a table, reflects functional limitations in fine motor skills. For the present, we ignore the possible hierarchical nature of these items.

Persons are defined as "inactive" if they report that *three or more* of these activities are:
AHEAD survey: a little difficulty/very difficult/cannot do
HRS survey: some difficulty/a lot of difficulty.

The "inactive" status thus references the confluence of multiple functional limitations, tapping into the idea of severity. We chose this approach rather than focusing on the severity levels defined by the specific item responses because of comparability problems across the surveys and over time. Although we have not conducted an exhaustive assessment of the shortfalls of our measurement approach, our results are highly consistent with results from prior research using a variety of measurement approaches (Rogers, Rogers, and Belanger 1989, 1990; Crimmins, Hayward, and Saito 1994).

Education is measured in terms of years of completed schooling. Education is categorized into 0–6 years, 7–8 years, 9–11 years, 12 years, 13–15 years, and 16 or more years of schooling. As noted above, we are especially interested in the lowest educational group and those persons who have completed 12 years of education. These groups roughly approximate the educational levels pre and post the spread of universal education.

As described later, educational disparities in morbidity and mortality experiences are directly estimated using the HRS/AHEAD data. Mortality rates for earlier historical periods, 1900 and 1940, are estimated by increasing the age-specific mortality rates by 200 percent and 225 percent. This produces life expectancies for persons aged 50 that closely approximate published life expectancies for 1900 and 1940 produced by the Social Security Administration (Bell, Wade, and Goss 1992).

3.2.1. Estimation of Transition Rates and the Multistate Life Tables

We estimate the transition rates for the multistate life table model using a hazard modelling approach (Hayward and Grady, 1990; Land, Guralnik, and Blazer, 1994). In this approach, the instantaneous transition rate, $\mu_{ij}(x)$, is the force of transition from state i to state j. The rate is defined as:

$$\mu_{ij}(x) = \lim_{\Delta x \to 0} \frac{p_{ij}(x, x + \Delta x)}{\Delta x} = \mu_{ijx} \qquad (0.1)$$

Note that the transition rate is specified to be equal to a constant quantity, μ_{ijx}, for individuals aged x to $x + n$, but may vary across different ages x. This is a piece-wise exponential transition rate model.

We estimate the transition rates using a log-linear modelling approach, i.e.,

$$\ln \mu_{ijx} = \beta_0 + \beta_1 \text{AGE}_x + \beta_2 \text{Male} + \beta_3 \text{Educ}_{0-6} + \ldots + \beta_7 \text{Educ}_{13-15} \qquad (0.2)$$

where the education group of 16 years or more of completed schooling is treated as the reference group. The parameter estimates of the statistical model are then used to calculate predicted transition rates, $m^*(x)$, based on an exponential function of the log-linear equation. The predicted rates serve as the inputs for the multistate life table model.

It is possible to mimic a life table by estimating $m^*(x)$ for single years of age. This would entail creating age-specific dummy variables for each single year of age (minus 1). Exponentiating the parameter estimates would yield age-specific transition rates. Here, we take a somewhat different approach. Age is assumed to be continuous, and (in the above equation) the log of the rate is specified to be a linear function of age. This is similar to the exponential smoothing techniques used in life table construction, only now applied to individual level data.

We tested a variety of functional forms of age-dependency in the transition rates (e.g., polynomial and Weibull specifications) to determine the "best fitting" model for each transition rate. We also tested for nonproportionality in the effects of sex by interacting sex and age. We found no statistical evidence that the effects of sex on the transition rates differed significantly by age. The Gompertz specification shown in equation 1.2 consistently provided the best fitting model for all of the morbid and mortality transitions.

All of the final hazard models used to calculate the transition rates for the multistate life tables include the effects of sex and education, regardless of whether the effects are statistically significant. We based our decision on nonstatistical reasons. First, we were concerned that nonsignificant effects could have a substantive impact, if the effect was persistent over a long period of time (e.g., we consider 50 years of life in this analysis). Second, we wanted to allow for the possibility of reinforcing nonsignificant effects across multiple transitions. Presently, we are engaged in methodological work using a bootstrapping approach to develop estimates of the standard errors for the multistate life table state probabilities and life expectancies based on the variance–covariance matrix of the hazard model parameter estimates.

Using the transition rates as inputs, we first calculated sex-specific life tables for persons with 0–6 years of education and persons with 12 years of education. For each education–sex group, the life table cohort (100,000) was allocated to the functioning states at the radix age 50 according to the prevalence observed for persons aged 45–54 (centred on age 50) at baseline. Persons in each functioning state at age 50 were then subjected to the transition rates. Next, we recalculated the life tables using mortality rates that were adjusted to produce life expectancies approximating those observed in 1940 and 1900 (by sex).

We rely primarily on the years of life a life table cohort can expect to spend in each functioning state as our metric for the life-cycle burden of disease in a population. That is, what is the expected number of years the average person at a given age could expect to live with and without functional limitations, given the mortal and morbid conditions defined by the hazard rates. We also show the corresponding age distribution of functional limitations for each of the life table cohorts to gauge the societal level of burden of disease, i.e., the health conditions of a population that a society must grapple with at a particular time point.

3.3. FINDINGS

3.3.1. Educational Disparities in the Burden of Disease Process

The hazard model results shown in Table 1 provide strong evidence of the associations between the level of education, functioning changes, and mortality. Relative to persons with 16 years of education or more (i.e., the reference category), persons with 0–6 years of education have higher rates of functional loss (2.67 times higher) and higher rates of death among persons having no major functional problems (1.63 times higher). Not surprisingly, the gap between the lowest education group and persons completing high school is not as great. Persons with the lowest education have a rate of functional loss that is 1.64 times greater than persons who completed high school (this effect is statistically significant). Among active persons, persons with the lowest level of education face a risk of death that is 1.15 times higher than persons completing high school.[2]

The results also illustrate that education affects the burden of disease primarily through its association with the onset of functional problems and mortality. Once functional problems occur, however, education does not appear to influence subsequent improvements in functioning or mortality. The education effect, in a sense, is played out relatively early in the disease process (a similar finding was reported by Zimmer et al. 1998). This pattern differs from that defined by sex, where sex effects operate in a synergistic fashion throughout the process. Females are more likely to experience the onset of functional limitations and less likely to recover or experience death, pointing to a lengthy period of functional limitations for females compared to males.

3.3.2. Active Life Expectancy Under Conditions of Improving Mortality and Education

The right-hand side of Table 2 illustrates the *current* disparity in active and inactive life expectancy separating persons with only a primary education from those completing high school. Note that the life table results substantively reflect the statistical associations shown in Table 1. Men aged 50 years, for example, with only a primary education can expect to live 2 years less than high school educated men. Primary school men are even more disadvantaged in terms of active life—they can expect more than 3 years *less* of active life than high school men. Inactive life is 1.5 years longer for primary school men compared to high school men even though primary school men at this age have a 2-year shorter

[2] The pattern of results shows differing and (for some events) non-linear associations between the risk of an event and education. For example, persons with 9–11 years of education have the lowest rate of functional recovery and the highest rate of death among persons who are active. We do not focus on the substantive interpretation of these patterns given that the focus of our thought experiment is based on a contrast of high and low educational groups. Nonetheless, these results point to potential negative health consequences for Americans of not completing high school.

Table 1. Hazard models of changes in functional status and mortality.

Transition	Intercept	Age	Female	Educ 0–6	Educ 7–8	Educ 9–11	Educ 12	Educ 13–15
Active to inactive	−7.1681	0.047**	0.3896**	0.9834**	0.9419**	0.9474**	0.4913**	0.4445**
	(0.2027)	(0.0026)	(0.0593)	(0.1325)	(0.1264)	(0.1184)	(0.1143)	(0.1263)
Inactive to active	−0.7672	−0.0066**	−0.183**	−0.126	−0.188	−0.2069*	−0.0983	0.0212
	(0.1464)	(0.0018)	(0.0466)	(0.0944)	(0.0968)	(0.0919)	(0.0873)	(0.0986)
Active to death	−10.3593	0.0865**	−0.5384**	0.4882*	0.4963*	0.6618**	0.3504*	0.0776
	(0.3737)	(0.0047)	(0.0999)	(0.2073)	(0.1957)	(0.1830)	(0.1747)	(0.2070)
Inactive to death	−6.1551	0.0522**	−0.6298**	−0.0523	−0.0646	0.0060	0.0161	0.1142
	(0.3136)	(0.0036)	(0.0854)	(0.1770)	(0.1793)	(0.1771)	(0.1720)	(0.1929)

**p < 0.01 *p < 0.05

Table 2. Simulations of active life expectancy in the United States under conditions of improving mortality and educational attainment: civilian non-institutionalized population.

Age	1900 mortality with 0–6 years of education			1940 mortality with 0–6 years of education			1992 mortality with 0–6 years of education			1992 mortality with 12 years of education		
Males	Total	Active	Inactive	Total	Active	Inactive	Total	Active	Inactive	Total	Active	Inactive
50	20.60	18.17	2.43	22.65	19.90	2.76	28.32	24.12	4.19	30.41	27.72	2.69
60	14.14	12.45	1.68	14.98	13.11	1.87	20.59	17.28	3.31	22.22	19.93	2.29
70	8.70	7.39	1.31	9.34	7.86	1.48	13.76	10.99	2.78	15.06	13.11	1.95
80	4.91	3.95	0.97	5.34	4.24	1.10	8.51	6.30	2.21	9.42	7.86	1.56
90	2.58	1.89	0.69	2.84	2.04	0.79	4.67	3.03	1.65	5.51	4.31	1.21
Females												
50	24.73	19.67	5.06	25.89	21.29	5.60	33.32	24.82	8.50	35.24	29.14	6.09
60	17.52	13.53	3.99	18.50	14.13	4.37	25.06	17.82	7.24	26.70	21.40	6.30
70	11.37	8.14	3.23	12.15	8.58	3.57	17.68	11.45	6.23	19.00	14.37	4.63
80	6.85	4.37	2.48	7.04	4.34	2.69	11.77	6.65	5.11	12.70	8.82	3.88
90	3.91	2.07	1.83	4.30	2.23	2.07	7.63	3.56	4.08	8.20	5.04	3.15

life expectancy. Education under current conditions both extends total life, active life, and compresses the period of life with functional problems.

The expectancies based on the 1900 mortality schedule for a population with primary schooling or less (see the results in the far left of Table 2) present the situation currently faced by many developing countries. In most of these countries, older populations have a low level of educational attainment; and life expectancy is low compared to that for developed nations. Indeed, the life expectancy estimates using the 1900 mortality schedule are comparable to those for same-aged persons in the North African countries of Algeria and Egypt; the South Asian countries of Bangladesh, Indonesia, and Nepal; the Latin American countries of Bolivia, Ecuador, and Paraguay; and the sub-Saharan African countries of Gambia, Senegal, and Nigeria.[3]

We compare the expectancies based on the 1900 mortality and 0–6 years of schooling to the expectancies based on 1992 mortality and 12 years of schooling, to evaluate how the burden of disease shifts as a population's social capacity for health improves. Note first that the expected number of years with a functional problem increases slightly over the 92-year period. A man aged 50 in 1900 with 0–6 years of education could expect to live 2.4 years with a major functional problem. In 1992, the 50-year-old who completed high school could expect to live 2.7 years with a functional problem. Making this comparison among women, inactive life expectancy rose from 5 years to 6 years.

The major change, however, is in terms of active life expectancy. The simulated population's transformation from low levels of educational attainment and high mortality to high levels of achievement and low mortality resulted in a 9.5-year gain in active life among men aged 50 and almost a 10-year gain in men's total life expectancy. In relative terms, 97 percent of the gain in life expectancy among men aged 50 was in the years of life *without* a major functional limitation. Gauged in this sense, a major transformation in a population's social capacity for health appears to have profound and beneficial effects on the burden of disease. In our simulation, the burden of disease clearly fell in terms of the relative number of years a person can expect to have a functional problem. Given that statistical models were the basis on which the life expectancy estimates were calculated, the results for women parallel those for men.

The prevalence of functional limitations also fell for the 1900–1992 comparison. This is shown in Figure 1 that presents the prevalence rates for males implied by the incidence rates governing the mortality and morbidity experiences of the life table cohorts. The bottom line represents the implied prevalence of functional limitations for the population defined by the 1992 mortality rates and 12 years of education. The middle line represents the prevalence rates for the population defined by the 1900 mortality rates and 0–6 years of education. One way of gauging the magnitude of the fall in prevalence rates is to compare the populations in terms of the ages at which certain levels of functional problems are

[3] The estimates for life expectancy and educational attainment were obtained from the International Data Base developed by the U.S. Census Bureau. The web address for the IDB is http://www.census.gov/ipc/www/idbnew.html.

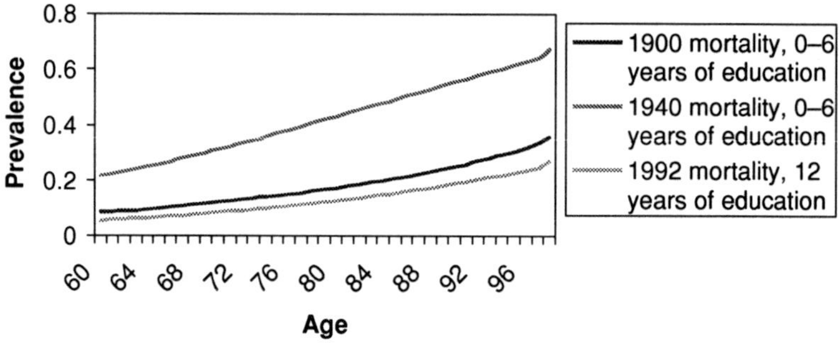

Figure 1. Prevalence of functional limitations, meals: life table cohorts.

reached. The prevalence rate for men aged 70 in 1900, for example, is 12 percent, and this level of functional limitations is not reached until age 78 in the 1992 population. Roughly 8 years separates the two populations in terms of when the prevalence of functional problems in the posttransition population is equivalent to that in the pretransition population.

Holding education and morbidity transition rates in the population constant, we also calculated a life table in which mortality was presumed to decline by 25 percent. In the United States, this approximates the mortality change that occurred from 1900 to 1940. We held education constant in order to approximate what might happen to the burden of disease when mortality is reduced in a population but improvements in education lag behind. As one can observe in Table 2, the pattern in terms of the relative number of years lived with a functional problem is roughly the same as that shown for 1900. Men aged 50 years, for example, can expect to live 12 percent of their total life expectancy (2.76 years out of 22.65) with a functional limitation. A mortality increase of 25 percent, given the base mortality shown for 1900, apparently has little consequence for the burden of disease in terms of disease experience over the life-cycle.

Turning back to Figure 1, however, we can see that the change in mortality, although not substantially altering health expectancies, has a strong effect on the prevalence of functional problems. The prevalence is dramatically shifted upward when mortality declines in a poorly educated population, under the assumption that an equitable decline in mortality occurs across the functioning states. Although the simulated mortality change appears to have had little consequence for the burden of disease experience over the life-cycle, the societal burden of disease is increased dramatically.

Our simulations are intended to reflect a plausible scenario of societal and demographic change. During the length of time spanning our simulations, we assume it is unlikely that medical progress in disease diagnosis and treatment or the lifestyle effects of education produce smooth and evolutionary improvements in the burden of disease. Progress across various diseases is likely to be uneven given the societal decisions about the allocation of health care resources, the nature of scientific discovery, and changes in population composition (Crimmins 1996; Hayward, Crimmins, and Saito 1998). Verbrugge (1989) also notes that in the early stages of fatal chronic disease fighting, medical care is often aimed at managing the fatal consequences of a disease. That is, the death of a person with a major health condition is postponed. This has the effect of extending the period of inactive life (Crimmins, Hayward, and Saito 1994; Hayward et al. 1998). As diseases become understood better, later stages of disease fighting focus on prevention and the postponement of disease onset. As prevention becomes the dominant focus in disease fighting, active life is extended and inactive life is compressed. Over a long-term period involving dramatic changes in a population's education and mortality, our basic point is straightforward.

At any one time we are likely to see improvements in some indicators of health and not others, and improvements in some age groups and not others (Crimmins 1996).

Here, we have shown that changes in the life-cycle burden of disease need not parallel changes in the societal disease burden.

4. Conclusions

The pace of fertility and mortality declines and the growth in educational attainment in most developing countries has been extraordinary since the end of World War II, and the trends show signs of continuing well into the present century (Ahuja and Filmer 1995; Easterlin 1997; Wils and Goujon 1998; Hayward and Zhang 2001). These trends will lead to significant population ageing on a global scale, an increase in the world's stock of human capital, and an increase in the social capacity for population health. Although this represents the world's demographic coming of age and an expansion of capabilities in economic development undertakings, many developing nations will nonetheless face the pressing dilemma of coping with the immediate demands of a rapidly growing older population while grappling with the necessity of increasing the stock of human capital among its younger population. Within the United States, this has led to age-targeted public policies pitting in a zero sum fashion the needs of the nation's younger population against those of its elderly citizens (Preston 1984).

There is growing recognition, however, that the education programs benefiting children have far-reaching multiplier effects and that some of these effects are manifested decades later in the reduction of major chronic diseases such as cardiovascular diseases or diabetes. Scientific evidence on this association has been growing rapidly over the past decade (Elo and Preston 1992; Kuh and Wadsworth 1993; Kuh and Davey Smith 1997; Kuh et al. 1997); and this evidence has fostered the development of policy related health programs such as WHO's Ageing and Life Course Program in the Department of Noncommunicable Disease Prevention and Health Promotion. Although public policy programs in the United States are still largely age-targeted, there is growing recognition that health care policies benefiting the elderly are inextricably tied to policies benefiting children. One of the mission's of WHO's Ageing and Life Course Program is to provide guidance for the development of these types of public policies.

Our analysis showcases the implications of the long-term investment in a population's social capacity for the burden of disease in the older population. As we have shown, populations that undergo a major increase in social capacity are likely to experience a substantial decline in the burden of disease. Many years of life are added to the average life-cycle and the vast majority of these years appear to be without major functional limitations. Prevalence rates of functional problems also drop substantially, reducing the short-term demand for protracted health care.

We have also cautioned against inferring that the changes in population health are necessarily linear *during* the transition in social capacity. In a crude sense the 1900–1992 simulation represents the beginning and ending equilibrium states, with disequilibrium defining much of the period in between. This point was exemplified by our 1940 simulation showing that reduced mortality without an improvement in the level of education led to a significant upswing in the prevalence of functional limitations in the population. Although the long-term consequence of the social capacity transition appears to be more years of life in good health and lower levels of functioning problems in the population, the march toward the reduced burden of disease need not be smooth and unidirectional. This reinforces the need to carefully lay out the scientific basis for long-term public policies aimed at reducing the burden of disease in the population, and it heightens the importance of a commitment to a long-term statistical monitoring of population health.

Preston (1984) argued that society as a whole gains more from a life-course perspective than a generational perspective. We concur and view this as a philosophical cornerstone in formulating public policies aimed at reducing the burden of disease in the population. Investing in a population's human capital through education is not only sound policy for economic development, but is also sound policy, both directly and indirectly, for the health of individuals and societies. Individuals gain longer, healthier lives while the collective costs of health care to future generations of elderly are reduced. This is particularly important given that as future cohorts of children become better educated, they may expect to live to older, and still older, ages.

5. Acknowledgements

Support for this research was provided by grants from the National Institute on Ageing (R01 AG11235 and R01 AG11758) and the National Institute on Child Health and Human Development (1 R24 HD41025).

References

Ahuja, V. and Filmer, D. (1995), "Educational attainment in developing countries: New estimates and projections disaggregated by gender." A Background Paper for the World Development Report 1995.

Behrman, J., Sickles, R., Taubman, P. and Yazbeck, A. (1991), "Inequalities in black-white mortality." *Journal of Econometrics*, 50:183–204.

Bell, F.C., Wade, A.H. and Goss, S.C. (1992), *Life Tables for the United States Social Security Area 1900–2080*, Actuarial Study No. 107. Social Security Administration Office of the Actuary: Baltimore, MD.

Brunner, E., Smith, G.D., Marmot, M., Canner, R., Beksinska, M. and O'Brien, J. (1996), "Childhood social circumstances and psychosocial and behavioural factors as determinants of plasma fibrinogen." *Lancet*, 347:1008–1013.

Caldwell, J.C. (1979), "Education as a factor in mortality decline: An examination of Nigerian data." *Population Studies*, 33:395–413.

Costa, D.L. (1999), "Understanding the Twentieth Century decline in chronic conditions among older men." *Demography*, 37:53–71.

Crimmins, E.M. (1996), "Mixed trends in population health among older adults." *Journal of Gerontology: Social Sciences*, 51B:S223–S225.

Crimmins, E.M., Hayward, M.D. and Saito, Y. (1994), "Changing mortality and morbidity rates and the health status and life expectancy of the older population." *Demography*, 31:159–175.

Crimmins, E.M., Hayward, M.D. and Saito, Y. (1996), "Differentials in active life expectancy in the older population." *Journal of Gerontology: Social Sciences*, 51B:S111–S120.

Day, J.C. and Bauman, K.J. (2000), "Have we reached the top? Educational attainment projections of the U.S. population." Working Paper Series No. 43. Poulation Division, U.S. Census Bureau. Washington, DC.

Easterlin, R.A. (1997), *Growth Triumphant: The Twenty-first Century in Historical Perspective*, The University of Michigan Press: Ann Arbor.

Elo, I.T. and Preston, S.H. (1992), "Effects of early-life conditions on adult mortality: A review." *Population Index*, 58:186–212.

Feinstein, J.S. (1993), "The relationship between socioeconomic status and health: A review of the literature." *The Milbank Quarterly*, 71:279–322.

Freedman, V.A. and Martin, L.G. (1999), "The role of education in explaining and forecasting trends in functional limitations among older Americans." *Demography*, 36:461–473.

Gustavus, S.O. and Nam, C.B. (1968), "Estimates of the 'true' educational distribution of the adult population of the United States from 1910 to 1960." *Demography*, 5:410–421.

Hayward, M.D., Crimmins, E.M., Miles, T.P. and Yu, Y. (2000), "The significance of socioeconomic status in explaining the racial gap in chronic health conditions." *American Sociological Review*, 65:910–930.

Hayward, M.D., Crimmins, E.M. and Saito, Y. (1998), "Cause of death and active life expectancy in the older population of the United States." *Journal of Aging and Health*, 10:192–213.

Hayward, M.D. and Zhang, Z. (2001), "The demographic revolution in population aging: A century of change, 1950–2050." In: *Handbook of Aging and the Social Sciences, Fifth Edition*, R.K. Binstock and L.K. George, eds. pp. 69–85. Academic Press: New York.

House, J.S., Lepkowski, J.M., Kinney, A.M., Mero, R.P., Kessler, R.C. and Herzog, R.A. (1994). "The social stratification of aging and health." *Journal of Health and Social Behavior*, 35:213–234.

Kuh, D. and Smith G.D. (1997), "The life course and adult chronic disease: A historical perspective with particular reference to coronary heart disease." In: *A Life Course Approach to Chronic Disease Epidemiology*, D. Kuh and B. Ben-Shlomo, eds. pp. 15–41. Oxford University Press: New York.

Kuh, D., Power, C., Blane, D. and Bartley, M. (1997), "Social pathways between childhood and adult health." In: *A Life Course Approach to Chronic Disease Epidemiology*, D. Kuh and Y. Ben-Shlomo, eds. pp. 169–198. Oxford University Press: New York.

Kuh, D. and Wadsworth, M.E.J. (1993), "Physical health status at 36 years in a British national birth cohort." *Social Science and Medicine*, 37:905–916.

Lynch, J.W., Kaplan, G.A. and Salonen, J.T. (1997), "Why do poor people behave poorly? Variation in adult health behaviours and psychosocial characteristics by stages of the socioeconomic lifecourse." *Social Science and Medicine*, 44:809–819.

Manton, K.G. and Stallard, E. (1997), "Health and disability differences among racial and ethnic groups." In: *Racial and Ethnic Differences in the Health of Older Americans*, L.G. Martin and B. Soldo, eds. pp. 285–300. National Academy Press: Washington, DC.

Palloni, A. (1981), "Mortality in Latin America: Emerging patterns." *Population and Development Review*, 7:623–649.

Preston, S.H. (1975), "The changing relation between mortality and level of economic development." *Population Studies*, 29:231–248.

———. (1984), "Children and the elderly: Divergent paths for America's dependents." *Demography*, 21:435–457.

Preston, S.H. and Taubman, P. (1994), "Socioeconomic differences in adult mortality and health status." In: *Demography of Aging*, L.G. Martin and S.H. Preston, eds. pp. 279–318. National Academy Press: Washington, DC.

Rogers, A., Rogers, R.G. and Belanger, A. (1990), "Longer life but worse health? Measurement and dynamics." *The Gerontologist*, 30:640–649.

Rogers, R.G., Rogers, A., and Belanger, A. (1989), "Active life among the elderly in the United States: Multistate life-table estimates and population projections." *The Milbank Quarterly*, 67:370–411.

Ross, C.E. and Mirowsky, J. (1999), "Refining the association between education and health: The effects of quantity, credential, and selectivity." *Demography*, 36:445–460.

Ross, C.E. and Wu, C.L. (1996), "Education, age, and the cumulative advantage in health." *Journal of Health and Social Behavior*, 37:104-20.

Verbrugge, L.M. (1989), "Recent, present, and future health of American adults." *Annual Review of Public Health*, 10:333–361.

Wils, A. and Goujon, A. (1998), "Diffusion of education in six world regions, 1960–1990." *Population and Development Review*, 24:357–368.

Winkleby, M.A., Jatulis, D.E., Frank, E. and Fortmann, S.P. (1992), "Socioeconomic status and health: How education, income, and occupation contribute to risk factors for cardiovascular disease" [see comments]. *American Journal of Public Health*, 82:816–820.

Zimmer, Z., Hermalin, A.I. and Lin, H.-S. (2002), "Whose education counts? The added impact of adult-child education on physical functioning of older Taiwanese." *Journal of Gerontology: Social Sciences*, 57:S23–S32.

Zimmer, Z., Liu, X., Hermalin, A. and Chuang, Y.L. (1998), "Educational attainment and transitions in functional status among older Taiwanese." *Demography*, 35:361–375.

CHAPTER 10. EFFECTS OF CHANGING AGE STRUCTURE AND INTERGENERATIONAL TRANSFERS ON PATTERNS OF CONSUMPTION AND SAVING

DIANE J. MACUNOVICH
Department of Economics, University of Redlands, California

1. Introduction

This study examines the issue of age-structure transitions and intergenerational transfers from a macro rather than a micro perspective: to what extent might changing age structure, brought about by the demographic transition and subsequent baby booms and busts, have contributed to economic fluctuations experienced in countries around the globe in the last century? And to what extent might intergenerational and interhousehold transfers have confounded attempts to quantify this relationship? These issues are explored here using data for the United States throughout the 20th century, as the United States moved from "developing" to "developed" status.

The post World War II baby boom—and then bust—in many Western nations, as illustrated in Figure 1, provides a natural experiment for examining the effects of a significant population "bulge" as it moves through the life-cycle. A simple life-cycle model suggests that there is a marked age-related fluctuation in the proportions of income consumed and saved over the life-cycle, as illustrated in Figure 2, which when superimposed on Figure 1 would suggest major shifts in patterns of expenditure and savings. Individuals are thought to overspend relative to their incomes (dis-save) at younger ages and again in retirement, with a period of saving during the prime years. If such a pattern exists at the micro level, does it carry through when the data are aggregated over individuals, and if so how significant is it in affecting macro level economic variables?

Similar issues have been addressed in a wide range of studies focused on adults aged 15 or 20 and older, like those of Angus Deaton and Christina Paxson. (See, for example, Deaton and Paxson 1997.) But another strand of the literature on these types of age structure effects has focused on the impact of children on savings rates and economic growth in less developed countries (LDC): what is referred to as the "dependency effect" (Leff 1969; Mason 1981, 1988; Fry and Mason 1982; Collins 1991; Kelley and Schmidt 1995). There are no clear answers there, however: it appears that the literature has highlighted two puzzles in recent years, both of them apparent reversals in the sign of the impact. The first is a very well documented reversal in the estimated effect of youth dependency on savings and

Anne H. Gauthier et al. (eds.), Allocating Public and Private Resources across Generations, 243–277.

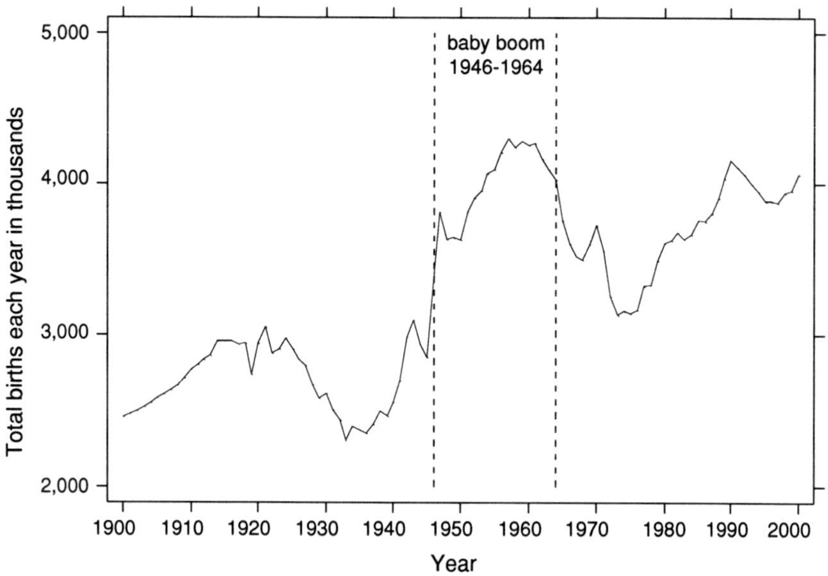

Figure 1. Pattern of births in the United States in the 20th century.

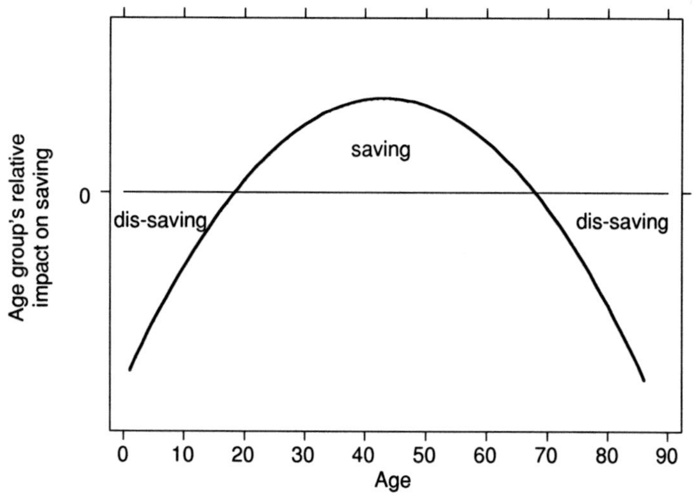

Figure 2. The hypothesized "life-cycle" pattern of consumption and saving, as a share income.

economic growth worldwide: what was a negligible or even benign effect in the 1960s and 1970s seems to have turned negative in the 1980s (Kelley and Schmidt 1994). The second is an apparent variation in the sign of the effect, in moving from LDCs to DCs (developed countries). That is, while these studies in general support the idea of a significant negative (short-term) effect of high birth rates on savings rates and economic growth rates in lower income countries, they suggest that this negative effect may be ameliorated as income levels rise.[1] Does this mean that little can be learned from the study of dependency rates in LDCs, to assist in estimating age structure effects in Western nations, and vice versa?

Not necessarily. Given the right data set it is possible to control for—and even quantify—age-specific effects that change with level of development, although this has not, to the author's knowledge, been done in any studies to date. Inconsistent estimates of the dependency effect may be due to inadequate measures of population age structure. Until very recently the standard measure in the dependency literature was simply the population share aged 0–14—or that share relative to the share aged 15–64—and there seems reason to suspect that measure might be capturing something other than, or in addition to, youth dependency. Why? Because the sizes of various age groups tend to be highly correlated in any population.[2] Parallel movements in multiple age groups make it impossible to determine if low savings rates are actually caused by the dependents themselves, or by adult age groups with whom their size is correlated, without adequate controls for the relative sizes of those adult age groups.

In addition, recent research suggests that even the 0–14 age group cannot be treated as homogeneous, although this is a standard approach in the dependency literature. The ratio of expenditure to income is significantly affected by the age of children in a family, even holding constant the age and real income of adults in that family (Lazear and Michael 1988). Holding income and other factors constant, parents in the United States are found to spend about 15 percent more on teens as opposed to younger children; this result is supported by data reported in Lino (1998). Thus, in addition to controlling for many heterogeneous adult age groups in dependency studies, it may be necessary to recognize not just the presence of children, but also their age distribution.

None of this is to say that there is no negative youth dependency effect. The point is that it seems doubtful that we can even know if there is any effect at all, much less its direction, until or unless we can control for more than just one or two age groups. Fair and Dominguez (1991) introduced a unique way of doing just that, and it seems pertinent that two recent studies making use of their technique did not seem to find "inconsistent" results between LDCs and DCs, but rather variations in effects as children aged. Higgins and Williamson (1997) examine what they term "population dynamics" using the full age distribution, examining the effects of a young population as it passes from childhood to working age. In doing so they identified what appear to be very significant economic effects: an initial period of what they term a "demographic burden" (when the new population is very young) followed by a "demographic gift" as the additions to the population age into productive

[1] Kelley and Schmidt (1995:544), in their extensive analyses of 99 DCs and LDCs between 1960 and 1990, found that the *overall net effect* of population growth on economic growth sometimes appears to move from negative to positive as per capita income rises—at least prior to the 1980s. They used current and lagged birth and death rates to attempt to capture dynamic age structure effects that they felt were missed by more static equilibrium analyses based on more traditional measures of dependency, like the share of population aged 0–14. But even some of the more traditional analyses seemed to find differential effects, for example Collins (1991); Taylor (1995); and Taylor and Williamson (1994).

[2] Examples of such correlations in the U.S. population, and their changes over time are presented in Macunovich (2001).

maturity. As in Higgins (1998), an initial period of rapid increase in investment seems to be generated by population growth, creating a substantial negative current account balance, but this is followed by a period of strong labour force growth accompanied by an increase in domestic savings and a transition to positive current account balances. These findings are consistent with Barlow (1994) and Kelley and Schmidt (1994, 1995), who find that long- and short-term effects of population growth differ, moving from negative to positive as a population bulge ages.

The purpose of the present analysis is to identify any effects of changing age structure on patterns of consumption and savings, while controlling for the sizes of all age groups in the population using the Fair and Dominguez technique, and permitting the age-specific effects to vary by level of development (per capita income). This has not been done in previous studies, possibly because in cross-national comparisons variations in per capita income tend to be confounded with a host of institutional and cultural differences. This study attempts to sidestep that problem by using cross sections of U.S. state-level data throughout the 20th century, tracing the United States from its "less" to its "more" developed status in the personal consumption expenditures (PCE) of up to 51 states (including the District of Columbia) with a greater degree of internal consistency in culture and institutions, than can be found in international cross sections.

In addition, the analysis presented here incorporates another aspect of changing age structure which has been omitted in all previous analyses: the potential effects of age structure—"relative cohort size"—on individuals' income relative to their material aspirations—their "relative income." These effects were first hypothesized by Richard Easterlin (1987), and are discussed in more detail in Section 4.

What this analysis finds is a more complex pattern of age structure effects than expected, a pattern that does indeed appear to vary both by level of income and by relative cohort size. The estimated effects are highly significant, suggesting that changes in age structure have caused PCE in the United States to vary by about 25 percent during the 20th century, holding constant all other factors including population size, real income, and its growth rate.

The next section addresses problems inherent in both micro and macro level analysis, while Section 3 explains the potential significance of intergenerational transfers and Section 4 explains the relative cohort size concept. Section 5 describes the data and method used in this study; Section 6 demonstrates the effect of using these data in a few model formulations from previous analyses in the dependency literature; and Section 7 presents results from applying a new method and model to the U.S. state-level data. The chapter concludes in Sections 8 and 9 with an application of these predicted age structure effects in a simulation of consumption behaviour in the United States given the population changes that occurred throughout the 20th century.

2. Micro versus Macro

It must be emphasized at the outset that any age structure effects identified at the aggregate level cannot be interpreted as support for, or refutation of, the standard life-cycle model of saving and consumption. We should neither assume nor expect to see the standard "hump-shaped" pattern of saving and spending in the aggregate (as emphasized in Higgins 1998, p. 351). That is, for example, the longevity of elderly parents—and therefore their continued presence in the age structure—might induce middle-aged adults to spend more in supporting them than they otherwise would have, or alternatively might induce those middle-aged adults to *save* more in the expectation of their own increased longevity. The presence of children in an extended family might induce additional saving, or spending, on the part of aunts, uncles, and grandparents. In this sense, the life-cycle model is simply a starting point generating the possibility of age structure effects, rather than the model being tested here. And conversely, this implies that any coefficient estimates of life-cycle spending/saving patterns based on individual household micro data cannot simply be aggregated to estimate macro level effects—although this has been a common practice in the savings literature for industrialized nations.[3]

Weil (1994) addresses this aggregation issue in explaining the differences between measured age patterns of savings at the micro and macro levels, that he feels result from intergenerational responses to the bequest motive. He posits that increased saving by elderly households planning bequests, as measured in micro level surveys, is masked at the macro level because of reduced saving by adult children expecting to receive those bequests. He concludes that "... one cannot use the mean saving of people at different ages (or any other coefficients that come from micro data that do not account for members of other generations) to forecast changes in the aggregate saving rate in response to changes in the age structure of the population" (p. 67).

Weil's findings in support of that hypothesis argue strongly for the use of macro level data to estimate effects of changing population age structure. Another study that supports this type of intergenerational effect is Attanasio (1998), where the savings profile of cohorts born between 1920 and 1939 appears to be "shifted down" relative to that of preceding and subsequent cohorts. Attanasio suggests that this shift might have resulted from baby boom-induced intergenerational transfers. In addition, members of the 1920–1939 birth cohort may have reduced their savings because as the "sandwich generation" they found themselves simultaneously caring for both children and ageing parents who may not reside in the same household.[4] The use of micro level data like the Consumer Expenditure Survey (CES) will miss much of this type of intergenerational effect, to the extent that it occurs across rather than within households.

But as Taylor and Williamson (1994) point out, macro data present problems as well: "The use of multi-country cross sections for short periods (or even a single year) raises the possibility of omitted variable bias... The use of long time series for a few countries (or only one country) raises the question of the structural stability of the savings equation and inclines a model to track poorly over the short to medium term." (1994:360 footnote 14). The state level cross sections used in the present analysis lie midway between micro and macro, capturing the inter-household effects otherwise available only in national macro

---

[3] See, for example, Auerbach, Cai, and Kotlikoff (1990) and Bosworth, Burtless, and Sabelhaus (1991).

[4] I thank my former colleague at Williams College, Roger Bolton, for this insight.

data, but using aggregations exhibiting more cultural and institutional similarity than can be found in international cross sections.

3. Inter-Household Transfers and Their Role

Intergenerational—and therefore usually interhousehold—transfers have been documented to occur among all income groups and to be particularly significant from older to younger generations. Sumon Baumik, in another chapter of this book, emphasizes the significance of such transfers, and states that "an event like child birth can induce transfers equal to about 30 percent of the average annual income of the recipient's household." (p. 112) Other examples of such interhousehold transfers, as pointed out by Baumik, might be gift giving at graduation or marriage. Even the expenditures of family and friends in dressing for and travelling to such functions would be expenditure induced by the age of the younger group, rather than by the ages of persons attending.

A childless adult might choose to assist financially a younger sibling with "start-up" costs for the younger sibling's children, or might set up a savings account in the children's names in anticipation of future educational expenses. Similarly parents might draw down on their savings in order to provide their adult children (no longer living at home) with cash for a car or a deposit on a house. This would be behaviour induced by the age and presence of the children, rather than by the age of the parents or other relatives; but the children's presence would not be detected in micro level surveys of the expenditures of individuals living outside the children's own households.

If, as suggested by Welch (1979) and Macunovich (1999), the incomes of young adults in large birth cohorts are adversely affected by cohort size, while at the same time the incomes of their parents are favourably affected, such cross-cohort giving might produce patterns of savings and consumption not predicted by models dealing with households in isolation—patterns related to the age distribution in the *total population*, rather than to the age of a household's own head. Similarly, when adult children contribute to the nursing home care of their elderly parents, that expenditure's relationship to the parents' age group would be missed in a household-level survey, but picked up in an analysis at a more aggregate level. Even expenditures of time on elderly parents—to the extent that they reduce paid work hours among the caregivers—could show up at the macro level as a change in the share of consumption out of income.

Why might these effects—and especially the effect of children—vary between lower and higher income countries? Given imperfect capital markets and an absence of government provisions for old age security in early stages of development, children themselves are effectively a form of saving on the part of their parents—their "old age security". Thus, it is perhaps to be expected in developing countries that high youth dependency rates will have an immediate negative effect on formal savings rates. But the effect of children in more developed economies is open to question. As material aspirations increase along with economic development, and parents begin to opt for "quality" over "quantity" in children, higher levels of education become mandatory, and parents begin to have more ambitious plans for their children as young adults. (See, for example, Kelley and Schmidt 1994: 24.)

It seems possible, then, that with increasing levels of development parents might be induced to save more, rather than less, when they have children, in anticipation of higher costs in the children's late teens, and when they set up on their own as young adults. Grandparents, as well, may save more, either to assist with the costs of higher education, or through an increased bequest motive as suggested by Collins (1991). Even college endowments are a form of saving by society in general, induced by the presence of children who will need education in the future. In addition, economists at least as far back as Malthus have recognized that children cause parents to work harder,[5] so that although absolute consumption expenditures may rise in the presence of children, there may be little increase in consumption relative to income. The net effect of children on aggregate patterns of consumption and savings is thus an empirical question.

4. An Additional Wrinkle: Relative Cohort Size Effects

There is an additional complication that arises in analyses of this type, because of changing age structure's potential to exert not only simple compositional effects on consumption (through life-cycle changes in behaviour), but also to be responsible for *changes in age-specific effects over time.* This phenomenon has not been addressed in the savings and consumption literature to date. It arises from the impact of *changing relative cohort size on young adults' relative income*, as first pointed out by Richard Easterlin (1987), and later substantiated in the United States by Welch (1979); Berger (1984, 1985, 1989); and Macunovich (1999), among others. Similar effects have been documented in other countries, both developed (Korenman and Neumark 2000) and developing (Higgins and Williamson 1999).[6]

The phrase "relative cohort size" refers to the size of a birth cohort of young adults relative to the size of their parents' birth cohort. "Relative income," in turn, refers to the earning potential of young adults relative to their material aspirations, which Easterlin hypothesized would be a function of the standard of living experienced while at home with their parents. The mechanism of transmission from relative cohort size to relative income is *imperfect substitutability in the labour market.* That is, increases in the size of entering cohorts in the labour market might be expected—through standard demand–supply effects—to reduce wages, but this effect does not occur uniformly throughout all experience groups. Because young inexperienced workers are very poor substitutes for older experienced workers, the increase in their own supply will depress their own wages, but may even increase older workers' wages. And because the older workers are, in the aggregate, the younger workers' own parents, this depresses the young workers' own relative wages and hence their earning potential relative to their material aspirations—their relative income, and thus the average relative income of all young adults at the aggregate level.

---

[5] Malthus wrote in his 1817 appendix to *An Essay on the Principle of Population*: "If it were possible for each married couple to limit by a wish the number of their children, there is certainly reason to fear that the indolence of the human race would be very greatly increased..." (Page 369 in the Cambridge University Press 1992 edition selected and introduced by Donald Winch.)

[6] I am indebted to members of this IUSSP workshop—especially to Nancy Folbre, for suggesting explicit controls for relative cohort size effects.

Easterlin hypothesized that this reduction in relative income would cause compensatory changes in age-specific behaviour in larger birth cohorts, adjustments such as reduced marriage and fertility and increased female labour-force participation. These changes in age-specific behaviour would in turn very likely affect patterns of age-specific consumption relative to income, as young people struggle to maintain their desired standard of living, and parents make transfers to young adult children to supplement their reduced wages. To the extent that this is the case, any study of age structure effects on consumption must control for changes in relative cohort size, as well as in the overall age structure.

Although these relative cohort size effects are difficult to study at the micro level, due to the absence of adequate income data describing the parents of both partners in a marriage, and the impact of other factors on material aspirations, they are much more straightforward in national- and state-level data, where it is not necessary to link specific parents with specific children. If the effect is there, it should emerge at the generational level when changes in the average income and cohort size of, say, prime age workers, are compared with changes in the average income and cohort size of young adults.

5. Data and Methods

This study addresses the issue of age-related patterns of consumption in a new way, using state-level cross sections of PCE developed for the United States by Lebergott (1996). For five dates in this century—1900, 1929, 1970, 1977, and 1982—he provides data that have three significant advantages over other data used in the past to analyse age structure effects on macroeconomic measures. First, they permit the analysis of 48 to 51 (including Alaska, Hawaii, and the District of Columbia) "sufficiently similar" areas as suggested by Taylor and Williamson (1994), in order to minimize bias from omitted variables. Second, they permit the examination of the United States over a long time span but with less danger of bias due to the autocorrelation that occurs in annual time-series data. Stoker (1986) has demonstrated that the most common method of dealing with autocorrelation—the inclusion of leads and lags of the dependent variable—produces misleading estimates because the leads and lags normally include information on the only slowly changing patterns of age structure, and thus tend to reduce the estimated significance of age structure variables. And thirdly, these state cross sections permit comparisons between recent U.S. experience as an industrialized nation coping with baby booms and busts, and the United States at the beginning of the 20th century when it was itself virtually a "developing economy" still experiencing its demographic transition.

Lebergott developed several hundred new series for components of PCE annually for the years 1900–1929—and at the state level for 5 years—that are directly comparable with official Bureau of Economic Analysis (BEA) series as revised in 1993, in both current and constant dollars. These data prepared by Lebergott are far more comprehensive than data provided by the CES. As he points out the latter are based on interviews with "less than one-thousandth of one percent of American 'consumer units'" in which "individual members of households try to remember expenditures in the prior year" (p. 130). The 1984 survey, for example, "understated U.S. food and clothing expenditures by $173 billion. Not to mention $33 billion for house furnishings, $28 billion for alcohol, and $46 billion for entertainment (pp. 129–130)," while the census rent sample "was one thousand times greater than that of the [Bureau of Labour Statistics] BLS" (p. 131).

Lebergott's data began with BEA national income account totals that were then allocated to states. It is important to note that in no case were his allocation methods based on age distributions within the population. Rather, they were derived from census data on production and expenditures, as well as (for 1900) distributions of workers by occupation and service income. Lebergott's data are described in more detail in the Appendix, and summary statistics are presented in Table 1.

These state and national expenditure figures are supplemented with detailed population data for states in each of these years, and for the United States in the entire century, provided by the Bureau of the Census. The analysis makes use of detailed age breakdowns of the total population, rather than simple dependency rates or birth rates, following the lead of Fair and Dominguez (1991), and recent analyses by Higgins and Williamson (1997), and Higgins (1998).[7] This minimizes problems associated with identifying appropriate age groupings. For example, including the 25–34 age group with assumed savers in the 35–44 age group would dilute the measured effect of that older group, if in fact the 25–34 year olds were dis-savers rather than savers.

Ideally a large number of age groups would be separately identified as independent variables in the analysis to avoid this age group identification problem. But a model that includes a large number of age groups in order to overcome the possibility of erroneous groupings, encounters a problem of severe multicollinearity that calls into question the accuracy of any individual coefficient estimates. And problems of multicollinearity are compounded by the marked loss of degrees of freedom in estimating those coefficients, as the number of age groups is increased: an important consideration in time series analyses. As observed by David (1962), "Age varies continuously and there are few convenient demarcations between age groups with significantly different behaviour patterns."

In order to circumvent these associated problems of age group identification and multicollinearity, the present analysis uses the Fair-Dominguez (1991) method of parameterizing multiple age groups, which is in turn based on Almon's (1965) distributed lag technique. This methodology has subsequently been adopted in more recent studies, as mentioned earlier, and is described in more detail in the Appendix. In general terms, it is one that permits the estimation of coefficients on single year population age shares by constraining those coefficients to lie along a polynomial. The coefficients φ_j on J population age shares p_j are assumed to enter the consumption equation in the form

$$\sum_{j=1}^{J} \varphi_j p_j \tag{1}$$

[7] It must be emphasized, in response to an earlier reviewer's comments, that the inclusion of the full age distribution in a more aggregated consumption/savings model does not in any way imply or necessitate decision-making on the part of children themselves, with regard to their patterns of consumption. It simply allows for the fact that children in one household might affect the spending patterns of individuals in other age groups and/or households.

Table 1. Summary statistics on state-level variables used in the analysis.

Variable (by Year)	Obs	Mean	Std. Dev.	Min	Max
		Personal consumption expenditures as a share of total personal income			
1900	48	1.09	0.19	0.76	1.77
1929	48	0.91	0.07	0.73	1.06
1970	51	0.82	0.07	0.66	1.04
1977	51	0.84	0.08	0.63	1.17
1982	51	0.79	0.09	0.61	1.20
		Real personal income (in millions of 1987 dollars)			
1900	48	4849.85	6608.08	257.88	35618.23
1929	48	12563.50	17961.98	558.05	99954.68
1970	51	43283.79	52840.65	3370.37	247811.91
1977	51	54474.04	62727.56	5117.60	317822.91
1982	51	62932.26	74108.20	6346.59	395150.09
		Total population (in thousands)			
1900	51	1486.67	1539.29	42.33	7268.89
1929	49	2505.61	2519.98	91.06	12588.06
1970	51	3985.32	4314.83	302.34	19964.68
1977	51	4309.02	4559.73	403.44	22352.41
1982	51	4542.44	4846.96	449.61	24820.02
		Growth rate of real GDP (percent)			
1900	47	3.98	3.14	−3.76	15.58
1929	49	0.72	1.69	−2.44	3.65
1970	51	3.71	1.35	0.99	7.46
1977	51	2.13	1.97	−4.43	10.05
1982	51	1.10	1.48	−1.92	4.45
		Percent urban			
1900	49	33.90	23.54	6.20	100.00
1929	49	46.92	21.01	16.60	100.00
1970	51	66.44	15.14	32.20	100.00
1977	51	67.02	14.99	33.00	100.00
1982	51	67.60	15.00	33.80	100.00
		Percent foreign-born			
1900	51	14.83	11.77	0.20	58.90
1929	49	8.97	7.41	0.28	25.35
1970	51	3.44	2.81	0.40	11.60
1977	51	4.34	3.56	0.90	15.10
1982	51	4.34	3.56	0.90	15.10
		Z1 (using logged population shares)			
1900	49	−2174.93	370.86	−3047.18	−1320.76
1929	49	−1818.45	353.85	−2720.06	−1200.02
1970	51	−1177.47	272.80	−2560.16	−766.52
1977	51	−1209.02	276.60	−2623.17	−695.44
1982	51	−997.92	254.79	−2256.73	−492.93

Table 1. ***(Continued)***

Variable (by Year)	Obs	Mean	Std. Dev.	Min	Max
U.S. averages:					
1900–1925	26	−1991.70	78.02	−2108.43	−1862.23
1926–1950	25	−1534.28	174.02	−1842.22	−1298.83
1951–1975	25	−1192.50	68.89	−1289.75	−1050.85
1976–1999	24	−877.30	84.41	−1030.34	−747.03
		Z2 (using logged population shares):			
1900	49	−205414.50	34096.00	−297895.81	−125640.50
1929	49	−175913.00	33640.09	−268351.50	−116720.30
1970	51	−109927.40	26520.66	−244218.50	−71987.44
1977	51	−115733.20	26706.89	−253996.20	−69216.45
1982	51	−96052.79	24458.84	−220786.41	−49608.36
U.S. averages:					
1900–1925	26	−189452.30	6344.23	−198650.50	−178894.70
1926–1950	25	−150154.80	15745.81	−177205.00	−126569.20
1951–1975	25	−112941.00	7332.17	−125312.90	−99988.94
1976–1999	24	−85372.21	7345.82	−98363.23	−74111.45

which is estimated as a polynomial

$$\sum_{j=1}^{J} \varphi_j p_j = \zeta_1 Z_1 + \zeta_2 Z_2 + \cdots + \zeta_n Z_n \tag{2}$$

in which n is the degree of the polynomial and Z_n is a weighted sum of individual population shares, defined as

$$Z_n = \sum_{j=1}^{J} p_j j^n - 1/J \sum_{j=1}^{J} p_j \sum_{j=1}^{J} j^n \tag{3}$$

Estimating the degree of that polynomial, n, appears in earlier studies to have been based largely on theory, assuming a quadratic at the aggregate level based on the hypothesized life-cycle "hump-shaped" pattern at the micro level. But as emphasized earlier in this chapter, because of interhousehold effects a different pattern may emerge at the aggregate level, and there is more danger in over- than in underestimating the degree of the polynomial. Judge et al. (1985:359–360) state that when fitting an Almon lag, overestimates of the true degree of the polynomial produce estimators that are unbiased although inefficient, whereas underestimates of the true degree produce estimates that are "always biased." For this reason they suggest starting with a higher n, than is assumed to apply in the true model, and stepping down, testing each additional restriction and finally accepting the level that "produces *the last acceptable hypothesis* [their italics]." That procedure has been adopted in the present study, and is documented in results not presented here, but available on request.

6. Reproducing Earlier Results, Using U.S. State-Level Data

The approach adopted in this analysis was first to attempt to reproduce (using U.S. 20th century data) the results of analyses of age structure effects that included countries at different income levels and stages of development, and then to expand on those models by including all age groups in unconstrained versions of the models (i.e., permitting age-specific effects to vary by income level and relative cohort size).

The studies addressed in this way were by Collins (1991); Taylor and Williamson (1994); Taylor (1995); Leff (1969); Fry and Mason (1982), and recent studies by Higgins and Williamson (1997); and Higgins (1998).[8] All of these analyses estimated similar models, except for the fact that the last two tested for effects of the full population age structure rather than just selected age groupings. None permitted the dependency effect to vary by level of income or relative cohort size, although some of the later ones, based on Fry and Mason's (1982) "variable rate-of-growth" model, tested for effects of the growth rate of GDP. They hypothesized that "[t]he dependency ratio [defined as the population aged 0–14 relative to the population aged 15–64] exerts both level and timing effects: a higher dependency ratio reduces the aggregate saving rate, the magnitude of the effect increasing with the rate of growth in real GNP" (p. 427). Fry and Mason's analysis was in turn an extension of earlier work by Leff (1969), who hypothesized (consistent with the analysis in Tobin 1967) that real GNP growth would increase the incomes of younger (saving) relative to older (dis-saving) households, thus increasing overall levels of saving. Thus, the Fry and Mason model of national saving, tested using panel data from seven Asian developing countries between 1962 and 1972,[9] included not just dependency and growth rates, but also their interaction, which was found to have the expected negative effect:[10]

$$\begin{aligned}
\ln[1/(1-s)] &= -\ln(c) \\
&= \underset{(0.066)}{0.147} - \underset{(1.841)}{3.769\,D} - \underset{(0.130)}{0.526\,f} + \underset{(0.410)}{1.644\,g} - \underset{(18.224)}{69.457\,Dg} + \underset{(1.159)}{4.822\,rg} \\
&\quad + \underset{(0.088)}{0.249\,\ln[1/(1-s)]_{t-1}}
\end{aligned} \tag{4}$$

where s is the aggregate rate of saving, the log-odds savings rate was approximated using the negative of the logged rate of consumption c, D is the population under age 15 divided by population aged 15–64, f is the foreign saving rate, g is the rate of growth in real GNP, r

---

[8] The Kelley and Schmidt (1994, 1995) studies are not included in this list because of their use of birth rates rather than specific measures of age structure, to explain effects on growth and savings.

[9] Fry and Mason's sample began with 10 countries, but dropped Pakistan, Sri Lanka, and Thailand because of binding foreign exchange constraints, ending with only seven: Burma, India, Korea, Malaysia, Philippines, Singapore, and Taiwan.

[10] Fry and Mason's full initial model was

$$\begin{aligned}
\ln[1/(1-s)] &= -\ln[c] \\
&= \alpha_0 + \alpha_1 D + \alpha_2 r + \alpha_3 f + \alpha_4 g + \alpha_5 Dg + \alpha_7 rg + \alpha_8 fg + \alpha_6 \ln[1/(1-s)]_{t-1}
\end{aligned} \tag{4a}$$

but they dropped r and fg after finding their estimated coefficients were not significant.

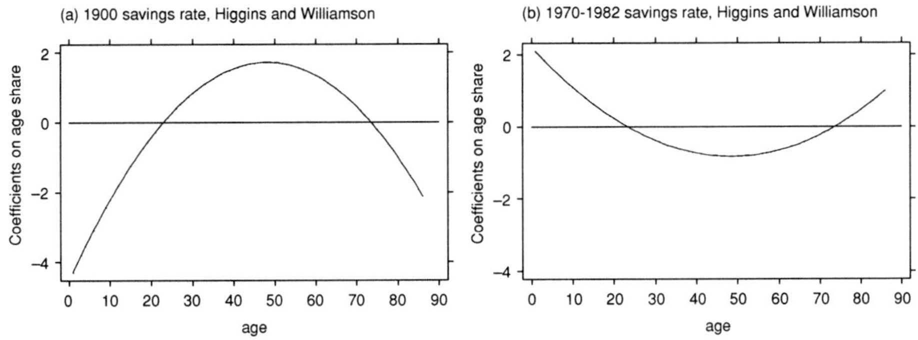

Figure 3. Estimating the Higgins and Williamson (1997) model presented in Table 2 using U.S. state-level data for 1900 and for 1970–1982, illustrating the apparent reversal that occurs in moving from low- to high-income economies.

is the real rate of return on financial assets and the adjusted R^2 was 0.87 (with t−statistics in parentheses).

It should be noted, however, that later models based on Fry and Mason have tended to find positive, rather than negative, effects of the interaction term between dependency and the growth rate. These models are described in more detail, and presented with their results, in Macunovich (2001). In addition, Macunovich (2001) presents detailed results from reestimating each of these models using Lebergott's state-level data for the United States at the five available dates (1900, 1929, 1970, 1977, and 1982) in place of the various groupings of LDCs and MDCs (more developed countries) used in the earlier studies. A common pattern that emerges in all of the reestimates is what appears to be a structural break between the earlier and later years, with coefficients supporting the dependency hypothesis in 1900, and "inconsistent" in more recent years. And contrary to Fry and Mason's results, in all of those reestimates the level and rate-of-growth effects of dependency tend to move in opposite directions.

This is perhaps best illustrated graphically, as in Figure 3 and Table 2, using the Higgins and Williamson (1997) model. A negative effect of children on savings in 1900 is demonstrated by the strong negative coefficients that were estimated for the lowest ages in Figure 3a (which are a result of the positive sign on the coefficient estimated for Z_1 in column 3 of Table 2). The hump shape in Figure 3a is the shape expected under the standard life-cycle model—higher levels of consumption out of income, and thus a tendency to "dis-save" at younger and older ages, with lower levels of consumption relative to income in the middle-age groups, the "saving" years. But those expectations are not met in Figure 3b, as was typical of all of the model re-estimates: younger children are estimated here to have a positive effect on savings in more recent years.[11]

[11] The age share coefficients presented in Figure 3b are based on column 6 in Table 2, but the pattern is virtually unaffected by a change to a fixed effects model (as in column 4) or when year dummies are excluded (as in column 7).

Table 2. Original Higgins and Williamson (1997) model estimates (column 2) and reestimates of their model using Legergott state-level U.S. data for 1900 and for 1970–1982, illustrating the apparent reversal in the effects of age structure in moving from the early to late 20th century U.S.—and by implication, from lower to higher income economies.

	Higgins and Williamson fixed effects results	Results using Lebergott U.S. state-level data				
		1900	1970–1982			
			Fixed effects		Random effects	
Z1	0.703**	0.2597***	−0.1217*	−0.1138**	−0.1258***	−0.0770**
		(0.0584)	(0.0687)	(0.0448)	(0.0392)	(0.0372)
Z2	−0.046**	−0.0027***	0.0012	0.0015***	0.0013***	0.0010**
		(0.0007)	(0.0008)	(0.0005)	(0.0004)	(0.0004)
Z1 growth rate*	na†	−0.0219*	0.0232**	0.0376***	0.0239**	0.0232**
		(0.0109)	(0.0104)	(0.0113)	(0.0097)	(0.0110)
Z2 growth rate*	na†	0.0002*	−0.0002**	−0.0004***	−0.0003**	−0.0003**
		(0.0001)	(0.0001)	(0.0001)	(0.0001)	(0.0001)
growth rate	−0.035	−3.3455	1.4504*	0.8461	0.7717	−0.6956
		(4.8952)	(0.8011)	(0.8585)	(0.7493)	(0.8098)
Rel. Price of Investment Goods	−0.046**					
Year = 1977?			−0.5088		−0.5661	
			(2.5124)		(1.5158)	
Year = 1982?			8.1445***		6.7653***	
			(2.3096)		(1.7831)	
No. of obs	916	46	153	153	153	153
Chi^2					72.92	18.51
F statistic	22.55	10.28	10.84	7.63		
Prob > Chi^2 (or F)	<0.001	0.0000	0.0000	0.0000	0.0000	0.0024

Notes: Standard errors in parentheses. Dependent variable is savings as a percent of aggregate real income in column 2, and 100 minus personal consumption expenditures as a percent of total personal income in columns 3–7. Estimates for constant term not reported. Growth rate is annual percent change in aggregate GDP. The original Higgins and Williamson model included a control for the relative price of investment goods, which is proxied in columns 5 and 7 using dummy variables for year under the assumption that there would be little cross-sectional variation in this measure in the U.S. data. The Higgins and Williamson model also included a lagged dependent variable to correct for serial correlation, which is not required in the U.S. data. Model in column 3, for 1900, estimated using robust regression and GLS. Z1 and Z2 are calculated using logged population shares, as in the original Higgins and Williamson (1997) model.

*** $p < 0.01$; ** $p < 0.05$; * $p < 0.10$. † These coefficient estimates for the interaction terms were not provided in Higgins and Williamson (1997).

7. Resolving the Paradox: Are Different Models Needed for DCs and LDCs?

Is there a model that bridges the apparent gap between early and late 20th century U.S. experience (and, by analogy, between age structure effects in LDCs and DCs)? Has behaviour changed, or only appeared to change because of the effects of other factors? And why do various models produce "inconsistent" results for the interaction term between dependency and the growth rate?

If in fact age structure effects vary by level of income, and relative cohort size, the models in the previous section cannot capture those changes since none control for relative cohort size and most exclude income as an independent variable, including it only as the denominator in the dependent variable and thus constraining its coefficient to unity. The one exception appears in Collins (1991)—but Collins does not include an interaction term between dependency and income, nor does she include any control for changing relative cohort size. It is significant, though, that her estimated (positive) coefficient on the dependency rate increases by 50 percent and becomes statistically significant when low-income countries are excluded from her full model (with the savings rate as dependent variable).

A more appropriate model, then, would seem to be one that includes, along with the variables used in previous models, a control for relative cohort size, and some form of income on the right-hand side, together with its interaction with the age structure variables. If the model is estimated in logged form the exact income specification does not really matter, as long as income and total population are included on the right-hand side as separate variables: all forms are mathematically equivalent[12]. Total population is used in the model, although virtually identical results are produced using just the adult population, or the working-age population.

Relative cohort size is specified here using the age groups hypothesized to be least substitutable in the labour force; those aged 20–24 (inexperienced workers) relative to those aged 45–49 (prime-age workers). The results presented in the remainder of this chapter were found to be fairly insensitive to changes in this measure, however.[13] Table 3 presents summary statistics on a range of relative cohort size measures.

Using the Lebergott data, the following basic model (equation (5)) has been estimated:

$$
\begin{aligned}
-\ln \mathrm{PCE}_{S_t} = {} & \varphi_0 + \sum_{j=1}^{J} \varphi_{0j} \ln(p_j) + \varphi_1 \ln \mathrm{RCS} + \varphi_2 \ln y_{St} + \ln y \sum_{j=1}^{J} \varphi_{2_j} \ln(p_j) \\
& + \varphi_3 \ln P_{S_t} + \varphi_4 g_{S_t} + v_S + u_{S_t} \qquad (5)
\end{aligned}
$$

---

12 Although the form of the income variable that is interacted with the age structure variables will change the *estimated coefficient on the interaction term*, it will have no effect on the estimated *total* effect of each age share on the dependent variable, at various levels of income, when the model is estimated in logged form.

13 That is, although the level of statistical significance varies among the different measures, the estimates of overall effect of changing age structure, as presented in Figures 4–6, are virtually unaffected by changes in the ratio used. These additional results are available from the author on request.

Table 3. Summary statistics on a range of relative cohort size variables, at the state and national levels. The ratio of 20–24/45–49 year olds was used in this analysis.

Variable (by Year)	Obs	Mean	Std. Dev.	Min	Max	U. S. actual levels
			Ratio 15–24/total population			
1900	49	0.20	0.02	0.17	0.23	0.20
1929	49	0.19	0.02	0.16	0.22	0.18
1970	51	0.19	0.01	0.16	0.21	0.18
1977	51	0.20	0.01	0.17	0.23	0.19
1982	51	0.18	0.01	0.16	0.19	0.18
U.S. averages:						
1900–1925	26	0.19	0.01	0.18	0.20	
1926–1950	25	0.18	0.01	0.15	0.18	
1951–1975	25	0.16	0.02	0.13	0.19	
1976–1999	24	0.16	0.02	0.14	0.19	
			Ratio 15–24/25–59			
1900	49	0.49	0.10	0.32	0.71	0.48
1929	49	0.43	0.08	0.30	0.63	0.41
1970	51	0.47	0.04	0.37	0.58	0.46
1977	51	0.47	0.04	0.39	0.59	0.45
1982	51	0.40	0.03	0.34	0.47	0.40
U.S. averages:						
1900–1925	26	0.44	0.03	0.40	0.48	
1926–1950	25	0.38	0.03	0.31	0.41	
1951–1975	25	0.38	0.07	0.29	0.46	
1976–1999	24	0.35	0.06	0.28	0.45	
			Ratio 15–24/45–54			
1900	49	2.27	0.46	1.50	3.32	2.25
1929	49	1.75	0.37	1.11	2.63	1.70
1970	51	1.67	0.20	1.27	2.30	1.60
1977	51	1.94	0.25	1.47	2.72	1.82
1982	51	1.88	0.14	1.57	2.35	1.83
U.S. averages:						
1900–1925	26	1.98	0.19	1.73	2.25	
1926–1950	25	1.53	0.15	1.24	1.75	
1951–1975	25	1.38	0.21	1.13	1.75	
1976–1999	24	1.50	0.32	1.02	1.87	
			Ratio 20–24/45–49			
1900	49	2.05	0.38	1.53	2.98	2.06
1929	49	1.57	0.30	1.04	2.22	1.52
1970	51	1.52	0.21	1.11	2.35	1.49
1977	51	1.92	0.25	1.46	2.74	1.85
1982	51	1.99	0.18	1.60	2.45	1.95

Table 3. ***(Continued)***

Variable (by Year)	Obs	Mean	Std. Dev.	Min	Max	U. S. actual levels
U.S. averages:						
1900–1925	26	1.82	0.21	1.52	2.06	
1926–1950	25	1.41	0.09	1.23	1.53	
1951–1975	25	1.24	0.23	1.00	1.72	
1976–1999	24	1.49	0.40	0.93	1.99	
			Ratio 20–22/45–49			
1900	49	1.23	0.24	0.90	1.82	1.25
1929	49	0.95	0.19	0.62	1.37	0.92
1970	51	0.92	0.13	0.65	1.38	0.90
1977	51	1.17	0.15	0.89	1.70	1.14
1982	51	1.21	0.11	0.95	1.46	1.18
U.S. averages:						
1900–1925	26	1.10	0.12	0.92	1.25	
1926–1950	25	0.85	0.06	0.72	0.93	
1951–1975	25	0.77	0.15	0.60	1.06	
1976–1999	24	0.90	0.24	0.58	1.21	

where S is state (from 1 to 51); t is year (1900, 1929, 1970, 1977, and 1982); PCE is personal consumption expenditure as a share of total income, with its inverse treated as savings; RCS is relative cohort size, the ratio of population aged 20–24 relative to those aged 45–49; p_j is the age group j share of total population; y is real per capita personal income; P is total population; and g is the growth rate of real GDP.[14] In addition, four alternative formulations were tested, including controls for:

1. the percent of a state's population that is foreign-born, and the percent living in urban areas, both of which can be expected to increase consumption's share of income [equation (6)];
2. interactions between the growth rate and age structure variables, as in the Fry and Mason formulation [equation (7)];
3. dummy variables for year, to capture any period effects [equation (8)]; and
4. a model containing all of the above variables [equation (9)].

[14] Total population is also needed in the model to control for the fact that the denominator is lost in calculating Zs with logged population shares. A reformulation of equation (3), substituting s_j/P for p_j, where s_j is the absolute number in age group j and P is total population, indicates that if $\ln(s_j/P)$ is used in this equation in place of s_j/P, the total population terms cancel, leaving a variation on Z_n based on absolute age group sizes, rather than age group shares:

$$X_{i,t} = \gamma_{0,i} + \gamma_1 X_{i,t-1} + \gamma_2 g_{i,t} + \gamma_3 RPI_{i,t} + \gamma_4 Z_{1_{i,t}} + \gamma_5 Z_{2_{i,t}} + \gamma_6 g_{i,t} Z_{1_{i,t}} + \gamma_7 g_{i,t} Z_{2_{i,t}} + u_{i,t}$$

Apparently the Higgins (1998) and Higgins and Williamson (1997) formulations both used logged population shares. Higgins (p. 348, footnote 7) notes that "the (coefficients on the age shares) must be restricted to sum to zero because the population age shares sum to unity, and thus are collinear with the intercept term", suggesting that he is working with unlogged population shares. However, in all of his table notes (as, for example, on p. 357) he explains that "the age distribution coefficients represent the change in the dependent variable associated with a unit change in the corresponding log population age shares."

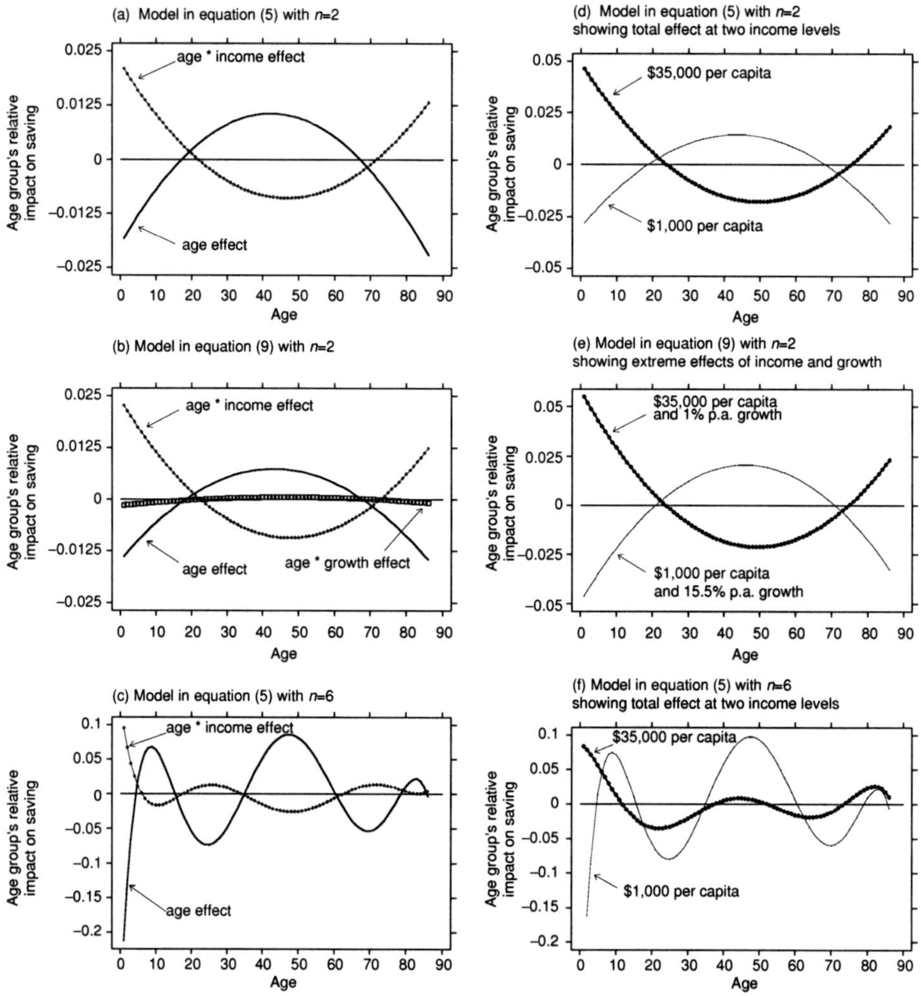

Figure 4. Estimating age-share coefficients, using models in columns 2 and 6 of Table 4.
Left: Panels 4a–c presents estimated effects of each age group on the intercept (coefficients on the age shares themselves) and slope (coefficients on the interaction terms between age shares and income and growth) of an equation estimating savings as a share of total personal income using Lebergott (1997) state-level U.S. data for 1900, 1929, 1970, 1977, and 1982.
Right: Panels 4d–f present estimates of the net effect of age structure at high and low levels of per capita income ($1,000 and $35,000, in year 1999 dollars, using the minimum and maximum values in the data). Panel 4e also incorporates the esitimated effects of growth on the age structure coefficients, and thus might be thought to present "extreme" cases: a low-income country ($1,000 per capita) experiencing rapid growth (15.5 percent based on the maximum in the data) and a high-income country ($35,000 per capita) experiencing low growth (1 percent p. a.). The coefficients in these three graphs represent elasticities. For example, Figure 4e suggests that a 10 percent

Although each of these additional controls produces statistically significant coefficients, and increases the overall explanatory power of the model, none have any substantial effect on the estimated level and pattern of age structure effects over time, as illustrated in Figures 4–6, and explained in the next section of this chapter.[15]

But what should be the degree of the polynomial used to represent age structure effects? Higgins and Williamson (1997) and Higgins (1998) assumed quadratic and cubic forms, no doubt with the life-cycle model in mind. Similarly Fair and Dominguez (1991) used a second-degree polynomial, expressly in the expectation of a life-cycle pattern of effects. But as emphasized earlier there is no theoretical reason to expect the life-cycle pattern to dominate at the aggregate level: once interhousehold effects are included we are no longer estimating a behavioural model.

The testing procedure adopted here, with the models indicated above, is that recommended in Judge (1985), i.e., starting from a higher degree polynomial than might be thought appropriate ($n = 8$) and sequentially adding restrictions. Although this procedure has been shown to be problematic in time series data when *both* the degree of the polynomial (n) and number of lags must be estimated (Terasvita 1976; Schmidt and Mann 1977; and Schmidt and Waud 1973), it is reliable when the number of lags (in this case, the maximum age) is known (Godfrey and Poskitt 1975; Harper 1977; Frost 1975; Schmidt and Sickles 1975). The results of these tests are presented in Table A.1, where it can be seen that only the quadratic form ($n = 2$) is unambiguously indicated. Only at that level are the t and Wald chi-squared statistics significant in all tests performed. The t-statistics on the Z's and their interactions with income are significant for $n = 6$ in equation (5) and $n = 5$ in equation (9), as well—but Wald tests of the highest coefficients in the fifth- and sixth-degree polynomials are singular, suggesting that the model is over fitted when $n > 2$.[16] However, because the overall explanatory power of the sixth-degree polynomial (as indicated by R^2 and Mean Squared Error (MSE)) is in general greater than that of the quadratic, the results presented here will first examine the quadratic model, and then the effect of adopting the sixth-degree model, on the pattern of estimated age share coefficients and their ultimate effect in simulations at the national level.

Figure 4. (*continued*)
increase in the population share of 1-year-olds would generate a 0.5 percent decline in the savings rate in a low-income high-growth economy, whereas the same increase would generate a 0.6 percent increase in the savings rate in a high-income low-growth economy.
The top four panels illustrate the patterns of age share coefficients obtained using a quadratic polynomial ($n = 2$), while the bottom two illustrate the patterns obtained using a sixth-degree polynomial.

[15] Similarly, even changes in the growth rate and total population size have virtually no effect on the estimated pattern of age structure effects, although excluding the growth rate altogether reduces the estimated level of age structure effects by about one-third.

[16] Results from earlier estimates of a model without any control for relative cohort size are presented in Macunovich (2001). Without that additional control, fourth- (and sixth-) degree polynomials are required to fit the data.

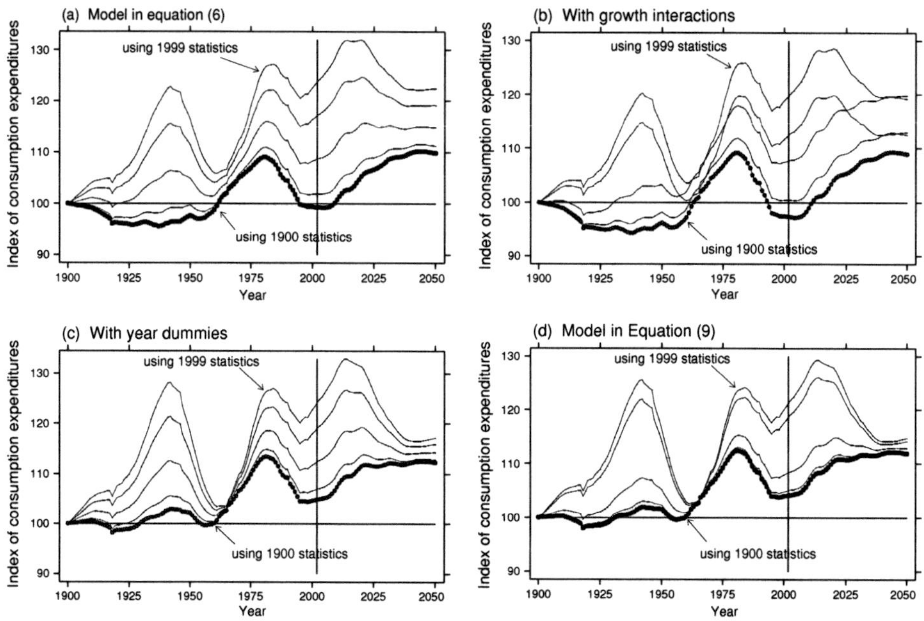

Figure 5. Simulated effect of changing age structure on absolute personal consumption expenditures in the United States, holding everything except age structure constant. Each curve holds real income, total population size, and GDP growth constant at the levels experienced in either 1900, 1925, 1950, 1975, or 1999. The four models, used to prepare the simulations are presented in columns 3–6, respectively, in Table 4. There are strong similarities among the results from all models, especially in the marked declines from 1930 to 1960, increase from 1960 to 1980, and declines again between 1980 and the mid 1990. Rates of change in these indexes are presented in Figure 6.

Including income and its interaction with the population shares in this way, together with a relative cohort size measure, creates a model that explains the apparent reversal in age structure effects between low- and high-income economies produced in earlier models, as demonstrated in Tables 4 and 5, and in Figure 4. Column 2 in Table 4 presents the estimates using $n = 2$ in the basic model set out in equation (5), and columns 3–6 illustrate the effect of successively adding controls for percents urban- and foreign-born, interaction terms between age structure and the growth rate, and year dummies. Table 5 presents results of model estimates designed to illustrate the significance of the age structure variables in alternative formulations, again in the quadratic model; the coefficients on the age structure variables, their interaction with income, and on relative cohort size remain highly significant in all formulations.

The top four panels in Figure 4 illustrate the patterns of age structure coefficients estimated in models (5) and (9), with $n = 2$, i.e., with age share coefficients constrained to lie along

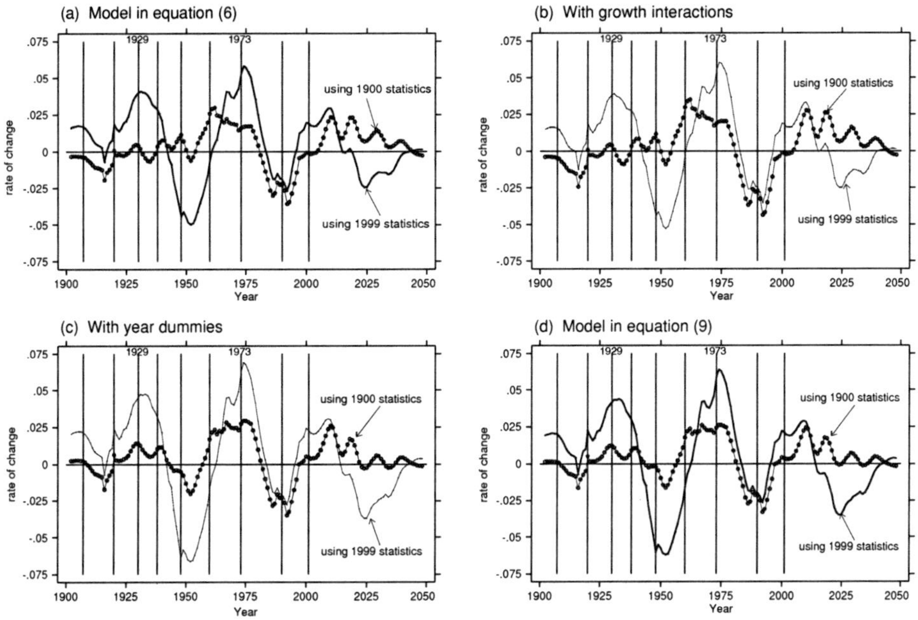

Figure 6. Rates of change in indexes of simulated absolute personal consumption expenditures generated by age structure changes in the United States as presented in Figure 5. The four models used to prepare the simulations are presented in columns 3–6, respectively, in Table 4. There are strong similarities among the results from all models, in terms of the timing of peaks and troughs. The vertical lines indicate peaks in consumption expenditure growth that coincide with business cycle peaks. These account for 9 of the 21 20th century business cycle peaks identified by the NBER, including some of the most significant, such as those in 1929 and 1973.

a quadratic polynomial. Figures 4a and 4b illustrate the patterns of age share coefficients estimated in the two models presented in columns 2 and 6 (equations (5) and (9)), respectively, of Table 4. Each graph illustrates two separate sets of age share coefficients—the effect on the intercept (coefficients on the age shares themselves) and slope (coefficients on the interactions between age shares and per capita income) of a savings equation. The simple age share effects conform to the standard "hump-shaped" life-cycle pattern of saving and dis-saving, but the hump is inverted for the coefficients on the interaction terms with income, suggesting a shift in the total effect of age structure as real per capita income rises.

It is difficult to conceptualize the total effect of age structure from the panels on the left of Figure 4, however, since the total effect depends on the magnitude of the interaction effects, which in turn depends on level of real per capita income. Thus, the panels on the right in Figure 4 illustrate the combined effects of age structure in a savings equation, at two different levels of per capita income—$1,000 and $35,000 in year 1999 dollars. Figure 4e

Table 4. Full regression results for the savings rate as a function of age structure variables, using Lebergott state-level data for the U.S. in 1900, 1929, 1970, 1977, and 1982. Columns 3–6 in this table correspond to the four models used in Figures 5 and 6. Columns 2 and 6 in this table were used as the basis for simulations reported in Figure 4.

	Equation (5)	Equation (6)	w/growth interactions	w/year dummies	Equation (9)
Relative cohort size	−0.125***	−0.134***	−0.161***	−0.140**	−0.132**
	(0.043)	(0.040)	(0.041)	(0.063)	(0.063)
Z1	0.00143***	0.00193***	0.00172***	0.00116***	0.00103***
	(0.0005)	(0.0004)	(0.0004)	(0.0004)	(0.0004)
Z2	−1.70e-05***	−2.11e-05***	−1.88e-05***	−1.33e-05***	−1.20e-05***
	(4.90e-06)	(4.60e-06)	(4.70e-06)	(4.30e-06)	(4.30e-06)
Z1 * income/capita	−0.00133***	−0.00141***	−0.00141***	−0.00141***	−0.00140***
	(0.0002)	(0.0002)	(0.0002)	(0.0002)	(0.0002)
Z2 * income/capita	1.43e-05***	1.45e-05***	1.46e-05***	1.47e-05***	1.47e-05***
	(2.50e-06)	(2.30e-06)	(2.40e-06)	(2.20e-06)	(2.20e-06)
Z1 * growth rate			9.38e-05**		8.75e-05**
			(4.59e-05)		(4.11e-05)
Z2 * growth rate			−1.10e-06**		−9.00e-07**
			(5.00e-07)		(4.00e-07)
real income/capita	0.272***	0.226***	0.227***	0.428***	0.428***
	(0.05)	(0.048)	(0.048)	(0.065)	(0.067)
growth rate of GDP	−0.015***	−0.014***	−0.026***	−0.008***	−0.009
	(0.003)	(0.003)	(0.008)	(0.003)	(0.007)
total population	0.0095	0.025***	0.027***	0.026***	0.027***
	(0.007)	(0.007)	(0.007)	(0.007)	(0.007)
% urban		−0.118***	−0.115***	−0.116***	−0.126***
		(0.023)	(0.024)	(0.022)	(0.023)
% foreign born		−0.018***	−0.019***	−0.038***	−0.034***
		(0.007)	(0.007)	(0.009)	(0.009)
Year = 1929?				0.010	0.020
				(0.023)	(0.024)
Year = 1970?				−0.252***	−0.232***
				(0.060)	(0.063)
Year = 1977?				−0.276***	−0.263***
				(0.065)	(0.066)
Year = 1982?				−0.209***	−0.197***
				(0.072)	(0.073)
Constant	−0.564***	−0.698***	−0.666***	−1.025***	−1.025***
	(0.127)	(0.120)	(0.122)	(0.133)	(0.138)
sigma_u	0.04029	0.035102	0.03435	0.0384	0.037827
sigma_e	0.075900	0.074200	0.073396	0.062970	0.062216
rho	0.219913	0.182874	0.179700	0.271119	0.269896
R^2: within	0.7599	0.7702	0.7782	0.8319	0.8363
between	0.1843	0.4130	0.4324	0.4622	0.4631
overall	0.6320	0.6996	0.7082	0.7569	0.7597
Wald chi^2	546.11	650.90	673.23	941.78	955.92
Prob > chi^2	0.0000	0.0000	0.0000	0.0000	0.0000

Notes: Dependent variable is − ln (personal consumption expenditures/total personal income). 248 observations. All variables except growth rate used in logged form. Standard errors in parentheses. Estimated as random effects models using GLS to control for heteroscedasticity.

$^{***}p < 0.01$; $^{**}p < 0.05$; $^{*}p < 0.10$.

Table 5. Partial models of the savings rate as a function of age structure variables, using the Legbergott state-level U.S. data. Here it can be seen that the interaction terms between age structure and the growth rate are not significant (column 6) until a control for relative cohort size is introduced (column 7), and even then these interactions are significant only at the 10 percent level. Columns 4 and 6 in Table 4 indicate that these variables become much more significant when income interactions are added to the model. The relative cohort size variable remains highly significant (in results not presented here) even when no other variables are included in the model.

	Equation (5)					
Relative cohort size		−0.129***		−0.125***		−0.235***
		(0.033)		(0.043)		(0.040)
Z1			0.00205***	0.00143***	−0.00041*	−0.00115***
			(0.0004)	(0.0005)	(0.0002)	(0.0002)
Z2			−2.26e-05***	−1.70e-05***	3.66e-06	1.09e-05***
			(4.64e-06)	(4.91e-06)	(2.28e-06)	(2.45e-06)
*Z1 * income/capita*			−0.00154***	−0.00133***		
			(0.0002)	(0.0002)		
*Z2 * income/capita*			1.61e-05***	0.0000143***		
			(2.52e-06)	(2.55e-06)		
*Z1 * growth rate*					8.02e-05	9.17e-05*
					(5.39e-05)	(5.08e-05)
*Z2 * growth rate*					−8.28e-07	−1.04e-06*
					(5.67e-07)	(5.35e-07)
real income/capita	0.170***	0.168***	0.199***	0.272***	0.205***	0.244***
	(0.010)	(0.010)	(0.044)	(0.050)	(0.021)	(0.020)
growth rate of GDP	−0.007***	−0.007***	−0.012***	−0.015***	−0.008	−0.028***
	(0.003)	(0.003)	(0.003)	(0.003)	(0.008)	(0.008)
total population	−0.004	−0.006	0.010	0.010	0.003	0.009
	(0.008)	(0.008)	(0.008)	(0.007)	(0.008)	(0.007)
constant	−0.156***	−0.064	−0.444***	−0.564***	−0.372***	−0.413***
	(0.057)	(0.061)	(0.122)	(0.127)	(0.103)	(0.096)
sigma_u	0.051234	0.053423	0.043497	0.04029	0.040087	0.034738
sigma_e	0.087571	0.083229	0.078360	0.075900	0.086738	0.080427
rho	0.255003	0.291787	0.235552	0.219913	0.176003	0.157225
R^2: within	0.6700	0.7045	0.7457	0.7599	0.6716	0.7216
between	0.0583	0.0519	0.1766	0.1843	0.1451	0.2400
overall	0.5323	0.5388	0.6238	0.6320	0.5555	0.6031
Wald chi^2	371.24	418.65	527.44	546.11	363.67	441.95
Prob > chi^2	0.0000	0.0000	0.0000	0.0000	0.0000	0.0000

Notes. Dependent variable is − ln (personal consumption expenditures/total personal income). 248 observations. All variables except growth rate used in logged form. Standard errors in parentheses. Estimated as random effects models using GLS to control for heteroscedasticity.

*** $p < 0.01$; ** $p < 0.05$; * $p < 0.10$.

compares a low-income ($1,000 per capita) but high-growth (15.5 percent p.a.) economy, and a high-income ($35,000 per capita) but low-growth (1 percent p.a.) economy, consistent with the standard model of economic convergence. In Figures 4d and 4e the pattern of age structure effects at low incomes, as in LDCs, is "hump" shaped as in the life-cycle model, but at levels of income typical of those in industrialized nations *the pattern is inverted*. This is the same effect obtained in Table 2 with the Higgins and Williamson (1997) model—but there it was necessary to estimate two separate models to fit the two sets of data, for low- and high-income periods.

The coefficients presented in Figure 4 represent elasticities. For example, Figure 4d suggests that a 10 percent increase in the population share of 1-year-olds would produce a 0.25 percent drop in the savings rate at the aggregate level in LDCs, but a 0.5 percent *increase* in the savings rate in high-income economies. Figure 4e indicates that although the *overall pattern* of age share coefficients is not changed in moving from equation (5) to equation (9), the *magnitude* of the total effect increases. In Figure 4e a 10 percent increase in the population share of 1-year-olds produces a 0.5 percent drop in the savings rate at the aggregate level in the low-income high-growth economy, but a 0.6 percent *increase* in the savings rate in the high-income low-growth economy.

For those who find a complete inversion of the life-cycle savings pattern at higher levels of income difficult to accept, it is perhaps helpful to consider how the pattern of coefficients changes as the degree of the polynomial increases. As mentioned earlier, there is some support in the statistical tests for the sixth-degree polynomial (in terms of higher R^2 and significant t-statistics on the highest Z's, although Wald tests are singular). What is the pattern of age share coefficients produced using a sixth-degree polynomial? This is demonstrated in the bottom two panels of Figure 4, where (in panel 4f) the standard hump-shaped life-cycle pattern is retained for ages 15–75 even at the higher income level, although with a considerably reduced amplitude.

Although the pattern of effects estimated here is more complex than the standard life-cycle model, it is not inconsistent: it simply adds a new wrinkle, with respect to the effect of children once we have controlled for the ages and spending patterns of their parents. In addition, these parameters do not necessarily reflect individual behaviour as it might be observed at the household level; rather they estimate the behaviour induced by various age groups at the societal level, including all interhousehold and intergenerational spending. Given the complexity of the effects estimated here, it is hardly surprising that studies over the years have produced so many different estimates of the effect of dependency on savings and growth.

8. Might These Effects Be Significant in Terms of Macroeconomic Fluctuations?

Returning to the question at the opening of this chapter, what impact has changing age structure had on actual patterns of consumption in the United States throughout the 20th century, as estimated by this model? The model results presented in the previous section are all highly significant *statistically*, but how substantively significant are they in *real world* terms? Whereas for the reasons discussed earlier it would be inappropriate to attempt to answer that question at the national level using parameters obtained in a micro level analysis, this analysis has provided macro level parameters, measured at the state level that *can* be applied at the national level. The goal is to use them to simulate the pattern of PCE over time in the United States at the national level that would have resulted if nothing other than age structure had changed: i.e., holding constant the total population size, income level, and GDP growth rate.

But holding income, population size and growth constant makes it very difficult to illustrate the full effect of changing age structure, both directly and as it operates through income and the growth rate. In order to estimate the full effect of changing age structure, each of the graphs in Figure 5 (which are based on the models in columns 3–6 of Table 4, moving from equations (5) to (9) with $n = 2$) presents the results of five different simulations for the United States. In each one, all variables except age structure are held constant, but at a different level—one, at the levels observed in 1900, another at 1925 levels, and three others at 1950, 1975, and 1999 levels, respectively. They trace out undulating "ribbons" of effects over the century on total PCE, with a maximum range of about 25 percent using the 1999-level characteristics, but applied to age structure changes that actually occurred in the 1960 to 1980 period.

It is worth emphasizing the meaning of this figure. If, in 1999, the age structure had been that observed in 1980, when the largest baby boom cohorts were about 20 years old—but with all other factors including income and population size held constant—the level of PCE would have been *25 percent higher* than the expenditures that would have occurred with a 1960 U.S. age structure, when birth rates were at their peak and the share of 20-year-olds was very low. This swing is virtually unaffected by the different formulations of the model that are presented in columns 3–6 of Table 4.

Despite the difference in levels among the five curves in each graph in Figure 5, they all have in common extremely strong growth between 1960 and 1980, and then decline after 1980, as a result of the baby boom's passing from childhood to young adulthood. It is outside the scope of this study to say whether this is a "good" or "bad" effect; it implies strong growth in aggregate demand, but could obviously result in declining savings rates unless per capita incomes rise correspondingly. And, of course, they did rise dramatically in the 1960s and early 1970s—assisted at least in part by the rising domestic aggregate demand. As Higgins and Williamson (1997) and Higgins (1998) emphasize, in open economies an increase in the share of domestic income devoted to consumption need not result in a decline in overall investment and hence a decline in productivity and real growth. These researchers trace a pattern of increasing age-structure-induced investment supported by an increasingly negative current account balance, culminating in rising productivity and growth in real per capita output (and a reversal to a positive current account balance) as children become productive young adults.

The similarity in the estimated effects in the four panels of Figure 5 is emphasized if we examine the *rates of change* in consumption expenditures, as in Figure 6, rather than the levels shown in Figure 5. Each of the four panels in Figure 6 illustrate rates of change in consumption expenditures using the 1900 and 1999 values for income, total population size, and growth rate. The similarities in the four panels are striking, with peaks occurring

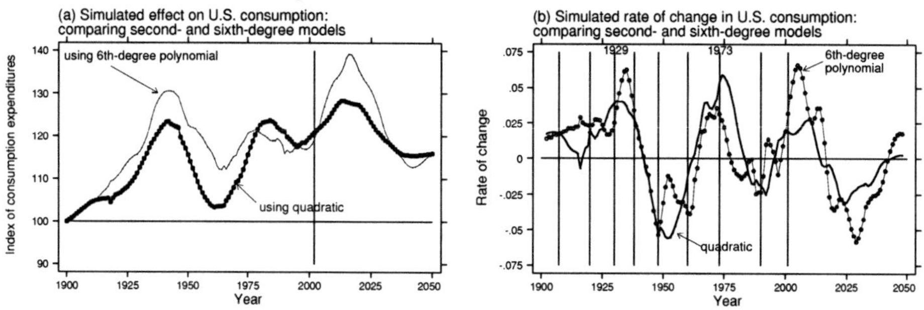

Figure 7. The effect of moving from a quadratic to a sixth-degree polynomial, on simulated impact of changing age structure on U.S. personal consumption expenditures.

in virtually the same years regardless of model formulation or levels of nonage-structure variables. The vertical lines in Figure 6 mark peaks in the growth rate of age-structure-induced consumption expenditures that coincide with business cycle peaks during the 20th century. This coincidence of age-structure-induced peaks and business cycle peaks occurs for at least 9 of the 21 business cycle peaks identified in the 20th century by the NBER, including several of the most significant, such as those in 1929 and 1973.

And finally, there is the question of sensitivity to model formulation. Although statistical tests indicate that anything higher than a quadratic polynomial for the age structure coefficients probably overfits the data, it is nevertheless worthwhile to examine the effect of moving from a quadratic to a sixth-degree polynomial in estimating age structure effects at the aggregate level. Figure 7 presents a comparison, using 1999 levels of income, total population size, and growth rate, of simulated effects using the age share coefficients estimated with two different polynomials in equation (5): $n = 2$ and $n = 6$. The two patterns are remarkably similar, in terms of the overall sinusoidal pattern, although differences between the two appear to be greatest in the most recent period when age structure changes have been most dramatic, with considerably wider swings over time estimated using the sixth-degree polynomial.

9. Conclusions

As expected, the results of this analysis support the hypothesis that the dependency effect changes with income level; children in lower income economies have an overall positive (negative) effect on the consumption (savings) rate, but as per capita income rises that effect is reversed. Children in high-income areas appear to motivate increased saving—or at least relatively lower levels of consumption—on the part of other members of the population, whether through precautionary motives or to finance bequests. Perhaps when infant and child mortality rates are low, and we turn from a desire for "quantity" to "quality" in our offspring, children make us more forward-looking: they provide a reason to place increased weight on future consumption. If so, and if indeed, the savings rate has declined in the United States over the past two decades, perhaps that decline occurred because there are relatively fewer children to save *for*, which would be ironic given the negative emphasis often placed on children in the dependency literature.

But despite this shifting effect, the estimates presented here suggest that changes in age structure—especially those due to the baby boom—have had a major impact on the U.S. economy, probably contributing a huge boost in the 1960–1980 period, but then a considerably weakened effect after 1980. Boomers and their children are likely to exert another strong positive effect in the first part of the 21st century, based on the parameters in the model and current Census Bureau population projections.

In addition, exploratory work on nations around the world from 1950 to 2000 indicate a strong correlation between changing age structure and the incidence of "financial crises" over the past 20 years in developing nations, as a type of delayed fallout from the demographic transition.[17] Perhaps the boom in consumption expenditures generated by increasing young adult cohorts contributes to "irrational exuberance" on the part of investors and lenders—which then collapses when growth in the size of those cohorts switches to decline.[18]

But obviously this analysis is only a beginning, given this "new" technique for estimating age structure effects that was introduced by Fair and Dominguez (1991). It suggests a very complex pattern of age structure effects that is consistent with the life-cycle model, but at the same time generates a much wider range of dependency effects than has been contemplated up to now in the literature. It is hoped that the results here will generate additional interest in using full age distributions to examine aggregate age structure effects, and allowing for the effects of interhousehold transfers on aggregate patterns of age-specific effects.

Appendix: Data Sources

The data on consumption expenditures which were used in the analysis were prepared by Stanley Lebergott (1997), in a painstaking and well-documented effort which produced PCE at the state level, broken down into the 100+ sub-categories itemized in the national income accounts, for the years 1900, 1929, 1970, 1977, and 1982, as well as annual data at the national level for the years 1900–1993. The same source provides personal income figures at the state and national levels, for the same dates.

As Lebergott describes, his data were developed chiefly from

> the Censuses of Retail Trade, Services, Housing, Government and Population. Each Census drew on an enormous sample of knowledgeable respondents. For the most part the reports rest on detailed and original records rather than fleeting and vagrant consumer memories. Thus the 1977 Census of Retail Trade collected data from firms with over 85% of all retail sales. The

---

17 A discussion of this potential relationship is presented in Macunovich (2002).

18 This pattern is supported by findings in Higgins and Williamson (1997) and Higgins (1998), who find that GDP growth and investment spending in Pacific Rim nations was substantially increased as postdemographic transition population "bulges" entered productive adult life.

> 1977 Census of Services relied on direct reports for about 70% of all services in scope. The 1980 Census of Housing collected rent and value data from over 95% of the population. (p. 71)

His state estimates

> uniformly began with estimates, by specific item, for the United States as a whole... Each U.S. total was distributed among the states by an allocator, and usually checked against another allocator. Had estimates been directly made for each state, they would not necessarily add to an adequate U.S. total. More important, comparisons against the per capita average for the U.S. as well as nearby states, permit some judgment as to whether a state estimate falls outside reasonable limits.

It is important to note that in no case were the allocators used by Lebergott based on age distributions within the population. Rather, they were derived from various census data on production and expenditures, as well as distributions of workers by occupation and service income. The following passages, taken from his documentation for 1900, illustrate the methodologies employed by Lebergott in allocating expenditures at the state level.

> For such major items as food, clothing, furniture, and lighting we utilize the 1901 expenditure survey of 25,440 families by the U.S. Commissioner of Labour. Our individual state averages for these individual items were checked by regressing them against relevant occupation counts times average nonagricultural service income. (p. 92) For food off-premise the result was then checked against the Population Census count of persons engaged in food retailing: merchants and dealers (excl. wholesale) in groceries and produce, hucksters and peddlers, butchers, bakers, and confectioners. (p. 93) A U.S. total of meals and beverages "was allocated by the number of persons in specified occupations (hotel keepers, bartenders, restaurant keepers, saloon keepers, and waiters) times the average service income per worker in the state." (p. 93) Food furnished to employees was allocated using two series: "One was the aggregate expenditures on farm labour reported by farmers. The other was (a) average monthly wages without board, minus average wages with board, divided by (b) farm wages without board." (p. 94)
>
> The value of dairy products consumed on farms was used to allocate the U.S. total for food produced and consumed on farms by farm operators. (p. 94)
>
> Clothing expenditures per capita in thirty-three states can be derived from the survey by the Commissioner of Labour. The intra-regional variation shown by these figures seemed unreasonably great, and was probably a reflection of sampling variability. We therefore averaged the per capita figures within each of eight regions. These averages were then tested by correlating them with per capita expenditures given by multiplying the occupation count for two groups of merchants and dealers (clothing and men's furnishings plus dry goods, fancy goods, and notions) times the service income per worker... (p. 95)

POPULATION DATA

The population data were taken from hard copy and electronic files provided by the Bureau of the Census. These data were on occasion available only by 5-year age group: in such cases, single years of age were estimated as one-fifth of each corresponding 5-year age group (although econometric results were found to be unchanged using simple 5-year age groups).

1900
Data at the state level on total population and population by 5-year age group through age 34, and 10-year age group 35–64, were taken from the 1900 Census of Population, Volume 1, Part 1, Table 3, pp. 110–111.

These data were supplemented by percentage distributions of state populations by 5-year age group through age 84, provided for 1900 in the 1930 Census of Population, Volume II: General Report, Statistics by Subject, Table 25, pp. 660–668.

1929
Data for the year 1930, as provided in the 1930 Census of Population, were used for 1929. Data at the state level on total population and population by 5-year age group through age 84 were taken from the 1930 Census of Population, Volume II, General Report, Statistics by Subject, Table 24, pp. 610–658.

1970
Data at the state level on population aged 0–2, 3–4, 5–13, by single year of age from 14–24, and for age groups 25–29, 30–34, 35–44, 45–54, 55–59, and 65+ were available in electronic form for the year 1970 in file e7080sta.txt from the Bureau of the Census website *http://www.census.gov/population/estimates/state*

The 1970 estimates in this file are consistent with those published in Current Population Reports Series P-25, No. 998. These electronic data were supplemented using total population and population by 5-year age group through age 84 taken from the 1970 Census of Population, General Population Characteristics, Volume I, Section I, Part 1, Table 62, pp. 297–309.

1977
Two alternative methods were used to prepare population by single year of age for 1977:

1. Data for 1980 are available at the state level by single year of age to 84 (based on the 1980 Census), and there are significant discrepancies between the Census Bureau's population estimates for the late 1970s, and the 1980 census. These discrepancies, taken together with the fact that the 1977 data are too aggregated in certain age ranges (see point 2 following), led to an effort to "backdate" the 1980 data by 3 years, to be used in place of the questionable 1977 figures. These backdated 1980 population figures were assumed to be preferable for calculating population age shares.
2. Data at the state level on population aged 0–2, 3–4, 5–13, by single year of age from 14 to 24, and for age groups 25–29, 30–34, 35–44, 45–54, 55–59, and 65+ were available in electronic form for the year 1977 in file e7080sta.txt from the Bureau of the Census website
   http://www.census.gov/population/estimates/state
   The 1977 estimates in this file are consistent with those published in Current Population Reports Series P-25, No. 998. Because there is no additional source of data for 1977, as there is for 1970, to break down the larger age aggregates (35–39 from 35–44, 45–49 from 45–54, 5–9 from 5–13, and 5-year groups above 64), national patterns

Table A.1. Coverage in each decade.

	1900–1929	1930–39	1940–49	1950–59	1960–79
Age	**0–75+**	**0–75+**	**0–85+**	**0–85+**	**0–85+**
Resident population	Yes	Yes	Yes	Yes	Yes
AK and HI?	excluded	excluded	excluded	included	included
AF overseas?	excluded	excluded	included	included	included
Reference CPR P-25	#311	#11	#311	#311	#519 and #917

for the year 1977 were used within these age groups. These estimated 1977 figures were used as a check on the backdated 1980 figures: it was found that there is a close correspondence.

1982
Data at the state level on total population and population by single year of age through age 84 on July 1, 1982 were taken from computer files downloaded from the Bureau of the Census website
http://www.census.gov/population/www/estimates/st_stiag.html

1900–1979 US data
These data were taken from Lotus files of population by single year of age, provided on diskette by the Bureau of the Census (Kevin Deardorff, Population Division, at 301-763-7950).

1980–1989 US data
(Resident population plus Armed Forces overseas, including AK and HI, age 0–100+ and total, as of July in each year): Files e8081pqi.txt through e8990pqi.txt from Bureau of the Census website
http://www.census.gov/population/www/estimates/nat_80s_detail.html

1990–1998 US data
(Resident population plus Armed Forces overseas, including AK and HI, age 0–100+ and total, as of July in each year): Files e9090pmp.txt through e9898pmp.txt from Bureau of the Census website
http://www.census.gov/population/www/estimates/nat_90s_2.html

Method Used to Estimate Coefficients on Population Age Shares

In an unconstrained model, population age shares—if assumed to affect the intercept of the consumption equation—would enter the equation for personal consumption expenditures as

$$\sum_{j=1}^{J} \varphi_j p_j \tag{A.1}$$

where p_j is the share of total population represented by age group j; and φ_j is the coefficient to be estimated. Since the population data used here are available in single

Table A.2. Statistics used to determine appropriate degree of polynomial (*n*).

Degree of polynomial:	**8**	**7**	**6**	**5**	**4**	**3**	**2**
A. Equation (5)							
R-sq: within	0.8142	0.8147	0.8170	0.8034	0.7972	0.7867	0.7599
between	0.2404	0.2409	0.2363	0.1858	0.1750	0.1802	0.1843
overall	0.6913	0.6910	0.6842	0.6457	0.6378	0.6362	0.6320
overall Wald chi^2	85.72***	91.53***	94.31***	113.13***	105.16***	607.07***	546.11***
t on relative cohort	2.4**	2.5**	1.8*	2.8***	1.7*	4.6***	2.9***
t on highest Z	0.2	1.0	3.4***	1.2	1.7*	1.0	3.5****
t on highest Zy (Z * income/cap)	0.0	1.5	2.0*	1.8*	1.1	2.1**	5.6***
combined Wald chi^2 on all Z	na	na	na	na	na	30.82***	17.39***
combined Wald chi^2 on all Zy	na	na	na	na	na	48.45***	33.35***
Wald chi^2 on highest Z & Zy	na	na	na	na	4.95*	13.60***	46.06***
Wald chi^2 on highest Z	na	na	na	1.46	3.01*	1.02	12.00***
Wald chi^2 on highest Zy	na	na	na	3.22*	1.14	4.38**	31.30***
B. Equation (6): including percent foreign-born and percent urban in the basic model							
R-sq: within	0.8295	0.8293	0.8244	0.8111	0.7992	0.7909	0.7702
between	0.4300	0.4331	0.4436	0.4028	0.3976	0.3949	0.4130
overall	0.7522	0.7525	0.7504	0.7274	0.7187	0.7122	0.6996
overall Wald chi^2	151.27***	158.42***	164.21***	179.92***	164.13***	719.76***	650.90***
t on relative cohort size	1.9*	1.8*	1.5	2.9***	1.4	4.8***	3.3***
t on highest Z	1.4	1.2	7.9***	1.1	1.6	0.1	4.5***
t on highest Zy (Z * income/cap)	0.5	0.7	1.6	0.3	0.7	1.2	6.21***
combined Wald chi^2 on all Z	na	na	na	na	na	34.81***	20.97***
combined Wald chi^2 on all Zy	na	na	na	na	na	64.37***	48.77***
Wald chi^2 on highest Z & Zy	na	na	na	na	7.15***	15.83***	40.38***
Wald chi^2 on highest Z	na	na	na	1.18	2.59*	0.01	20.69***
Wald chi^2 on highest Zy	na	na	na	0.08	0.52	1.54	36.93***
C. Equation (6) plus growth rate interactions with age structure variables							
R-sq: within	0.8278	0.8282	0.8250	0.8110	0.7988	0.7916	0.7782
between	0.4495	0.4483	0.4546	0.4150	0.4132	0.4123	0.4324
overall	0.7543	0.7541	0.7524	0.7292	0.7209	0.7149	0.7082

Table A.2. (*Continued*)

Degree of polynomial:	**8**	**7**	**6**	**5**	**4**	**3**	**2**
overall Wald chi^2	160.13***	156.91***	164.92***	178.29***	163.58***	705.88***	673.23***
t on relative cohort size	1.9*	1.8*	1.5	2.9***	1.5	4.7***	3.9***
t on highest Z	1.1	0.7	2.9***	0.9	1.5	0.1	4.0***
t on highest Zy (Z * income/cap)	0.0	0.4	1.5	0.1	0.7	1.2	6.2***
combined Wald chi^2 on all Z	na	na	na	na	na	59.21***	48.89***
Wald chi^2 on highest Z & Zy	na	na	na	na	6.08**	9.06**	46.04***
Wald chi^2 on highest Z	na	na	na	0.80	2.19	0.01	16.33***
Wald chi^2 on highest Zy	na	na	na	0.01	0.44	1.46	38.56***
D. Equation (9): basic model plus % foreign-born, % urban, growth interactions, and year dummies							
R-sq: within	0.8437	0.8452	0.8445	0.8437	0.8429	0.8365	0.8363
between	0.7061	0.6949	0.6840	0.5697	0.4552	0.4694	0.4631
overall	0.8133	0.8126	0.8101	0.7895	0.7644	0.7605	0.7597
overall Wald chi^2	246.98***	260.83***	257.18***	278.68***	249.58***	924.45***	955.92***
t on relative cohort size	1.1	1.1	2.1**	2.2**	3.1***	2.3**	2.1**
t on highest Z	0.6	1.0	2.7***	2.1**	0.5	0.4	2.8***
t on highest Zy (Z * income/cap)	0.6	0.4	0.9	3.3***	1.5	0.6	6.8***
combined Wald chi^2 on all Z	na	na	na	na	na	8.77**	10.39***
combined Wald chi^2 on all Zy	na	na	na	na	na	46.19***	51.84***
Wald chi^2 on highest Z & Zy	na	na	na	na	5.69**	0.51	67.65***
Wald chi^2 on highest Z	na	na	na	4.49**	0.25	0.13	7.70***
Wald chi^2 on highest Zy	na	na	na	11.21***	0.14	0.34	45.87***

The standard test for the appropriate n starts from a high-degree polynomial ($n = 8$) and then successively adds restrictions until the last restriction fails. In the cases above, although t-statistics for the highest Z's are significant in a sixth-degree polynomial in Model A, and in a fifth-degree polynomial in model D (indicated in bold), Wald tests are singular in these cases, suggesting an over-fitted model. The only D supported by all test statistics is $n = 2$ (a quadratic polynomial, indicated in bold in all models).

Dependent variable is −ln (personal consumption expenditure/total personal income).

248 observations.

All variables except growth rate used in logged form.

Standard errors in parentheses. ***: $p < .01$; **: $p < .05$; *: $p < .10$.

A designation na indicates cases in which the Wald test is singular.

Estimated as random effects models using GLS to control for heteroscedasticity.

years of age 0–84 and 85+, J—the total number of age groups—is 86 in this analysis. The φ_j are constrained to sum to zero in order present comparable results across models.

However, it would be impractical to attempt to estimate 86 separate coefficients. As an alternative this analysis has adopted a method suggested by Fair and Dominguez (1991), which is in turn similar to Almon's (1965) distributed lag technique. The φ_j are constrained to sum to zero, and constrained to lie on a polynomial of degree n (where n is to be determined in fitting the model) such that

$$\varphi_j = \zeta_0 + \zeta_1 j + \zeta_2 j^2 + \zeta_3 j^3 + \cdots + \zeta_n j^n \qquad j = 1 \ldots J \tag{A.2}$$

and the following constraint has been imposed:

$$\sum_{j=1}^{J} \varphi_j = 0 \tag{A.3}$$

Substituting equation (3) into equation (2) produces

$$\zeta_0 = -\zeta_1(1/J)\sum_{j=1}^{J} j - \zeta_2(1/J)\sum_{j=1}^{J} j^2 - \cdots - \zeta_n(1/J)\sum_{j=1}^{J} j^n \tag{A.4}$$

and thus

$$\sum_{j=1}^{J} \varphi_j p_j = \zeta_1 Z_1 + \zeta_2 Z_2 + \cdots + \zeta_n Z_n \tag{A.5}$$

where

$$Z_n = \sum_{j=1}^{J} p_j j^n - \sum_{j=1}^{J} p_j \sum_{j=1}^{J} j^n \tag{A.6}$$

Acknowledgements

The work presented here has evolved over a 4-year period, as the author has gradually and gratefully incorporated the comments of reviewers and seminar and conference participants. A more general discussion of earlier findings is presented in Macunovich (2002), but none of the formal analysis has been published previously. Reference is made in this chapter to an earlier working paper (Macunovich 2001) containing a more comprehensive discussion of certain issues than could be accommodated here.

References

Almon, S. (1965), "The distributed lag between capital appropriations and expenditures." *Econometrica*, 33(January):178–196.

Attanasio, O.P. (1998), "Cohort analysis of saving behavior by U.S. households." *Journal of Human Resources*, 33(3):575–560.

Auerbach, A.J., Cai, J. and Kotlikoff, L.J. (1990), "U.S. demographics and saving: Predictions of three saving models." *NBER Working Paper 3404*, Cambridge, MA.

Barlow, R. (1994), "Population growth and economic growth: Some more correlations." *Population and Development Review*, 20(1):153–165.

Berger, M.C. (1984), "Cohort size and the earnings growth of young workers." *Industrial and Labor Relations Review*, 37(4):582–591.

Berger, M.C. (1985), "The effect of cohort size on earnings growth: A re-examination of the evidence." *Journal of Political Economy*, 93:561–573.

Berger, M.C. (1989), "Demographic cycles, cohort size, and earnings." *Demography*, 26(2):311–321.

Bosworth, B., Burtless, G. and Sabelhaus, J. (1991), "The decline in saving: Evidence from household surveys." *Brookings Papers on Economic Activity*, 1991(1):183–241.

Collins, S.M. (1991), "Saving behavior in ten developing countries." In: *National Saving and Economic Performance*, J.B. Shoven, ed. pp. 349–375. University of Chicago Press: Chicago.

David, M.H. (1962), *Family Composition and Consumption*. North-Holland Publishing Co.: Amsterdam.

Deaton, A. and Paxson, C. (1997), "The effects of economic and population growth on national saving and inequality." *Demography*, 34(1):97–114.

Easterlin, R.A. (1987), *Birth and Fortune*, second edition. University of Chicago Press: Chicago.

Fair, R.C. and Dominguez, K. (1991), "Effects of the changing U.S. age distribution on macroeconomic equations." *American Economic Review*, 81(5):1276–1294.

Frost, P. (1975), "Some properties of the Almon lag technique when one searches for degree of polynomial and lag (in Theory and Methods)." *Journal of the American Statistical Association*, 70(351):606–612.

Fry, M.J. and Mason, A. (1982), "The variable rate-of-growth effect in the life-cycle saving model." *Economic Inquiry*, 20:426–442.

Godfrey, L.G. and Poskitt, D.S. (1975), "Testing the restrictions of the Almon lag technique (in Theory and Methods)." *Journal of the American Statistical Association*, 70(349):105–108.

Harper, C.P. (1977), "Testing for the existence of a lagged relationship within Almon's method." *Review of Economics and Statistics*, 59(2):204–210.

Higgins, M. (1998), "Demography, national savings, and international capital flows." *International Economic Review*, 39(2):343–369.

Higgins, M. and Williamson, J.G. (1997), "Age structure dynamics in Asia and dependence on foreign capital." *Population and Development Review*, 23(2):262–293.

Higgins, M. and Williamson, J.G. (1999), "Explaining inequality the world round: Cohort size, Kuznets curves, and openness." *NBER Working Paper 7224*, Cambridge, MA. http://www.nber.org/papers/w7224

Judge, G.G., Griffiths, W.E., Carter-Hill, R., Lütkepohl, H. and Lee, Tsoung-Chao. (1985), *The Theory and Practice of Econometrics*, second edition. John Wiley and Sons: New York.

Kelley, A.C. and Schmidt, R.M. (1994), "Population and income change: Recent evidence." World Bank Discussion Paper 249, August 1994, Washington, DC.

Kelley, A.C. and Schmidt, R.M. (1995), "Aggregate population and economic growth correlations: The role of the components of demographic change."*Demography*, 32(4):543–555.

Korenman, S. and Neumark, D. (2000), "Cohort crowding and youth labor markets: A cross-national analysis." In: *Youth Employment and Joblessness in Advanced Countries*, D.G. Blanchflower and R.B. Freeman, eds. pp. 57–105. University of Chicago Press: Chicago.

Lazear, E.P. and Michael, R.T. (1988), *Allocation of Income within the Household*. Univeristy of Chicago Press: Chicago.

Lebergott, S. (1996), *Consumer Expenditures: New Measures and Old Motives*. Princeton University Press: Princeton, NJ.

Leff, N.H. (1969), "Dependency rates and savings rates." *American Economic Review* 59(5):886–896.

Lino, M. (1998), *Expenditures on Children by Families:1997 Annual Report*. United States Department of Agriculture, Center for Nutrition Policy and Promotion, Miscellaneous Publication Number 1528–1997, Washington, DC. Available at http://www.usda.gov/fcs/cnpp/using2.htm

Macunovich, D.J. (1999), "The fortunes of one's birth: Relative cohort size and the youth labor market in the U.S." *Journal of Population Economics*, 12:215–272.

Macunovich, D.J. (2001), "Lessons from the baby boom: Delayed effects of fertility on patterns of consumption." http://newton.uor.edu/Departments&Programs/EconomicDept/macunovich/agdemand.pdf

Malthus, T.R. (1817), *An Essay on the Principle of Population*, selected and introduced by Donald Winch, Cambridge University Press, 1992.

Mason, A. (1981), "An extension of the life-cycle model and its application to population growth and aggregate saving." *East-West Population Institute Working Papers 4* (January).

Mason, A. (1988), "Saving, economic growth, and demographic change." *Population and Development Review*, 14(1):113–144.

Mason, A. (1987), "National saving rates and population growth: A new model and new evidence." In: *Population Growth and Economic development: Issues and Evidence*, D.G. Johnson and R.D. Lee, eds. pp. 523–560. University of Wisconsin Press: Madison.

Schmidt, P. and Mann, W.R. (1977), "A note on the approximation of arbitrary distributed lag structures by a modified Almon lag (in Theory and Methods)." *Journal of the American Statistical Association*, 72(358):442–443.

Schmidt, P. and Sickles, R. (1975), "On the efficiency of the Almon lag technique." *International Economic Review*, 16(3):792–795.

Stoker, T. (1986), "Simple tests of distributional effects on macroeconomic equations." *Journal of Political Economy*, 94(4):763–795.

Taylor, A.M. (1995), "Debt, dependence and the demographic transition: Latin America into the next century." *World Development*, 23(5):869–879.

Taylor, A.M. and Williamson, J.G. (1994), "Capital flows to the new world as an intergenerational transfer." *The Journal of Political Economy*, 102(2):348–371.

Terasvita, T. (1976), "A note on bias in the Almon distributed lag estimator." *Econometrica*, 44(6):1317–1321.

Tobin, J. (1967), "Life cycle saving and balanced growth." In: *Ten Economics Studies in the Tradition of Irving Fisher*, W. Fellner, et al. eds. Wiley and Sons: New York.

Weil, D.N. (1994), "The saving of the elderly in micro and macro data." *The Quarterly Journal of Economics*, 109(1):55–81.

Welch, F. (1979), "Effects of cohort size on earnings: The baby boom babies' financial bust." *Journal of Political Economy*, 87(5, pt.2):S65–S97.

CHAPTER 11. SOME INTERGENERATIONAL TRANSFER IMPLICATIONS OF BIRTH FLUCTUATIONS

ROBERT SCHOEN AND STEFAN JONSSON

Department of Sociology, Pennsylvania State University

With the demographic transition from high to low rates of birth and death accomplished or underway in most of the world, some observers have predicted that humanity is on the road to zero population growth. In the long term, the destination of unchanging population size is probably an illusion. In the short term, the road must surmount population growth and population ageing.

A good deal of research has been done on population momentum, the increase in population size that occurs after fertility has fallen to replacement level (cf. Keyfitz 1971). Preston (1986) was the first to realize that the population growth associated with momentum occurs only at ages over 30 years. Kim and Schoen (1997) showed that momentum and population ageing were simply two aspects of the same phenomenon—the restabilization of the age structure from that of a growing population to that of a stationary one.

The economic and social impacts of recent demographic changes, and their likely future consequences, have received considerable attention. Preston (1984) saw a shift in transfer payments to the elderly and away from the young, concentrating poverty in single parents and their children. Menken (1985) observed that middle-aged persons may be caught in a dependency bind, as they face simultaneous demands from their teenage children and ageing parents. Keyfitz (1988) saw a rise in intercohort inequities in the receipt of pension benefits. In the course of the adjustment to zero growth, population ageing may be temporarily exaggerated (Schoen and Kim 1996), further straining mechanisms for intercohort equilibration and intergenerational transfer. The national pension system in the United States, and in a number of other countries, is facing serious financial pressures as the "baby boom" cohorts begin to retire and there are relatively fewer economically active persons contributing to the system.

More generally, a considerable literature (see Lee 1997) has examined population–economic interactions. In a classic article, Tobin (1967) examined how demographic change affected savings rates and capital formation. More recently, Lee (1994) examined the formal demography of ageing and its economic consequences. In the context of a stable population model with the growth rate equal to the interest rate, he estimated that an additional year of life expectancy would reduce individual consumption (or increase individual labour supply) by about 1 percent.

Anne H. Gauthier et al. (eds.), Allocating Public and Private Resources across Generations, 279–290.

An end to sustained population growth does not necessarily mean an end to differences between cohorts. Population behaviour can cycle, and cyclical models have long played a prominent role in theories of population growth and economic change. To Malthus (cited in Lee 1997:1092) the fact that populations would oscillate irregularly about their equilibrium in long, slow swings was something that "no reflecting man who considers the subject deeply can well doubt." In population mathematics, cyclically stable models date back to Skellam (1967), and have been examined by a number of authors (cf. Caswell 1989:198–202; Schoen and Kim 1994). In particular, Tuljapurkar (1986a,b) provided an in-depth treatment of the mathematical structure of nonlinear, age-structured cyclical models. Theoretically, cycles have been seen as arising from a number of mechanisms. Cycles one generation long can arise from random perturbations (Lee 1974). "Malthusian cycles" of about two generations in length can arise from economic–demographic interactions. Bourgeois-Pichat (1979), reflecting on the development of below replacement fertility in Europe, saw a tendency for persons born in small families to have large families, and vice versa. He argued that that pattern would give rise to self-generated wave movements in fertility.

A similar idea was given great prominence by Easterlin (1978, 1980). He sought to explain the pronounced 40-year demographic cycles that characterized the twentieth century United States (and a number of other countries) in terms of a self-renewing demographic–economic interaction. Specifically, he argued that both the postwar baby boom and the subsequent baby bust were brought about by swings in generation size that affected the economic circumstances of young adults. Those born in the small cohorts of the 1930s encountered favourable economic circumstances and produced the large cohorts of the 1950s. Those large cohorts, in no small part because of their size, encountered unfavourable circumstances in the 1970s, and had few children. Easterlin found that the ratio of men 15–29 to men 30–64 closely tracked many economic and demographic variables. Arguing that cohort size affected economic circumstances, which in turn influenced demographic behaviour, Easterlin concluded that small generations tend to produce large generations and create self-generating cycles of some 40 years.

Methodologically, Easterlin's model was criticized for its informality and strong assumptions (Lee 1997:1105–1106), and doubts were raised as to whether such regular waves could be sustained (Wachter 1991). Empirically, the failure of fertility to rise as predicted during the 1980s and 1990s largely discredited the thesis, despite arguments that migration introduced a new element, which disrupted the process. Nonetheless, Schoen and Kim (1997) did find a demographic model consistent with the Easterlin hypothesis, and the idea of repeating cycles in demographic behaviour is one that is both plausible and appealing. Population size and structure have always fluctuated. Changing climates brought good and bad harvests, epidemics took their toll, and economic cycles continue to affect the most advanced countries.

We are a long way from being able to predict future demographic behaviour, let alone accurately model the dynamics of economic–demographic interactions. However, the importance of changes in age structure for individual welfare and for the growing number of increasingly more comprehensive transfer systems has risen considerably. In that context, assuming that the population of the future will be stationary and have birth cohorts unchanging in size is simply untenable.

To move toward a more realistic long-term population model, we present a straightforward, baseline demographic analysis of how birth fluctuations affect age composition and can impact intergenerational transfers. We begin with a simplified model where birth cohort size varies sinusoidally in the absence of long term population growth. We then examine how the population's dependency burden varies over time, and how differences in size can generate transfer disparities across cohorts.

1. The Basic Model

The framework has been kept as simple as possible, given a model with changing vital rates. In part that serves to simplify the calculations, but mostly it is to make the results as interpretable and transparent as possible. The framework can readily be adapted to a variety of birth functions, mortality conditions, and transfer systems.

Our model is cyclically stationary, i.e., there is no long-term population growth but the vital rates repeat themselves every T years. Mortality does not change over time, with every person born surviving to exactly 90 years and dying at that point. Fertility is such as to produce birth cohorts whose size in year t, $b(t)$, is given by

$$b(t) = 1 + A \sin \omega t \tag{1}$$

where $-1 < A < 1$ represents the amplitude of the sine wave and $\omega = 2\pi/T$ is its frequency. (If $A = 0$, the population is stationary, with one birth every year.) Kim and Schoen (1996) analysed such birth trajectories, assuming that the associated net maternity function was normally distributed. They found that the net reproduction rate function ($R(t)$) oscillated with the same cycle length as the birth function, but was not sinusoidal. The $R(t)$ curve could be as much as a quarter cycle ahead of or behind $b(t)$. The size of $R(t)$ relative to $b(t)$ varied substantially, and was sensitive to the ratio of cycle length T to the mean age at net maternity.

At time t, the total number of persons in the population between the ages of x and y ($x < y$), $P(x, y, t)$, is given by

$$P(x, y, t) = \int_{t-y}^{t-x} b(\tau)\, d\tau = (y - x) + [A/\omega][\cos \omega(t - y) - \cos \omega(t - x)] \tag{2}$$

When $x = 0$ and $y = 90$, equation (2) indicates that the total population size, $P(t)$, is

$$P(t) = 90 + [A/\omega][\cos \omega(t - 90) - \cos \omega t] \tag{3}$$

Since the cosine ranges between -1 and $+1$, the size of the population is always bounded by $90 \pm 2A/\omega$ (or by $90 \pm AT/\pi$).

We want to distinguish between the economically active and inactive (or dependent) components of the population. Any age criterion is bound to be arbitrary and subject to some error, but we believe that it is reasonable to consider ages 20 through 65 as the economically active years and ages 0 through 20 and 65 and over as the years of dependency. It follows that we can define the population dependency ratio at time t, $D(t)$, as

$$D(t) = [P(0, 20, t) + P(65, 90, t)]/P(20, 65, t) \tag{4}$$

or the ratio of the number in the dependent ages over the number economically active. In our model, the dependency ratio can be written as

$$D(t) = \frac{45 + [A/\omega][\cos\omega(t-90) + \cos\omega(t-20) - \cos\omega(t-65) - \cos\omega t]}{45 + [A/\omega][\cos\omega(t-65) - \cos\omega(t-20)]} \tag{5}$$

Here our focus is on the total population dependency burden. We do not distinguish between formal (e.g., pension system) or informal (e.g., familial) transfers. Moreover, we assume that, at the societal level, the burden of one younger person is the same as that of one older person. That is something of a heroic assumption, though Leff (1969) found that the effects on savings of a marginal child or marginal elder were approximately the same. In any event, that assumption can readily be modified as needed, and it facilitates the inclusion of dependency at both the oldest and youngest ages.

2. Changes in Aggregate Dependency

Our cyclically stationary model can describe a range of behaviours as its two parameters, cycle length (T) and maximum amplitude (A), vary. Here we emphasize cycle lengths from 15 to 105 years. Cycles of less than 15 years have little effect, while cycle lengths much beyond individual life expectancy are not predicted by theory and may be indistinguishable from long-term trends. We also limit A to values between 0 and 0.3. It is plausible to assume that reproduction might vary cyclically from a net reproduction rate (NRR) of about 0.7 to an NRR of 1.3, or that the total fertility rate (TFR) could range from 1.4 to 2.6. Larger fluctuations, while certainly possible, seem less likely, especially on a sustained basis.

Our initial focus is on fluctuations in the population dependency ratio. Since the model is specified with half of the lifespan active and half inactive, the base value of the dependency ratio is 1. Figure 1 shows how the maximum dependency ratio varies with T when $A = 0.3$. The minimum dependency is not shown as it is about equal to the reciprocal of the maximum D. We only show values for $A = 0.3$ in Figure 1, as the amplitude effect is essentially linear (e.g., when $A = 0.15$, the maximum D's deviation from 1 is approximately half of what it is when $A = 0.3$).

The dependency ratio always equals 1 whenever T is an integral divisor of 45: i.e., for graphed values of $T = 15$, 22.5, and 45 years. That occurs because the model has a fixed lifespan of 90 years, half of which is active and half dependent. Thus, with $T = 45$, the experience of each period captures stages of the life span across exactly two cycles of fertility, half of those years being active and half dependent. The pattern of resonances in Figure 1 would almost certainly be attenuated in more realistic models, which relax our strict assumptions.

For $15 < T < 22.5$, the maximum D is less than 1.1. Departures from unity are more substantial for $22.5 < T < 45$, with D reaching a maximum of 1.195 when $T = 35$. The

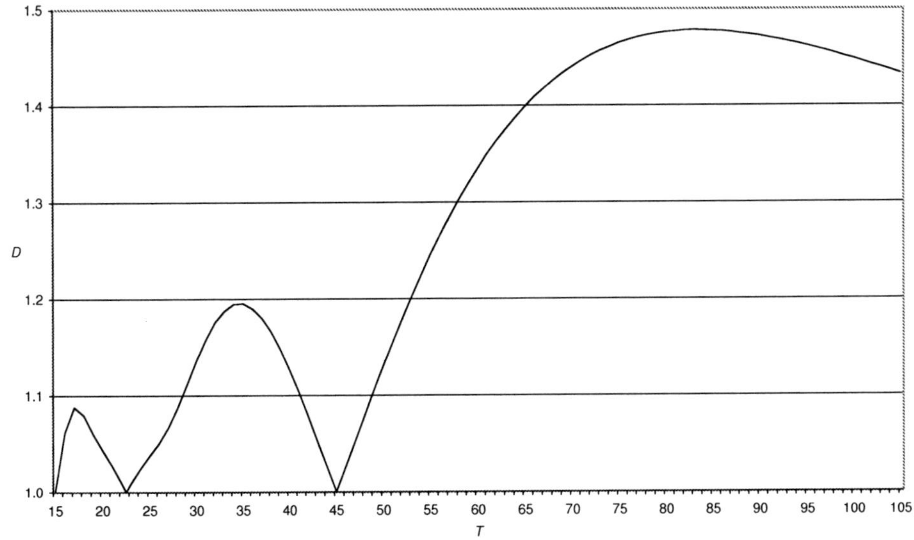

Figure 1. Maximum dependency ratio (D) when amplitude (A) is 0.3, for values of cycle length (T) from 15 to 105 years.

largest values of D come when $T > 45$. Values of D rise sharply after that point, reaching a maximum of 1.478 when $T = 83$. At that cycle length, the dependency ratio rises 0.814 from a low of 0.664, or more than doubles during the course of a cycle. For $T > 83$, the maximum D declines gradually, eventually approaching 1 when T is very large.

One might think that the maximum (and minimum) of D would occur when $T = 90$. At one point in the $T = 90$ cycle, all of the smaller (i.e., size less than 1) cohorts are between the ages of 20 and 65, while all of the larger cohorts are in the dependent ages. At that point, however, $D = 1.472$, slightly below the maximum value. As might be expected given the derivative of equation (5), the extrema of D are given by a complicated combination of trigonometric terms.

It is clear from Figure 1 that a cyclically stationary population's dependency burden varies markedly with T, and that it can depart substantially from 1. Although the mean age of net maternity is not a factor, a local maximum occurs with cycle lengths of approximately one generation. The value of D is rising at "Malthusian" cycle lengths of about two generations, while the global maximum occurs around the three-generation mark. At many cycle lengths, when A is sizeable, the variation in dependency over the course of a cycle can be quite pronounced. We now consider how those variations can affect persons born at different points in a cycle.

3. Individual Outcomes under Two Systems of Transfer Payments

Aggregate, or period, experiences can be quite different from individual, or life-cycle, experiences (cf. Preston 1982). To explore cohort differences, we examine how individuals born at different stages of a cycle, and thus into birth cohorts of different sizes, fare under two common types of transfer systems. They are known in the pension literature as defined contribution and defined benefit plans. Basically, a defined contribution plan is one where the contributions are specified. Here, we consider a "Pay \$45" plan, where each individual between the ages of 20 and 65 pays \$1 each year. As is the case with most governmental plans (and with most informal plans as well), we view the system as being "pay as you go"—each year's benefits are paid from that year's receipts. Hence with the Pay \$45 plan, each dependent in year t receives a benefit of $\$(1/D(t))$. In contrast, a defined benefit plan is one where the benefit to be received is specified. Here we examine a "Get \$45" plan, where each person in the dependent ages receives \$1. Under the "pay as you go" format, each person in the economically active ages is seen as contributing $\$D(t)$.

3.1. VARIATIONS UNDER A DEFINED CONTRIBUTION PLAN

Figure 2 shows the patterns of net transfer (i.e., lifetime benefits minus contributions). For a specified cycle length, each curve depicts the net lifetime transfers accruing to each person born in the indicated year of the cycle. The experience of only one cycle is shown; hence the curve for an N-year cycle ends in year N. The only amplitude shown is 0.3, as the effects of amplitude are again roughly linear. When $T = 60$, persons born in year 45 have a net gain of \$12.05, while persons born in year 14 have a net loss of \$8.36. For all cycle lengths, there is a strong pattern of net losses for those born in the first half of a cycle (where, following equation (1), cohort sizes are greater than 1), and net gains for those born in the

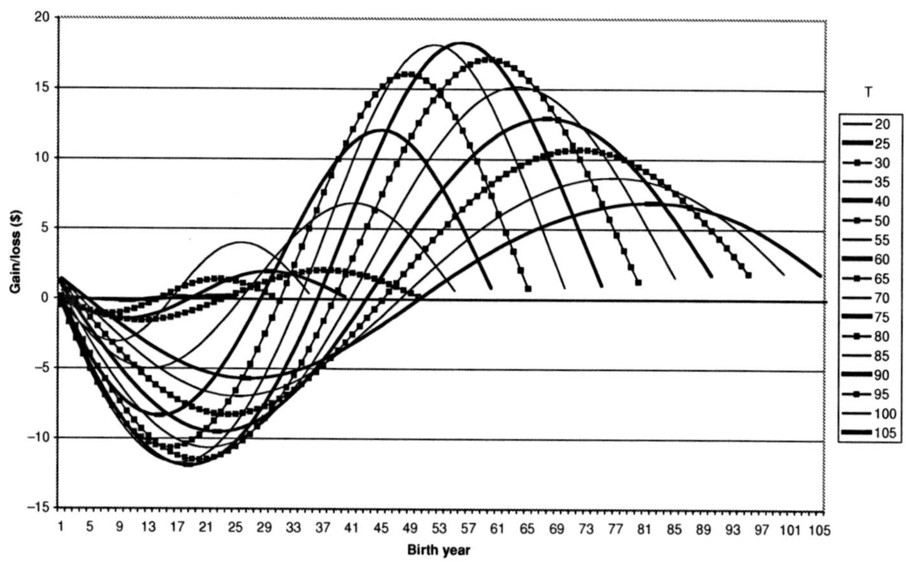

Figure 2. Lifetime net transfers (benefits minus contributions) under a "Pay \$45" plan when amplitude (A) is 0.3, for cycle length (T) of 20 to 105 years, and by year of birth in the cycle.

Table 1. Maximum dollar gain and loss values (benefits minus contributions) under Pay \$45 and Get \$45 plans, when amplitude (*A*) is 0.3, by cycle length (*T*).

	Pay \$45			Get \$45		
T	Maximum gain	Maximum loss	Gain minus loss	Maximum gain	Maximum loss	Gain minus loss
20	0.21	–0.17	0.38	0.17	–0.21	0.38
25	0.18	–0.16	0.34	0.16	–0.19	0.34
30	1.40	–1.03	2.43	1.03	–1.40	2.43
35	4.00	–3.09	7.09	1.34	–1.87	3.21
40	1.94	–1.42	3.36	0.50	–0.69	1.19
45	0	0	0	0	0	0
50	2.05	–1.55	3.60	0.54	–0.70	1.24
55	6.82	–5.01	11.83	1.92	–2.45	4.37
60	12.05	–8.36	20.41	3.66	–4.63	8.29
65	16.04	–10.62	26.66	5.37	–6.79	12.16
70	18.12	–11.73	29.85	6.83	–8.70	15.53
75	18.30	–11.93	30.27	7.95	–10.26	18.21
80	17.09	–11.50	28.59	8.74	–11.45	20.19
85	15.14	–10.64	25.78	9.25	–12.32	21.57
90	12.90	–9.52	22.42	9.52	–12.90	22.42
95	10.69	–8.28	18.97	9.61	–13.25	22.86
100	8.66	–6.99	15.65	9.57	–13.40	22.97
105	6.88	–5.72	12.60	9.42	–13.42	22.84

second half of the cycle (where cohort sizes are less than 1). The net transfer curve for $T = 35$ has the greatest amplitude for cycles where $T < 45$. For $T > 45$, the amplitudes of the net transfer curves continue to increase up to $T = 75$.

Table 1 shows the maximum dollar gains and losses for every fifth value of T from 20 to 105. There are no net gains or losses when $T = 45$ years, and the difference between the maximum gain and the maximum loss is less than 1 percent of the amount transferred (\$45) when T is 20 and 25. However, the difference between maximum gains and losses is greater than 10 percent of the amount exchanged when $T = 35$ and for $T = 55$ or more. When T is 75 years, that difference is more than two-thirds of the amount transferred.

Figure 3a, b shows how cohorts of different sizes fare under the Pay \$45 plan, showing cycle lengths of 20 and 70 years. The points indicating the first and last birth years of each cycle are always identified in the figure. The patterns resemble ellipses, although similar curves (not shown) for $T = 30$ and 90 collapse to yield approximately straight lines. It is evident that the larger the cohort the larger the loss (or the less the gain). For $T = 25$, losses are smaller when cohort size is increasing (birth years 1 through 5 and 16 through 20) than when cohort size is decreasing (years 6 through 15). The opposite is true for cycle lengths of 50 and 70. For cycle lengths of 30 or more, persons born in a small majority of the years in each cycle show a net gain, and the magnitude of the largest gain is consistently

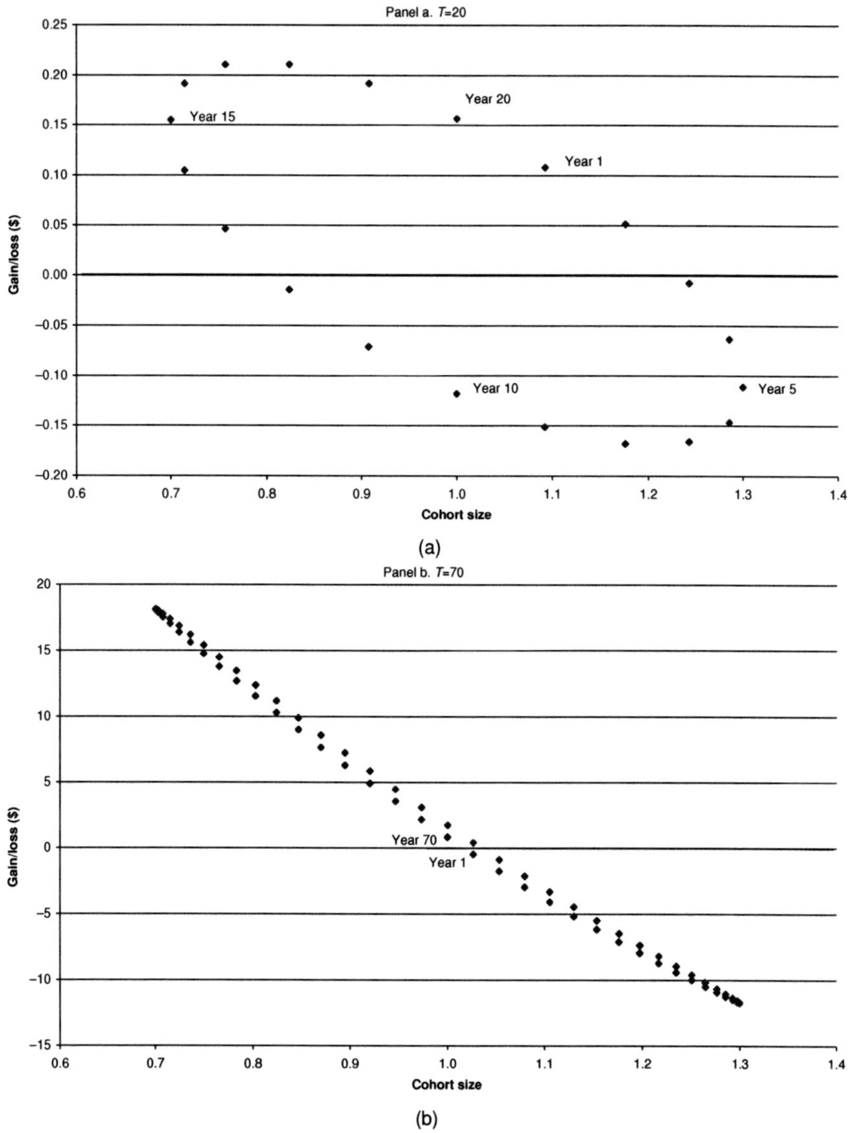

Figure 3. Lifetime gains and losses for persons born in different years of a cyclically stationary population with cycle length 20 and 70 years, assuming each active person pays $1 each year.

greater than that of the largest loss. However, when the gains and losses are weighted by cohort size, there is a net gain of zero across all cohorts in every cycle.

Year of birth thus plays an important role in how individuals fare under the Pay $45 plan. Although the magnitude of some of the differences may be surprisingly large, the pattern was predictable. All persons pay the same amount, but larger cohorts generally receive less

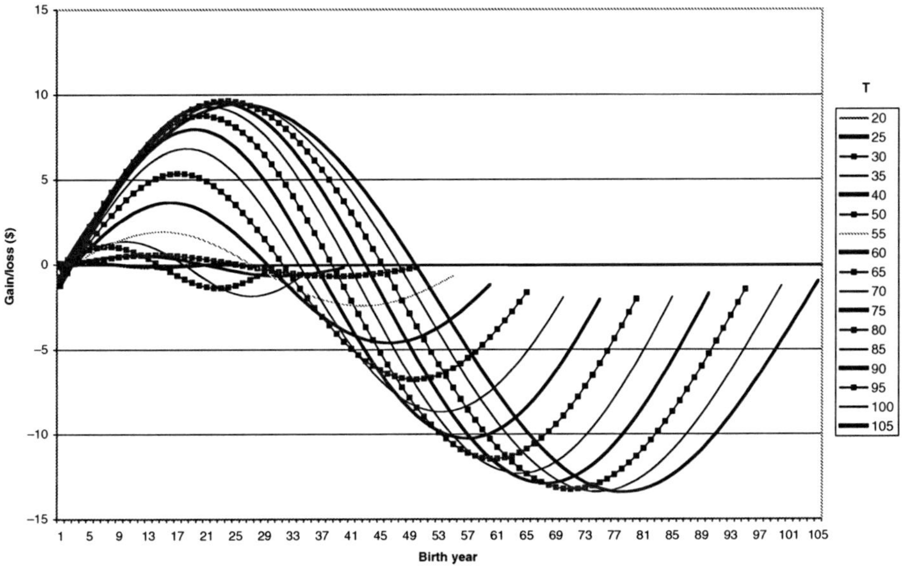

Figure 4. Lifetime net transfers (benefits minus contributions) under a "Get \$45" plan when amplitude (*A*) is 0.3, for cycle length (*T*) of 20 to 105 years, and by year of birth in the cycle.

because there are more of them to split the fixed per person contributions of the smaller cohorts.

3.2. VARIATIONS UNDER A DEFINED BENEFIT PLAN

Figure 4 shows the pattern of net transfers by cycle length and year of birth under a Get \$45 plan, with $A = 0.3$. The pattern is similar to that shown in Figure 2, but with a defined benefit plan the larger cohorts are the ones showing net gains. Looking more closely, the magnitude of the gains is somewhat smaller under the Get \$45 plan than under the defined contribution plan just examined. When T is 50 or less, the birth year curves depart only slightly from 0. However, the amplitude of the birth year curves continues to increase through $T = 90$.

The right-hand side of Table 1 shows the maximum gains and losses for every fifth value of T under the Get \$45 plan. Those maxima are never greater than the comparable values under the Pay \$45 plan. Especially for cycle lengths of 55 years or more, the difference between the maximum gain and loss is considerably smaller under the Get \$45 plan than under the Pay \$45 plan. Still, that difference exceeds 10 percent of the \$45 transferred when $T > 55$. The maximum difference, \$22.97 or 51 percent of the amount transferred, occurs when $T = 100$. (However, the maximum gain, \$9.62 occurs when $T = 96$, and the maximum loss, \$13.43, when $T = 102$.)

Figure 5a, b shows how cohorts of different sizes fare under the Get \$45 plan for cycle lengths of 50 and 90 years. Again, the patterns are similar to those shown in Figure 3, though

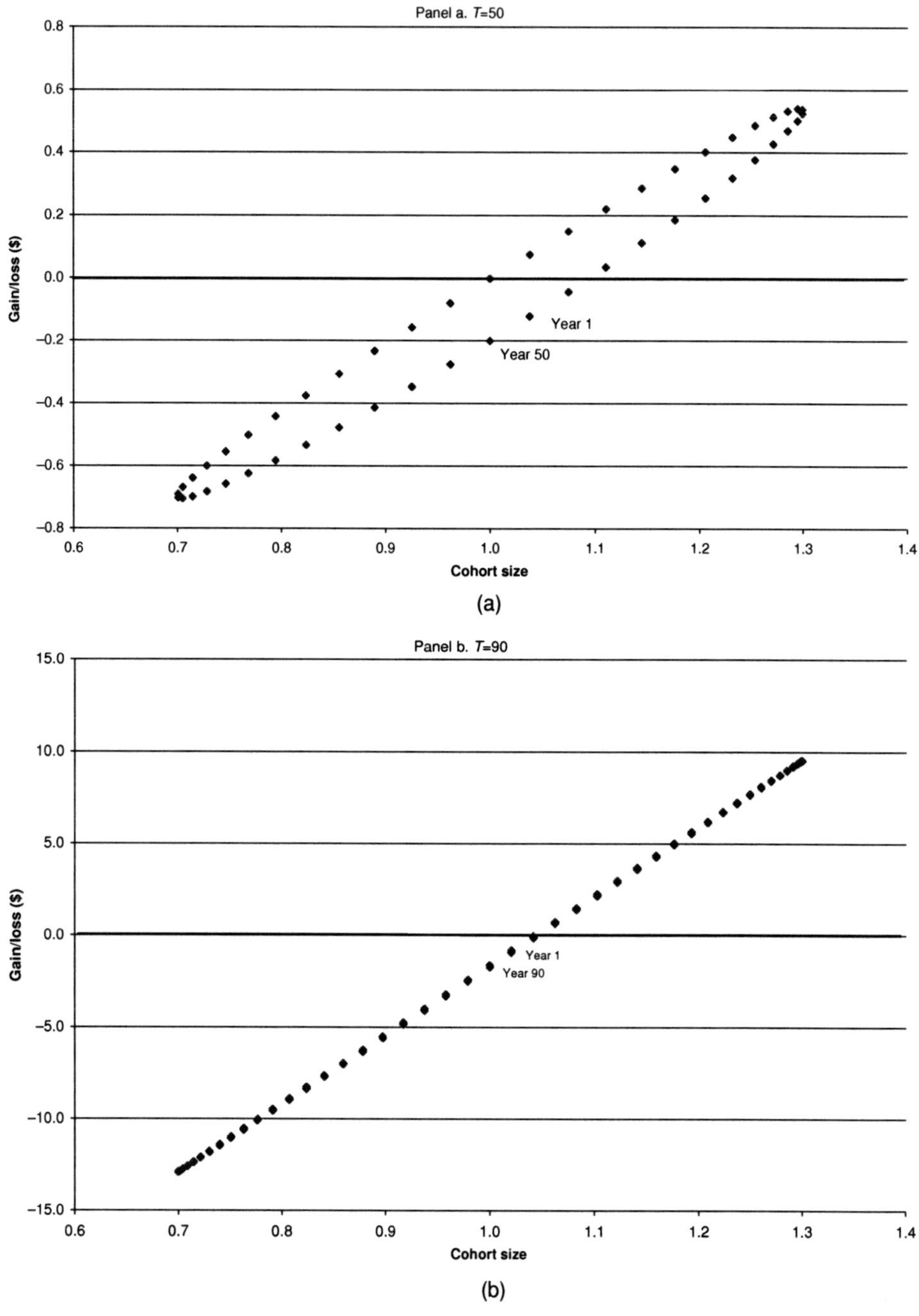

Figure 5. Lifetime gains and losses for persons born in different years of a cyclically stationary population with cycle length 50 and 90 years, assuming each dependent receives $1 each year.

the slant of the ellipses has changed so that now the smaller cohorts are disadvantaged. For $T = 30$ and 90, the values for each birth year are the same as under the Pay $45 plan, with the sign changed. The magnitude of the largest gains is less and, for T values of 30 or more, most birth years experience net losses.

4. Summary and Conclusions

The implications of birth fluctuations can be examined using a cyclically stationary population model with universal survival to age 90. The results show that population dependency burdens and individual gains and losses vary substantially with the amplitude of the fluctuations and the length of each cycle. There is no doubt that substantial temporal and intergenerational issues can arise in the absence of long-term growth when there are fluctuations in the number of births.

At the population level, dependency burdens can vary greatly over the course of a cycle. At the individual level, gains or losses from intergenerational transfers differ dramatically by year of birth. Under a defined contribution plan (Pay $45), large cohorts generally lose while small cohorts gain; the reverse is true under a defined benefit plan (Get $45).

With an amplitude of 0.3, cycles of around one generation (30–40 years) produce noticeable effects. Cycles of two generations produce even larger effects, though the maximum impact occurs in the neighbourhood of three generations. Although some qualifications are in order, this basic model finds substantial merit in Easterlin's argument that birth cohort size matters. Time of birth can greatly affect one's fortune, although the direction of that effect depends on the system of intergenerational transfers.

References

Bourgeois-Pichat, J. (1979), "La baisse actuelle de la fécondité en Europe s'inscrit-elle dans le modéle de la transition démographique?" *Population*, 34:267–306.

Caswell, H. (1989), *Matrix Population Models*. Sinauer: Sunderland, MA.

Easterlin, R.A. (1978), "What will 1984 be like? Socioeconomic implications of recent twists in age structure." *Demography*, 15:397–432.

Easterlin, R.A. (1980), *Birth and Fortune*. Basic Books: New York.

Keyfitz, N. (1971), "On the momentum of population growth." *Demography*, 8:71–80.

Keyfitz, N. (1988), "Some demographic properties of transfer schemes: How to achieve equity between the generations." In: *Economics of Changing Age Distributions in Developed Countries*, R.D. Lee, W.B. Arthur, and G. Rodgers, eds. pp. 92–105 Clarendon Press: Oxford.

Kim, Y.J. and Schoen, R. (1996), "Populations with sinusoidal birth trajectories." *Theoretical Population Biology*, 50:105–123.

Kim, Y.J. and Schoen, R. (1997), "Population momentum expresses population aging." *Demography*, 34:421–427.

Lee, R.D. (1974), "The formal dynamics of controlled populations and the echo, the boom, and the bust." *Demography*, 11:563–585.

Lee, R.D. (1994), "The formal demography of population aging, transfers, and the economic life cycle." In: *Demography of Aging*, L.G. Martin and S.H. Preston, eds. pp. 8–49. National Academy Press: Washington, DC.

Lee, R.D. (1997), "Population dynamics: Equilibrium, disequilibrium, and consequences of fluctuations." In: *Handbook of Population and Family Economics*, Vol. 1B. M.R. Rosenzweig and O. Stark, eds. pp. 1063–1115. Elsevier: Amsterdam.

Leff, N. (1969), "Dependency rates and savings rates." *American Economic Review*, 70:886–895.

Menken, J. (1985), "Age and fertility: How late can you wait?" *Demography*, 22:469–483.

Preston, S.H. (1982), "Relations between individual life cycles and population characteristics." *American Sociological Review*, 47:253–264.

Preston, S.H. (1984), "Children and the elderly: Divergent paths for America's dependents." *Demography*, 21:435–457.

Preston, S.H. (1986), "The relation between actual and intrinsic growth rates." *Population Studies*, 40:343–351.

Schoen, R. and Kim, Y.J. (1994), "Cyclically stable populations." *Mathematical Population Studies*, 4:283–295.

Schoen, R. and Kim, Y.J. (1996), "Stabilization, birth waves, and the surge in the elderly." *Mathematical Population Studies*, 6:35–53.

Schoen, R. and Kim, Y.J. (1997), "Exploring cyclic net reproduction." *Mathematical Population Studies*, 6:277–290.

Skellam, J.G. (1967), "Seasonal periodicity in theoretical population ecology." In: *Proceedings of the Fifth Berkeley Symposium on Mathematical Statistics and Probability*, Vol. IV, *Biology and Problems of Health*, L. Lecam and J. Neyman, eds. pp. 179–205. University of California: Berkeley.

Tobin, J. (1967), "Life cycle savings and balanced economic growth." In: *Ten Economic Studies in the Tradition of Irving Fisher*, W. Fellner, ed. pp. 231–256. Wiley: New York.

Tuljapurkar, S. (1986a), "Cycles in nonlinear age-structured models, I: Renewal equations." *Sloan-Berkeley Working Paper in Population Studies #1*. Institute of International Studies, University of California: Berkeley.

Tuljapurkar, S. (1986b), "Cycles in nonlinear age-structured models, II: McKendrick–Von Foerster equations." *Sloan-Berkeley Working Paper in Population Studies #2*. Institute of International Studies, University of California: Berkeley.

Wachter, K.W. (1991), "Elusive cycles: Are there dynamically possible Lee-Easterlin models for U.S. births?" *Population Studies*, 45:109–135.

CHAPTER 12. ON STOCHASTIC GENERATIONAL ACCOUNTING

JUHA M. ALHO
Department of Statistics, University of Joensuu

REIJO VANNE
Finnish Pension Alliance, Helsinki, Finland

1. Introduction

Part of the inheritance we leave to future generations is the (positive or negative) public net wealth. Depending on the country, the positive wealth may include roads, bridges, office buildings, hospitals, universities, natural parks etc. Such items resemble durable goods in that they depreciate at a relatively predictable rate.

In addition to state and municipalities, the public sector (or the *general* government) includes institutions that manage statutory pensions and unemployment and disability insurance. In Finland, most of the latter institutions are privately owned but they are included in the general government according to the national accounts standard. They own publicly traded stocks and other types of financial wealth. In some countries, such as Finland and Norway, the state also owns such assets. This part of the public wealth can be highly volatile (Vanne 2002).

On the other hand, negative wealth typically exists in the form of government bonds. They are sold to investors who appreciate low risk, low return assets, to cover government expenditures. The future real market value of such debt is relatively more predictable.

In addition to wealth, we leave to the future generations a set of economic rules including tax laws, entitlements to benefits, and rights to free or low price public services. Such rules are intended to apply to current and future generations, so any attempt at calculating the net bequest or burden we leave to future generations—in the sequel, we will speak of burden for short—should incorporate the rules. The so-called *generational accounts* (cf., Auerbach, Gokhale, and Kotlikoff 1991; Auerbach, Kotlikoff, and Leibfritz 1999) provide a possible system for the book-keeping of such future costs, and discounting their net effect to the present. In the usual approach, a deterministic view of the future population and future economy is adopted. Thus, the calculations are implicitly conditioned to the assumed future scenario. The goal of this paper is to show how uncertainty regarding the future can be incorporated into these calculations using stochastic methods. This will be

Anne H. Gauthier et al. (eds.), Allocating Public and Private Resources across Generations, 291–303.

expressed in terms of a predictive distribution. It indicates not only what the most likely burden will be, but also how uncertain it is.

A practical motivation for considering alternatives to the usual deterministic analyses is that their results can be highly volatile depending, e.g., on the current population projection and the current market value of the assets. This is illustrated with an example from Finland in Section 2. Section 3 reviews the sources of uncertainty that will be the basis of modelling in the sequel. In Section 4 some principles of standard generational accounts are reviewed. In Section 5 we will briefly describe the structure of Finnish public expenditures and the Finnish tax system, and present estimates of age-specific profiles for both. Section 6 reviews some of the empirical methods used in producing a stochastic population forecast of Finland, as well as the interpretation of the results. In Section 7 we describe statistical models for the relevant economic rates. Section 8 combines the results into a predictive distribution of the burden. Technical details about the analyses are reported elsewhere (Alho and Vanne 2006).

2. An Example of the High Volatility of the Deterministic Calculations

The so-called intertemporal public liabilities (Raffelhüschen 1999) can be defined as, IPL = (discounted future entitlements of all current and future generations) − (discounted future taxes of all current and future generations) − (current public net wealth). Attempts at estimating the IPL have lead to curious results in Finland. Estimates made in 1995 (Feist et al. 1999) showed the IPL as 253 percent of the gross domestic product (GDP), but estimates made in 2000 showed the IPL as −90 percent (Vanne 2002); i.e., a large implicit public debt was turned into a large implicit public wealth in a matter of 5 years. The seemingly incredible transformation was caused primarily by a decline in unemployment, which led to a decrease in projected future entitlements, by a rise of profits and capital income taxes, and by a phenomenal increase in the market value of information technology and other stocks held by the state and pension institutions. A year later, when the stocks had fallen in value, a similar calculation would have showed a lower—but still positive—net wealth.

Although the IPL calculations were, in both cases, arithmetically correct, the whole concept loses some of its attractiveness, if the results can suddenly change. A major reason for the volatility is that the calculations perpetuate the current conditions in a deterministic manner, ad infinitum. Since the volatility derives from the uncertainty of future developments, one should acknowledge that the results are inherently uncertain. This can be done in a probabilistic framework, in which current conditions are also perpetuated—but probabilistically, treating all future values as random variables. A realistic description of the future uncertainty is not as variable as the annual values of the relevant variables. We think of this as the major motivation of *stochastic generational accounting* to be described in this paper.

3. Sources of Uncertainty

Increases in productivity automatically increase tax revenues, earnings-related pensions, and standards of social services. Therefore, a major source of uncertainty in the IPL is the future development of productivity. A complication is that changes in tax rates and the costs of entitlements are, to some extent, predictable. In principle, this should be taken into account, if genuine predictive distributions are desired. For example, in the 1995 Finnish calculations (Feist et al. 1999) several alternative scenarios were produced that all lead to a smaller net liability than the 253 percent figure. In the most extreme scenario, it was assumed that levels of certain entitlements would be lowered, and unemployment would rapidly decrease, in the near future. In this case, the IPL was −10 percent. This still falls far short of the year 2000 figure −90 percent.

From a demographic point of view, generational accounts can be viewed as functionals of future population, as disaggregated by age and sex. Therefore, IPL calculations can be viewed as an exercise in the so-called functional forecasting (Bogue and Biehler 1979). The demographic aspect deserves emphasis, because the uncertainty of future population development is frequently underestimated. This has been emphasized by Lee and Tuljapurkar (2001), for example, in their analysis of the fiscal sustainability of the U.S. social security system. Here, we build on our earlier work on stochastic forecasting in Finland as discussed in Alho (1998). For the calculation of the functionals, we need estimates of age and sex-specific tax rates and entitlement rates. Here, we rely on our earlier work as described in Feist et al. (1999) and Vanne (2002). In addition, we will develop stochastic models for interest rates, stock returns, and labour productivity (Alho and Vanne 2006).

A topic of much debate in Finnish public life has been the government policy of reducing debt from the sale of stocks and from unexpected tax revenues when the economy has boomed. At the beginning of the year 2000, the value of government bonds (net of the bonds owned by insurance institutions that we include in general government) was 32.3 percent of the GDP of year 2000, and the value of the stocks held by the government was 91.3 percent of the GDP. A policy of reducing debt may increase the options of the government in an eventual downturn of the economy. This is not clear, however, since the decrease in debt is matched by a commensurate decrease in wealth. A point that has not been emphasized in the debate is that reducing debt in this way changes the stochastic characteristics of the government's portfolio. High volatility, high yield assets are traded to pay out low volatility debt of equal current value. We will compare the effect of these sources of uncertainty and the effect of productivity and demographics on the public liability.

4. Generational Accounting and Debt

As discussed by Kotlikoff (1999) fiscal book-keeping relies on labels that are conventional. Current spending can alternatively be viewed as decreasing the previously accumulated wealth of the government; as a "loan" that is expected to be paid back by the receiver or his/her descendants in the form of taxes, or as "borrowing" from future generations that is expected to be reimbursed in some form later. In a similar vein, family benefits, such as child allowances, could be thought of as either benefits to the child or benefits to the parent.

For an individual, an accounting of lifetime taxes and entitlements provides a net measure that is independent of labelling. For a government with a potentially infinite life span, one has to consider *all* future in- and out-payments simultaneously. Auerbach and Kotlikoff (1999) suggested that by making neutral assumptions about future tax rates and future entitlements, one could use discounting to achieve comparability of all public fiscal events. Implicit in the calculation is that the government must be able to service the possible debt at all times, but otherwise they make no assumption about the realism of the resulting economic and demographic paths. However, an assumption of intertemporal budget constraint is made. This says that, in present value, future net taxes must equal the current debt. Or, over the infinite horizon, the public sector does not make a surplus or a loss.

Although the generational accounts that are produced in this way go a long way toward avoiding the conventional aspects of current fiscal book-keeping, they do rely on a set of assumptions of their own. For example, there is no need to make comparisons of equity across current and future generations based on the intertemporal budget constraint of zero. Instead, one could determine what we would have to do to all future taxes, so that current wealth would be maintained. This could be sensible, if the public sector held a positive wealth at any given time. In general, some other level of net surplus could conceivably create more utility to the tax payers than the zero level.

5. Taxes and Public Expenditures

5.1. AGGREGATE STATISTICS

For the purposes of our analysis, the government includes the state, municipalities, and institutions managing statutory pension systems and other forms of social security, such as unemployment, disability, and health insurance. For the book-keeping of public finances, Finland applies the European System of Accounts (ESA95).

Based on the figures of the database of the Research Institute of the Finnish Economy (ETLA; http://www.etla.fi; October 9, 2001.), the public revenues in the year 2000 were, as percent of the GDP of year 2000,

Value added tax (VAT)	7.8 percent
Taxes on alcohol, cars, and tobacco	5.6 percent
Taxes on income and property	21.1 percent
Employers' social security contributions	8.9 percent
Employees' social security contributions	3.3 percent

The total revenues (excluding those deriving from wealth) were 46.7 percent. Therefore, this is also the tax rate. Note that since the GDP was 130 billion (or 130×10^9) euros, we have approximately that 1 percent is 1.3 billion euros, or 1.5 billion dollars (at the exchange rate of 2004).

The public expenditures were, correspondingly,

Business subsidies	1.5 percent
Pensions, unemployment etc.	16.5 percent
Education, social and health services	13.1 percent
Collective consumption of fixed capital	7.6 percent
Other, net of other revenues	1.8 percent

The total expenditures were 40.5 percent, so the primary balance was 46.7 percent − 40.5 percent = 6.2 percent of the GDP.

The capital income was 4.4 percent and the interest paid on public debt was 3.6 percent of the GDP, so the capital balance was approximately 0.7 percent. Therefore, the wealth increased due to the difference of revenues and expenditures, by 6.5% + 0.7% = 6.9% of the GDP. In addition to the conventional public surplus, the value of publicly owned stocks soared.

In our discussion of the IPL, the stochastic future taxes and expenditures are key processes. The capital budget enters via the initial net wealth, which is equal to the discounted value of the infinite streams of capital income and interest on bonds.

The Finnish surplus in the year 2000 was high in international comparison. Moreover, although Finland has a tradition of surpluses (due to the partial prefunding of statutory pensions), the figure in year 2000 was exceptionally high, from the Finnish perspective as well (Vanne 2002). Providing a description for the volatility of both primary and capital balances is a major motivation of our paper.

5.2. TAX, TRANSFER, AND PUBLIC SERVICE LEGISLATION

While taxes and public expenditures have a basis in legislation, they have to be suitably aggregated to allow for the statistical estimation of age–sex profiles from existing data. We provide here more details about the aggregates discussed above.

Finland has harmonized indirect taxation rules according to the European Union (EU) policies. The VAT rate varies by product, but 22 percent is a common rate. Despite reductions since 2000, taxes on alcohol, cars, and tobacco are also higher than the average in the other EU countries. (The reductions of taxes have not been taken into account in our calculations.) Indirect tax profiles by age and sex were estimated based on household surveys, the observed consumption patterns, and the indirect tax content of different products (Feist et al. 1999).

Taxes on wages and social transfers are progressive. Capital income is taxed at a flat rate of 29 percent. Personal wealth is taxed according to a low, but progressive schedule. Separate profiles for capital income and wages were estimated.

Employers and employees also contribute to a statutory earnings-related, defined benefit pension system, which is partially prefunded. They also pay for the national pension, universal health insurance, and statutory unemployment insurance that only run buffer funds.

The pension schemes provide old age pensions for all starting at the age of 65, as well as disability and survivors' benefits. There are complicated early retirement benefits that depend on the applicant's labour market status. These rules prevailing in 2000 have subsequently been modified, but this has not been taken into account in our calculations. The earnings-related pension system is mostly run by mutual pension companies that manage their assets like private investment funds.

In the ESA95 standard, education, social, and health services provided by the public sector are included under individual consumption. User fees are not included in the figures. Social services include an extensive childrens' day care system run by the municipalities.

5.3. THE AGE–SEX PROFILES OF TAXES AND EXPENDITURES

The primary balance derives from age–sex profiles of different taxes and expenditures. A multitude of statistics were used to derive the profiles by single year of age for the 10 primary balance variables mentioned in Section 5.1. They were from different calendar years during 1990–2000.

Twenty aggregates with an age–sex profile were available in the first phase of the calibration. For instance, the ESA95 category "social transfers" were divided into eight sub-aggregates, namely pensions, unemployment benefits, transfers related to children (family policy), daily allowances of health insurance, refunds of health insurance, rehabilitation benefits of health insurance, student benefits, and other social transfers.

The pension benefit profiles are from the Central Pension Security Institute and the Social Insurance Institution (2001); and all three health insurance profiles are from Social Insurance Institution (2001). The initial social and health service profiles are from the year 1998; and they were published by the Ministry of Social Affairs and Health (2001). All the other initial profiles are described in Feist et al. (1999).

The profiles for the year 2000 are given in Figure 1. The difference in taxes paid, between males and females, derives primarily from the higher labour-force participation and higher

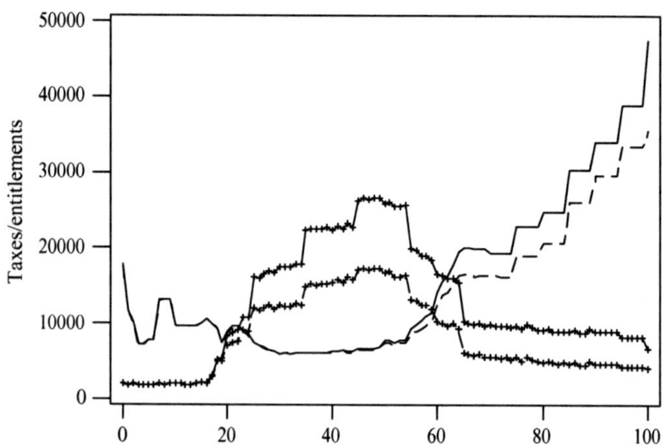

Figure 1. Taxes (+) and entitlements for males (solid) and females (dashed), in Finland in 2000.

wages of the males. This also explains the difference in entitlements, since the males receive higher earnings-based pensions.

6. A Stochastic Forecast of Finland

Population influences the value of IPL via the profiles of age-specific average rates of entitlements and taxes discussed above. Since the future population cannot be known accurately, neither can the IPL. A stochastic forecast is a method of accounting for uncertainty in a systematic way. In Alho (1998) we describe, in detail, the methods used to produce a predictive distribution of the Finnish population for 1999–2050. This forecast was extended 50 years for the present work. Here, we will summarize aspects that are relevant for the stochastic analysis of the IPL, and for the interpretation of the results.

We consider the population as disaggregated by sex and single years of age ($x = 0, 1, \ldots,$ 99, 100+). Age-specific fertility and mortality rates are forecast for each future year. Similarly, assumptions about the net number of migrants are made for every future year. The usual cohort-component book-keeping is used to derive the consequences for the population vector.

To describe the uncertainty of the future vital rates, a model of error introduced in Alho and Spencer (1997) was used. The calculations are carried out with the program PEP written at the University of Joensuu (for details, visit http://www.joensuu.fi/statistics/juha.html). The program stochastically simulates a full set of vital rates and calculates a cohort-component forecast for the population. The results were stored into a database for possible aggregation and statistical processing. In the application at hand, 1,500 simulations were used.

The point forecasts we used were essentially those of Statistics Finland. The age-specific fertility was assumed not to change over time with the total fertility rate at 1.80. Mortality was assumed to decline at a rate similar to that in the recent past so that, in 50 years, life expectancy would increase to 83.0 for males and 87.8 for females, and in 100 years it would increase to 88.2 and 91.0, respectively. Net-migration was assumed to stay at the recent past level of 4,000 per year.

The second moments for fertility and mortality were empirically determined, based on naive or baseline forecasts. In the case of total fertility, we calculated the empirical error of a forecast that assumes fertility does not change. This mimics closely the way actual forecasts of fertility have been made in the United States (Lee 1974). The data were from 1776–1995. A smoothed version of the error estimates gave the standard deviations. The estimates were adjusted for the first two decades to account for the recent low level of volatility. Cross-correlations across age were estimated using an AR(1) model. For long-term forecasting PEP has the option of keeping error variances at a fixed level from some year t^* on. In this application we used $t^* = 50$. At this point a 67 percent prediction interval for total fertility was approximately [1.3, 2.5].

In the case of mortality the procedure was the same, except that a constant rate-of-decline model was taken as the baseline forecast. We have shown earlier that this method is quite competitive with the more complex method used by the official forecasters (Alho 1990).

The data period was 1900–1994. After 50 years, a 67 percent prediction interval for life expectancy was [78.6, 87.3] for males and [84.5, 91.1] for females. After 100 years the intervals were [84.3, 91.9] and [87.7, 94.2], respectively.

The error model for migration was based on a time-series analysis of data from 1945–1995, but judgmentally calibrated. For the year 2030, a 67 percent interval was [−6,300, 14,300], for example.

For the results of the Finnish forecast, we refer the reader to Alho (1998). (An updated version appeared in Alho 2002.) Here we note merely that in the Appendix F (http://books.nap.edu/books/) of the National Research Council (2000) panel report *Beyond Six Billion*, a comparison was made between the forecast in Alho (1998) and a separate analysis, in which a prediction interval is calculated based on the analysis of the UN forecasts during 1970–1990. For the total population, the ratios of the upper end point of the 95 percent prediction interval to the median, for forecast years t, are 1.030 ($t = 10$), 1.153 ($t = 30$), and 1.402 ($t = 50$) in Alho (1998); and 1.032, 1.142, and 1.309 in the UN-forecast-based analysis. The results are close for the first 30 years. The discrepancy at $t = 50$ may be explained by the fact that the UN based analysis used recent past data for all lead times, whereas in Alho (1998) the higher volatility of the earlier years was used for longer lead times for fertility.

In summary, apart from recent fertility, the error structure we have assumed has been chosen to approximate the level of uncertainty of forecasting in the past. Its use in the IPL calculations assumes that the uncertainty of the future forecasts is of the same order of magnitude as in the past, or at least, that it is *not higher* than in the past.

7. Analysis of Interest Rates, Stock Returns, and Productivity

The uncertainty of the future economy influences the Finnish IPL via three processes. First, the value of government debt varies according to the real interest rates of government bonds. In the future, these interest rates are the same, with the exception of small liquidity related differences, all through the euro zone. Therefore, we used the German real interest from 1955–2000 as a basis for modelling. An AR(1) model was identified and estimated for the process (Alho and Vanne 2006). Second, the value of stocks held by the government varies like a large portfolio of publicly traded stocks. More than half of the stocks are foreign, and their share is expected to rise. To model this process in the short and medium term, we used the Dow–Jones index for industrial stocks from 1949–2000. We believe the U.S. index is a better guide to the future evolution of stock returns than the European or Finnish series that have developed in a more regulated environment. A random walk model with a drift (0.019) was estimated (Alho and Vanne 2006). Third, the tax rates and, hence, the entitlements should follow the productivity of labour. For this, we have a carefully constructed real GDP *per capita* growth series from Finland from 1860–2000 (Hjerppe 1989; Statistics Finland 2001). It reflects both productivity per person, labour-force participation, and unemployment rate. This series was modelled using an MA(1) process (Alho and Vanne 2006). In particular, the mean of the process was estimated as 0.022 with a standard error 0.005.

7.1. RANDOM RETURNS AND DISCOUNTING

The goal of the IPL calculations is to help anticipate the burden we leave to future generations. We incorporate uncertainty in four ways: the future stock prices are random; the future bond prices are random; the change in productivity is random; and the future population growth is random.

Our analysis tried to put the various money flows on an equal footing at 2000. The difficulty in assessing the future value of the financial assets derives from the infinite horizon we consider. A potentially useful resolution is obtained, when we realize that the results depend on the way the government portfolio is managed. We assumed that the government maintains a fixed value portfolio. That is, if the stocks increase in value (inflation adjusted), then enough are sold, so the value of the holdings does not change. Similarly, if the stocks decrease in value, more are bought to compensate for the loss in value. We assume that such purchases are financed by increasing taxes (over and beyond the tax schedules considered here). These effects on spending and taxes are excluded from the analysis.

Another problem in discounting is that, unlike a private consumer, the government does not have a particular time preference for money. In fact, by selling bonds the government provides a service to investors who want low yield, secure assets. If cash is needed, it can be raised via new taxes. On the other hand, the government is constrained by the political process not to raise taxes at will. As an alternative, by selling bonds the government can defer raising taxes indefinitely. In order to sell, the government must pay interest on the bonds to satisfy the time preference of the buyers. Therefore, a meaningful discount rate for any future income or out-payments is the expected real interest rate of bonds, in this case 0.038.

8. Predictive Distribution of the IPL

To define a random version of the IPL in a formal manner, we first represent current wealth W as the sum of the value of stocks held A, and bonds owed B, or $W = A + B$. Since both stocks and bonds are valued by discounting the money flow they generate, both A and B are random variables. Alho and Vanne (2006) derive the following expectations, $E[A] = 1.272$ and $E[B] = -0.323$, in the units of the GDP of year 2000. Note that the expected discounted income from the stocks is *higher* than their market value 0.90. Thus, $E[W] = 1.272 - 0.323 = 0.950$. Similarly, the standard deviation of W is estimated as 0.626 in the same units. The volatility of the stocks held produces 99 percent of the variance of W.

To define the discounted primary balances, let $V(x, s, t)$ be the number of people in age $x = 0, 1, \ldots, 99, 100+$), who are of sex s = F (female) or s = M (male), during year t. We assumed that the entitlements and taxes grow with productivity, from the values $S(x, s)$ and $T(x, s)$ they have at baseline $t = 2000$. For example, the discounted entitlements at time $t + u$ that are due to those of age x, who are of sex s, are of the form $V(x, s, t + u)S(x, s)\exp(-0.038u + 0.022u + Z(u))$, where 0.038 is the discount rate, 0.022 is the average growth rate of the entitlements, and $Z(u)$ is a once integrated MA(1) process of random variation of productivity around its mean. Since the mean of the productivity series had to be estimated, the mean was only assumed to be known up to the standard error (= 0.005). This induces a random component into the model, whose standard deviation at time u equals $0.005u$. Summing over x, s, and u yields the value of discounted future entitlements, or S. Replacing $S(x, s)$ by $T(x, s)$, we get the corresponding discounted taxes, or T.

Since we use a per capita measure for productivity, it is plausible that population development and productivity are approximately independent. Independence was assumed.

We can now define the random public liability as $L = S - T - W$. This random version of the IPL has a probability distribution that it inherits from S, T, and W. We can also think of this as the predictive distribution of the IPL.

We have simulated S and T based on the models described above. For technical reasons discussed in detail in Alho and Vanne (2006) we are able to report only on truncated calculations that involve the first 100 years only. We use $S^*(u)$ and $T^*(u)$ for S and T as truncated to the first u years.

In Alho and Vanne (2006) the following statistics were obtained for the evolution of the distribution of discounted primary balances $T^*(u) - S^*(u)$ at $u = 25, 50, 75, 100$,

u	Mean	Median	Q_1	Q_3	SD
25	0.70	0.69	0.59	0.80	0.16
50	0.16	0.22	–0.12	0.51	0.55
75	–0.46	–0.28	–0.93	0.19	1.03
100	–1.01	–0.70	–1.59	–0.08	1.50

Here, Q_1 and Q_3 are the first and third quartiles, respectively, and SD is the standard deviation. The distributions are skewed to the left. The medians are the most natural summary statistics of the predictive distributions. For the first quarter of the century the primary balances will increase the wealth; but after that the wealth will be depleted by approximately a half of the year 2000 GDP, each 25 years.

A more refined picture is obtained by looking at the whole predictive distribution. Consider the random public liability, as truncated at u, $L^*(u) = S^*(u) - T^*(u) - W$. The distribution is skew to the right, indicating that large liabilities are relatively more likely than corresponding levels of positive wealth. Alho and Vanne (2006) show numerically how the predictive distribution derives from its two independent components,

	Mean	Median	Q_1	Q_3	SD
$S^*(u) - T^*(u)$	1.01	0.70	0.08	1.59	1.50
$-W$	−0.95	−0.95	−1.36	0.53	0.62
$L^*(u)$	0.07	−0.18	−0.95	0.75	1.63

We see that, in terms of standard deviations, the uncertainty of the future primary balances is 2.5 times as high as the uncertainty of the current wealth, and the uncertainty of the future liability approximately equals that of the future primary balances.

Although the level of uncertainty is high, the chance that $L^*(u)$ is negative at $u = 100$, is 55 percent. In fact, the primary balances remain positive for the next 25 years, and then turn negative. At $u = 50$, the chance is still 91 percent that $L^*(u)$ is negative. That is, although the current entitlements are eventually too high for the current level of taxation, there is a good chance that it may not become a problem during the coming century, and it is not likely to become a problem in the next 50 years or so.

We have decomposed random public liability as $L^*(u) = S^*(u) - T^*(u) - W$. The wealth consists of stocks held and bonds owed, $W = A + B$. As noted above, 99 percent of the variance is related to stocks. On the other hand, under the fixed value portfolio policy, the stocks provide an expected increase of the government cash funds of 0.36. As before, the unit is the GDP of year 2000.

We have also shown that the uncertainty in discounted primary balance $T^*(u) - S^*(u)$ dominates that in W. Since we assume the independence of productivity per capita and population development, we can decompose the variance in $T^*(u) - S^*(u)$ into a component deriving from the uncertainty of future population development, and a component deriving from the uncertainty in future productivity. Alho and Vanne (2006) first estimate $\mathrm{Var}(T^*(u) - S^*(u)) = 0.90^2$. Therefore, the remaining variance due to productivity is $1.50^2 - 0.90^2 = 1.20^2$. The share of demographics out of $\mathrm{Var}(L^*(u))$ is $0.90^2/1.63^2$, or 30 percent. The rest is due to economics. The share of uncertain future taxes and entitlements is $1.20^2/1.63^2$, or 54 percent. The remainder, approximately 15 percent, is due to uncertainty in the initial wealth W, under the fixed value portfolio policy. (The percentages do not add up to 100 percent due to rounding and simulation error.)

9. Discussion

Our analysis of the intertemporal public liabilities shows that the current and past generations are not leaving the future generations an inheritance of debt and misery, in Finland. Current public wealth will be gradually consumed. This can be a moral problem, because the wealth has partly been generated by earlier generations, but it does not seem to be a similar problem relative to the future generations.

A central conclusion is that the primary balances offer both serious opportunities and risks for the public sector. This is simply the question of how tax policies and rules for entitlements will be formulated in the future. Adaptive mechanisms that react to demographic uncertainties (such as those suggested by Lassila and Valkonen (2001) for pension prefunding) deserve serious attention. Alho, Lassila, and Valkonen (2006) show that such policies can influence both the mean of the predictive distribution and decrease variance. Similarly, the dependence of the public portfolio on the volatility of international stock markets should be recognized, and the government should prepare for it. In contrast, with the partial exception of migration, demographics is largely outside the scope of governmental decision-making.

From the perspective of public liabilities the current policy of debt reduction has almost no effect on the uncertainty of the public liabilities. However, a reduction of debt will decrease the long-term earning opportunities of the government. Therefore, a more relevant discussion might involve investment strategies that would bring in a higher rate of return.

The analysis presented could be refined. For example, a key assumption was that the future taxes and entitlements grow in proportion to productivity. Both taxes and entitlements could be disaggregated into components for which a more refined specification would be feasible. This can involve elements of forecasting at jump-off. In addition, our analysis of the current wealth relies on a simple portfolio policy that is hardly optimal. Alternatives that optimize the portfolio subject to a low level of volatility could be considered.

Acknowledgement

This paper was prepared for presentation at the International Meeting on Age-Structure Transitions and Policy Dynamics: The Allocation of Public and Private Resources Across Generations, sponsored by IUSSP, in Taipei, December 6–8, 2001.

References

Alho, J. (1990), "Stochastic methods in population forecasting." *International Journal of Forecasting*, 6:521–530.

Alho, J. (1998), "A stochastic forecast of the population of Finland." *Reviews 1998/4*. Statistics Finland: Helsinki.

Alho, J.M. (2002), *The Population of Finland in 2050 and Beyond*, Discussion Paper 826. The Research Institute of Finnish Economy: Helsinki.

Alho, J.M., Lassila, J. and Valkonen, T. (2006), "Demographic uncertainty and evaluation of sustainability of pension systems." In: *Pension Reform: Issues and Prospects for Non-Financial Defined Contribution (NDC) Schemes*, R. Holtzmann and E. Palmer, eds. World Bank: Washington, DC.

Alho, J. and Spencer, B.D. (1997), "The practical specification of the expected error of population forecasts." *Journal of Official Statistics*, 13:203–225.

Alho, J. and Vanne, R. (2006), "On predictive distributions of public net liabilities." *International Journal of Forecasting*, 22 (to appear).

Auerbach, A.J., Gokhale, J. and Kotlikoff, L.J. (1991), "Generational accounts—A meaningful alternative to deficit accounting." *NBER Working Paper* No. 3589, January 1991.

Auerbach, A.J. and Kotlikoff, L.J. (1999), *The Methodology of Generational Accounting*, pp. 31–41 In: Auerbach, Kotlikoff and Leibfritz, Generational Accounting Around the World. The University of Chicago Press: Chicago.

Auerbach, A.J., Kotlikoff, L.J. and Leibfritz, W. (1999), *Generational Accounting Around the World*. The University of Chicago Press.

Bogue, D.J. and Biehler, S. (1979), *Techniques for Making Functional Population Projections*. Community and Family Study Center, University of Chicago.

Central Pension Security Institute and the Social Insurance Institution (2001), *Statistical Yearbook of Pensioners in Finland 2000*. Hakapaino, Oy: Helsinki.

Feist, K., Raffelhüschen, B., Sullström, R. and Vanne, R. (1999), "Finland: Macroeconomic turnabout and intergenerational redistribution." In: EU Generational Accounting in Europe, *European Economy, Reports and Studies No 1999:6*, Office for the Official Publications of the EC: Luxembourg.

Hjerppe, R. (1989), "The Finnish economy 1860–1985: Growth and structural change." Bank of Finland Publications, Studies on Finland's Economic Growth no. 13. Government Printing Centre: Helsinki.

Kotlikoff, L.J. (1999), "From deficit delusion to the fiscal balance rule: Looking for an economically meaningful way to assess fiscal policy." In: Auerbach, Kotlikoff and Leibfritz, eds. pp. 9–30.

Lassila, J. and Valkonen, T. (2001), "Pension prefunding, ageing, and demographic uncertainty." *International Tax and Public Finance*, 8:573–593.

Lee, R.D. (1974), Forecasting births in post-transition populations: stochastic renewal with serially correlated fertility. *Journal of the American Statistical Association,* 69:607–617.

Lee, R.D. and Tuljapurkar, S. (2001), "Population forecasting for fiscal planning: Issues and innovations." In: *Demographic Change and Fiscal Policy,* A. Auerbach and R.D. Lee, eds. pp. 7–57. Cambridge University Press: Cambridge.

Ministry of Social Affairs and Health (2001), *Strategies of Social and Health Care Policy 2010* (*Sosiaali- ja terveyspolitiikan strategiat 2010*). Ministry of Social Affairs and Health. Helsinki: Libris Oy.

National Research Council (2000), *Beyond Six Billion*, National Academy Press: Washington DC.

Raffelhüschen, B. (1999), "Generational accounting: Method, data and limitations," in *EU Generational Accounting in Europe*, (*European Economy, Reports and Studies* No. 1999:6), Office for the Official Publications of the EC: Luxembourg.

Social Insurance Institution (2001), *Statistics of the Health Insurance and Family Benefits 2000 (Kansaneläkelaitoksen sairausvakuutus- ja perhe-etuustilastot 2000)*, Publications of the Social Insurance Institution No. T11:12, Helsinki.

Statistics Finland (2001) Database.

Vanne, R. (2002), "The Finnish generational accounting revisited." *Nordic Journal of Political Economy*, 28: 13–25.

INDEX

Printed in the United States
87374LV00001B/109-120/A